57 Structure and Bonding

Editors:
M. J. Clarke, Chestnut Hill · J. B. Goodenough, Oxford
J. A. Ibers, Evanston · C. K. Jørgensen, Genève
D. M. P. Mingos, Oxford · J. B. Neilands, Berkeley
D. Reinen, Marburg · P. J. Sadler, London
R. Weiss, Strasbourg · R. J. P. Williams, Oxford

Complex Chemistry

With Contributions by
J. Emsley R. D. Ernst B. J. Hathaway K. D. Warren

With 85 Figures and 39 Tables

Springer-Verlag Berlin Heidelberg GmbH 1984

ISBN 978-3-662-15734-3 **ISBN 978-3-540-38948-4 (eBook)**

DOI 10.1007/978-3-540-38948-4

Library of Congress Catalog Card Number 67-11280

Originally published by Springer-Verlag Berlin Heidelberg New York Tokyo in 1984

2152/3140-543210

Table of Contents

Structure and Bonding in Metal-Pentadienyl and Related Compounds

Richard D. Ernst

Department of Chemistry, University of Utah, Salt Lake City, UT 84112, USA

Metal Complexes of the allyl (C_3H_5) and cyclopentadienyl (C_5H_5) ligands have played a major role in the development of organometallic chemistry. Metal allyl complexes are well known for the many intricate organic transformations which they may mediate, yet often possess limited stability. Metal cyclopentadienyl complexes on the other hand possess very high stabilities but only very limited catalytic chemistry. The pentadienyl ligand (C_5H_7 as well as various methylated derivatives) has similarities to both the allyl and cyclopentadienyl ligands, and it has now been demonstrated that metal pentadienyl complexes possess not only thermal stability but also chemical and catalytic reactivities. This favorable combination then will allow for detailed correlations to be made between chemical reactivities and the physical natures (structural, spectroscopic, etc.) of these compounds. In this article the presently available information on metal-pentadienyl compounds is discussed in order to gain some understanding of their nature compared to the related allyl and cyclopentadienyl systems, and to anticipate future applications of metal-pentadienyls. Particular emphasis is given to the bis(pentadienyl)metal complexes, which have been referred to as the "open metallocenes".

A. Introduction

The fields of Inorganic and Organometallic Chemistry experienced a great resurgence in interest some thirty years ago in large part due to the accidental discovery of the unusually stable compound ferrocene[1-3]. The subsequent emergence of the cyclopentadienyl group as one of the most important, versatile, and stabilizing ligands in all of inorganic chemistry is well known. Fairly detailed theoretical treatments have been published to account for the great stability of many of these compounds[4-7]. Similarly, the allyl group was also shown early to form stable π-complexes with transition metals[8], and once again the initial discovery was rapidly followed by the dedication of a great deal of effort to the study of such compounds[9-11]. Particularly interesting have been the fascinating applications of the allylic materials to a wide variety of organic transformations, including the legendary "naked metal" reactions[9,10]. A number of bonding schemes and theoretical treatments have also appeared regarding these systems[10-16]. It should be noted that while metal-cyclopentadienyl compounds often possess high stabilities, quite often their reactivities are rather uninteresting, at least when centered at the cyclopentadienyl ligand. Of course, a notable exception is the organic substitution chemistry of ferrocene and related molecules[17]. Conversely, the generally reactive metal allyl complexes often (especially in homoleptic first-row transition metal complexes) possess low thermal stabilities, which has provided a barrier to obtaining reasonably detailed physical data on these compounds[18]. Note, however, that despite the disadvantage of low thermal stability, a number of species such as $Co(C_3H_5)_3$ have proven to be useful in a large variety of interesting transformations[19].

While the first transition metal-pentadienyl compound, $Fe(C_5H_7)(CO)_3^+ClO_4^-$, was reported in 1962[20], this ligand has been greatly neglected in comparison to the cyclopentadienyl and allyl ligands. Only recently has it become clear that pentadienyl ligands offer substantial promise for the future of inorganic and organometallic chemistry. In part this apparent lack of interest may have been due to the greater difficulty associated with obtaining suitable starting materials for introducing these ligands into metal complexes. However, now that suitable materials and methods are becoming available (see Sect. C.I), the field of metal-pentadienyl chemistry is blossoming in many directions. It is the purpose of this treatment to present a perspective view of some of the more important physical, structural, and chemical aspects of pentadienyl and metal-pentadienyl chemistry as they now appear. Thus, the major focus of this article will be on acyclic pentadienyl compounds. However, as it is also desired to provide at least a glimpse of how pentadienyl systems compare to their allyl, butadiene, and various cyclic counterparts, a number of these other relatives will be included where appropriate.

B. General Electronic and Energetic Considerations

I. Physical and Theoretical Data

Before treating the structural and spectroscopic data pertaining to various pentadienyl entities, it is appropriate to examine some physical and theoretical aspects which should prove helpful to understanding and even predicting some of the features of metal-pen-

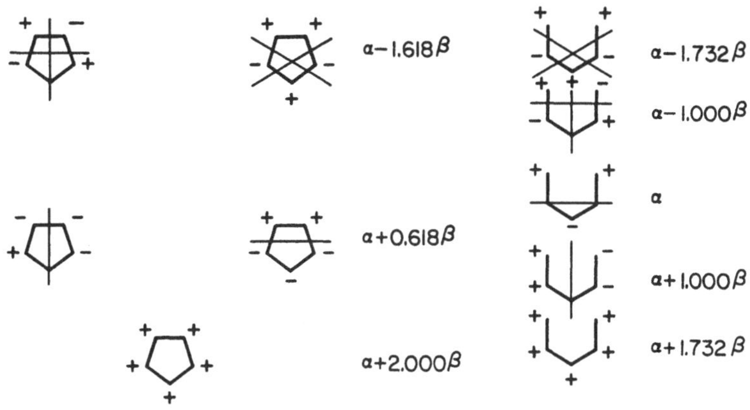

Fig. 1. Representations of the π-molecular orbitals of the cyclopentadienyl (*left*) and pentadienyl (*right*) groups

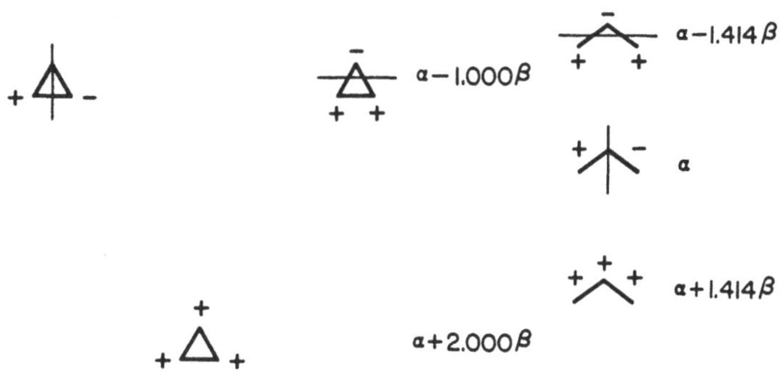

Fig. 2. Representations of the π-molecular orbitals of the cyclopropenyl (*left*) and allyl (*right*) groups

tadienyl chemistry. The pentadienyl group is a delocalized π system which commonly occurs as a cationic, radical, or anionic species. The π-molecular orbitals appropriate for this species are depicted in Fig. 1 along with the π molecular orbitals for the cyclopentadienyl group. It should be noted that the "U" pentadienyl conformation is presented here (and elsewhere) for convenience, although the more classic treatments have generally dealt with the "W" conformation[21], and an "S" (sickle) conformation also is possible. For comparison, the π molecular orbitals for the allyl and cyclopropenyl groups are

U W S

presented in Fig. 2. It can readily be seen from Fig. 1 that the pentadienyl nonbonding orbital (half-filled for the radical) is localized on the carbon atoms in positions 1, 3 and 5. Hence, any radical or charge character on this group will be found almost exclusively at these three positions, as can be seen in resonance hybrids *I*. In comparison, an allyl group can delocalize any charge or radical character on the 1 and 3 positions, while cyclopentadienyl can delocalize over all 5 positions. In general, therefore, resonance stabilization for these species will fall in the order cyclopentadienyl > pentadienyl > allyl, although

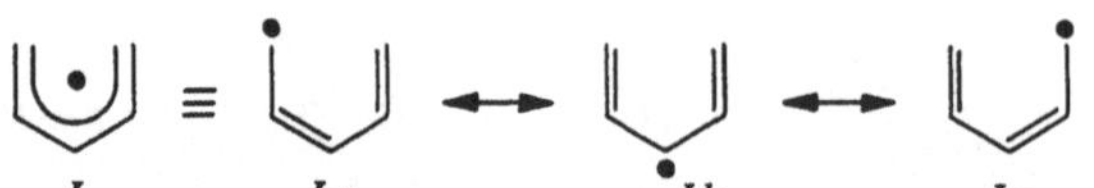

I *Ia* *Ib* *Ic*

the cyclopentadienyl cation is antiaromatic. As will be seen in Sect. B.II, the above MO diagrams are useful for comparing these three delocalized systems.

A number of pertinent physical parameters which relate to various valence states for these groups have been measured or estimated. In many cases, the actual values reported are heavily dependent on the definitions, assumptions, or techniques used, and therefore ranges of values often result. Nevertheless, these data are still of some use for purposes of comparison. Thus, allyl radical resonance or stabilization energies have been estimated as being anywhere from 9–25 kcal/mol[22–29], with a reasonable value probably being close to 13–14 kcal/mol[30]. For the pentadienyl radical, the resonance energy has been estimated as 15–24 kcal/mol, with a reasonable value probably being around 20 kcal/mol[27, 31–38].

From appropriate thermochemical data the standard heats of formation of the gas phase allyl[23, 24, 39–42], pentadienyl[37, 41, 43], and cyclopentadienyl[44, 45] radicals have been estimated as approximately 40, 48, and 62 kcal/mol, respectively. From these and other results the C–H bond dissociation energies of propene[23, 24, 29, 40–43, 46], pentadiene[37, 41, 43], and cyclopentadiene[44, 45] (to yield the allyl, pentadienyl, and cyclopentadienyl radicals), have been reported to be 88, 75–82, and 82 kcal/mol, respectively. Of course, the values for allyl and cyclopentadienyl are much better established than that for pentadienyl. Further, other complications arise for pentadienyl in that one could deal with either 1,3-pentadiene or 1,4-pentadiene. In any case, it should be clear that any real difference in these values could be important in the relative ease of formation or transformation of allyl, pentadienyl, and cyclopentadienyl ligands on a metal coordination sphere.

Other data have been obtained which reflect on transformations involving charged ligands. Thus, the electron affinities for allyl[28, 47], pentadienyl[48], and cyclopentadienyl[49, 50] are respectively 0.55, 0.90, and 1.79 eV. The ionization potentials are respectively 8.1, 7.8, and 8.7 eV[39, 51–55]. Of course, the electron affinity and ionization potential of C_5H_5 must be expected to be greater than those of C_3H_5 or C_5H_7 since in the latter two species the orbital involved is formally nonbonding, while it is bonding in the C_5H_5 case. It is particularly noteworthy that both the ionization potential and the electron affinity of pentadienyl are lower than the corresponding values for cyclopentadienyl. This suggests that C_5H_5 should be more electronegative than its open counterpart (C_5H_7), all other things being equal. In fact, relative gas phase acidities of propene, 1,4-pentadiene, and cyclopentadiene have been determined by competitive measurements[56–60]. Compared to cyclopentadiene, 1,4-pentadiene is less acidic by ca. 6.9 kcal/mol. The value for propene (relative to C_5H_6) is even larger at 34.3 kcal/mol. The approximate solution pK_a values for propene, 1,4-pentadiene, and cyclopentadiene are 40, 30, and >18, respectively[59, 61–65]

II. Prospects Regarding Metal Pentadienyl Complex Stability and Reactivity

Having a wide variety of pertinent data on hand such as the above, it is appropriate to attempt to anticipate the behavior of a pentadienyl ligand relative to that of the allyl and cyclopentadienyl ligands. First, the resonance hybrids of these three ligand systems can be compared. A η^3-allyl group can be considered to bond to a metal in two hybrid forms, *IIa* and *IIb*, each involving one sigma and one pi attachment. However, one can write

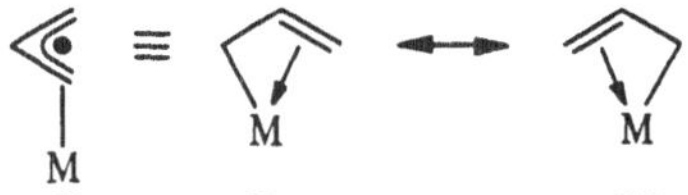

II *IIa* *IIb*

three resonance hybrids, *IIIa–IIIc*, to indicate the bonding of a η^5-pentadienyl ligand to a metal. It can readily be seen that a pentadienyl ligand forms an extra π-attachment to

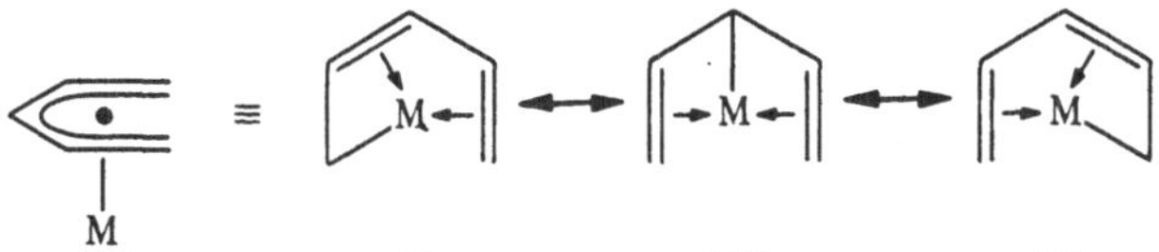

III *IIIa* *IIIb* *IIIc*

the metal as compared to the allyl ligand, and this allows an important expectation to be reached. Since the total metal-ligand interaction is greater for the η^5-pentadienyl ligand as compared to the η^3-allyl ligand, metal-pentadienyl compounds should possess greater thermal stability than the often unstable metal-allyl compounds, and this should greatly facilitate physical studies in this system. Other indications of stability can be gathered by comparing the π molecular orbitals of C_5H_7 with those of C_5H_5, each of which can be considered as 6π electron anions. Referring to Fig. 1, one can notice a very strong similarity of π molecular orbitals for the two 6π electron ligands. Thus, all of the bonding interactions which take place between C_5H_5 and a metal can also take place between C_5H_7 and a metal. Further, it can be seen that the energies of the three filled C_5H_7 π molecular orbitals are on the average substantially higher than those of the corresponding three filled C_5H_5 orbitals, and the energies of the two empty C_5H_7 π-molecular orbitals are on the average substantially lower than those of the corresponding two empty C_5H_5 orbitals. On an energetic basis, therefore, the C_5H_7 ligand could function as both a better donor and a better acceptor than the C_5H_5 ligand. Although other considerations such as overlap may also be important, at least under some conditions it might be expected that the pentadienyl ligands could actually be more strongly bound than even the cyclopentadienyl ligand.

Besides expectations of thermal stability, there are also reasons to believe that pentadienyl ligands should be capable of imparting chemical and catalytic reactivity into their metal complexes. An important reason for the high reactivity of metal allyl complexes has to do with the ability of the allyl ligand to bond in η^3 (*IVa*) as well as in η^1 (*IVb*) fashion. A pentadienyl ligand should be capable of bonding in η^5, η^3, and η^1 modes

IVa *IVb*

(*Va–c*, respectively), and therefore high reactivity should also be feasible. In fact, as a model for these important transformations one might compare the energetics of η^5-η^3 isomerizations for the C_5H_5 and C_5H_7 anions as well as the energetics of a η^3-η^1 isomerization of a C_3H_5 anion. In particular, the loss in π-delocalization energy might be one indicator of the barrier to such processes. Thus, in the conversion of the π-C_3H_5 anion

($\varepsilon = 4\alpha + 2.828\beta$) to a localized anion ($\varepsilon = 2\alpha$) with a free olefinic bond ($\varepsilon = 2\alpha + 2\beta$), a loss of 0.828β energy units occurs. Similarly, the conversion of a η^5-C_5H_7 anion ($\varepsilon = 6\alpha + 5.464\beta$) to a η^3-C_5H_7 anion ($\varepsilon = 4\alpha + 2.828\beta$) and a free olefinic bond results in a loss of 0.636β energy units – actually less than that for the allyl ligand. Thus, energeti-

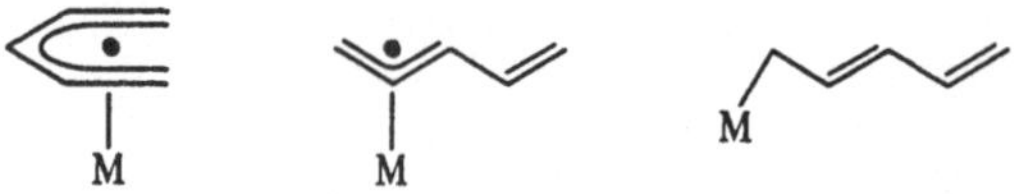

Va *Vb* *Vc*

cally the metal pentadienyl complexes could be even more reactive than metal allyl complexes, as long as such stepwise processes are involved. Of course, kinetic factors (e.g., steric) will also make some contribution. It is, however, interesting to note that the conversion of a η^5-C_5H_5 anion to an η^3-C_5H_5 anion requires a loss of 1.644β units of energy, which could easily account for the usual lack of such reactivity in these systems.

C. Experimental and Theoretical Studies of Pentadienyl Anions, Radicals, and Cations

In this section will be included the methods of preparations, selected reactions, spectroscopy, diffraction studies, and theoretical calculations on pentadienyl anions, radicals, and cations. Because the anionic species are more readily obtainable and have been greater utilized than the radical or cationic species, this treatment will naturally emphasize these anions. Of course, the anionic species generally have a metal counterion present, which can bring about some covalency due to metal-ligand interactions. Hence, it can be quite arbitrary as to which metal counterions constitute an ionic compound. This is particularly evident when one considers that even the replacement of lithium by potassium can bring about major changes in the nature of ionic interactions and pentadienyl conformation, and that the presence of coordinated Lewis bases (e.g., TMEDA) can greatly increase ionic character. Strictly for convenience, therefore, the alkali, alkaline earth, lanthanide and actinide metal ions will be treated here, along with zinc and aluminum. Transition metal complexes will be considered in Sect. D.

I. Preparative Methods

One of the most common routes to pentadienyl anions involves proton abstraction from a 1,4-diene. Lithium alkyls are useful reagents for these deprotonations, and a wide variety of alkylated, arylated, and siloxylated pentadienyl anions have been prepared in this fashion[66–71]. Alternatively, (trimethylsilylmethyl)potassium[72] and various metal amides[73] may be employed. The above reagents have also proven effective in the syntheses of resonance stabilized pentadienyl anions from 1,3-dienes such as 1,3,5-triphenyl-1,3-pentadiene[74], 1-trimethylsilyl-1,3-pentadiene[75, 76], 1,5-bis(trimethylsilyl)-1,3-pentadiene[76], and 1,3,5-tris(trimethylsilyl)-1,3-pentadiene[76]. However, if non-resonance stabilized 1,3-dienes are to be deprotonated, more reactive metallating agents are required. One particularly effective medium is a 1:1 mixture of butyllithium/potassium *t*-butoxide[72]. Also useful, especially for the preparation of crystalline anions, are alkali metal sands in the presence of triethylamine[77, 78], which inhibits anionic polymerization reactions[79]. However, in this method half of the diene is converted to a dimer.

In some cases, pentadienyl anions may result from transformations of other delocalized systems. Thus, deprotonation of various heptatrienes initially yields detectable heptatrienyl anion intermediates, which rapidly cyclize to cycloheptadienyl anions[80, 81]. Similarly, 3-vinylheptadienyl anions may also be readily prepared, but again cyclization results (following two sigmatropic proton shifts), leading to the 2-ethylcycloheptadienyl anion[82]. In some cases pentadienyl anions themselves may undergo cyclization or sigmatropic proton shift transformations, and these are discussed in Sect. C.II. Interestingly, 1,3,6,8-nonatetraene deprotonates readily to the blue-black nonatetraenyl anion (cf., allyl, colorless; pentadienyl, yellow; heptatrienyl, red), which is stable to cyclization[83].

For purposes of comparison it should be noted that very powerful metallating agents (e.g., butyllithium/TMEDA or butyllithium/potassium t-butoxide) are capable of abstracting two or even three protons from various alkylated dienes. Thus, various hexadienes and methyl pentadienes have been converted to dianionic species, while 2,4-dimethyl-1,4-pentadiene, 1,4-heptadiene, and 1,4-cycloheptadiene have all been converted to trianions[84–86].

II. Unimolecular Transformations

While it is beyond the scope of the present review to cover the very large body of reaction chemistry dealing with pentadienyl anions or metal-pentadienyl complexes, various spontaneous rearrangement processes may take place for isolated pentadienyl species, and these processes greatly affect the ability to isolate or utilize the anions. Hence these rearrangements will be covered here. One such rearrangement involves intramolecular 1,6-sigmatropic hydrogen atom shifts, which are possible whenever alkyl substituents are present on the terminal (1,5) carbon atoms[68, 87, 88]. The stereochemistry of these arrangements seems to follow Woodward-Hoffmann rules, being thermally antarafacial but photochemically suprafacial, and the rates of reaction correlate well with the total number of alkyl groups present on the 1 and 6 positions. In many cases the driving force appears to be the higher stability of an isomer with fewer alkyl groups in the 1, 3, and 5 positions. Thus, the 1,1-dimethylpentadienyl anion (1,1-C_7H_{11}) was observed to convert at 40° to a 48 : 52 mixture of 1,1-C_7H_{11} and 1,4-C_7H_{11}. That these products were present in an equilibrium ratio was demonstrated by obtaining the same product ratio from the 1,4-C_7H_{11} anion. Similarly, the 1,1,5,5-tetramethylpentadienyl anion is converted at 90° to a 10:90 mixture of 1,1,5,5-C_9H_{15} and the 1-isopropyl-4-methylpentadienyl anion. A related but more complex example is the following, whose equilibrium distribution was approached from all three constituents, *VI a–c:*

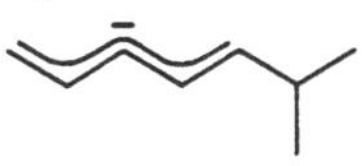

VI a, 10%

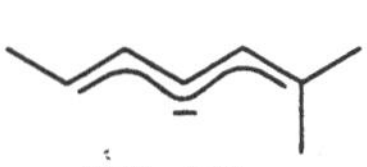

VI b, 0%

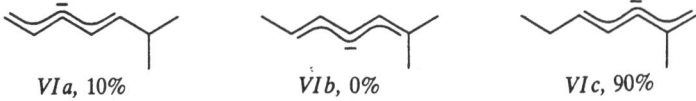

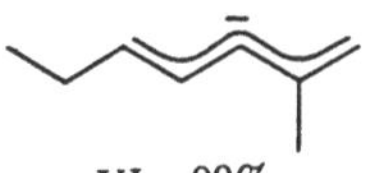

VI c, 90%

It should be noted that while both the 90% and the 10% components have only a single alkyl substituent at a 1, 3, or 5 position, the latter component possesses greater steric strain due to its trisubstituted sp^3 carbon atom. In another example, the rearrangement of linear *VII* to cyclic *VIII* was observed:

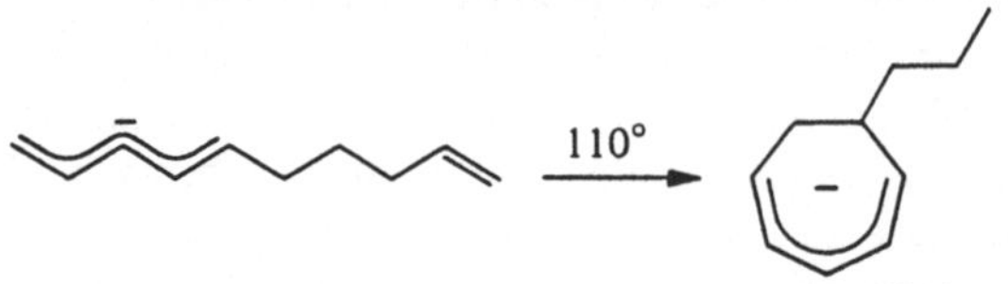

VII, 0% *VIII*, 100%

Apparently, the pentadienyl anion portion moved along the carbon atom backbone until a heptatrienyl anion was formed, which then underwent the expected cyclization (see Sect. C.I) to a heptadienyl anion, which is of greater stability due to its extra carbon-carbon single bond. Finally, while the 3-methylcycloheptadienyl anion was found to be thermally stable, it could be completely converted to the 2-methylcycloheptadienyl anion photochemically.

In a number of cases, pentadienyl species are subject to cyclization reactions[89–92] (cf., heptatrienyl, Sect. C.I). Thus, cyclooctadienyllithium undergoes a first-order conversion to a bicyclo(3.3.0)octenyl anion with a half-life of 80 min at 35°[93]. The 1,3,5-triphenyl-2,4-diazapentadienyl anion (1,3,5-$(C_6H_5)_3$-2,4-$C_3H_2N_2^-$) also undergoes very facile, stereospecific cyclization at a rate competitive with its formation even at − 78 °C[94]. The resulting 1,3-diazaallyl anion has negative charge localized on the nitrogen atoms, which was not the situation for the initial anion. The related 1,3,5-triphenyl-2-azapentadienyl anion is also subject to such cyclizations, but they are nonstereospecific and also much slower, which might be expected since the resulting azaallyl anion has only one nitrogen atom in a position to absorb the negative charge[95]. The 1,5-diphenylpentadienyl anion, when generated in situ at 225°, also undergoes an analogous reaction[96], as do various cyclic and heteroatom analogs[97,98]. Rather unusually, a 6-vinylcycloheptadienyl anion has been found to undergo a ring-opening reaction, leading to the previously mentioned nonatetraenyl anion[83].

Pentadienyl cations, generated by protonation of trienes or dienols, have generally been found to undergo much more facile cyclizations than do the pentadienyl anions[99–106]. It has been noted that the presence of a 3-methyl group tends to bring about the fastest cyclizations. Thus, in the case of the 1,1,5,5,-tetramethylpentadienyl cation the cyclization is rather slow, having a half-life of ca. 3 days at 25° and yielding the 1,2,3,4-tetramethylcyclopentenyl cation[99]. The 1,1,5-trimethylpentadienyl cation is reasonably stable at − 30°, but at − 10° is transformed to the 1,5,5-trimethylcyclopentenyl cation, and thereafter to the 1,2,3-trimethyl isomer[107]. The 1,1,3,5,5-pentamethyl and 1,5-dimethyl substituted cations could not be detected under normal conditions in concentrated H_2SO_4 (ca. − 30°)[101, 102], but were readily preparable and observable at ca. − 60° in HSO_3F/SbF_5[103]. On warming each initially formed a metastable cyclized intermediate (1,3,4,4,5-pentamethylcyclopentenyl and 1,5-dimethylcyclopentenyl cations, respectively), which then rearranged to the final products, the 1,2,3,4,4-pentamethylcyclopentenyl and 1,2-dimethylcyclopentenyl cations, respectively. However, the products were in some cases observed to be dependent on the choice of reaction medium. Initial attempts to prepare the 1,3,5-trimethylpentadienyl and 1,1,3,5-tetramethylpentadienyl cations led to the 1,3,4-trimethylcyclopentenyl and 1,3,4,4-tetramethylcyclopentenyl cations[100], although in a later study the 1,1,3,5-tetramethylpentadienyl cation, and its

(1,1,2,5), (1,1,4,5), (1,1,2,5,5), and (1,1,2,3,4,5,5) methylated analogs, were observed during a kinetic study of their cyclization reactions. Of these last four systems, only the first and fourth yielded single initial and final products. Thus, 1,1,2,5-tetramethylpentadienyl cation initially formed the 1,2,5,5-tetramethylcyclopentenyl cation, which then rearranged to its 1,2,3,4 isomer, and the heptamethylpentadienyl cation only led to the heptamethylcyclopentenyl cation. An interesting situation was observed during the protonation of various isomers (cis,cis-, trans,trans-, cis,trans-) of 3,4,5-trimethyl-2,5-heptadien-4-ols, wherein the cis,trans isomer led primarily to the cis-1,2,3,4,5-pentamethylcyclopentenyl cation, while the other two isomers led primarily to the trans-1,2,3,4,5-pentamethylcyclopentenyl cation[104)].

It should also be noted that even when coordinated to transition metals, formal pentadienyl cations can undergo ring closure reactions[108)], as can various heteroatom analogs[109)]. Finally, transformations involving pentadienyl and cyclopentenyl radicals have also been observed[110, 111)].

III. NMR and Conformational Studies

Both ^{1}H and ^{13}C nmr spectroscopy have provided important information regarding charge densities, molecular rearrangements, and molecular conformations of various pentadienyl and related species. Charge localization in particular is well illustrated by ^{13}C nmr data (see Table 1). Thus, for the cyclohexadienyl cation, $C_6H_7^+$, the formally charged 1, 3, 5 carbon atom positions are all characterized by higher (i.e., downfield) ^{13}C chemical shift values than the uncharged 2 and 4 positions[112)]. Conversely, in both the lithium and potassium salts of the pentadienyl anion, the ^{13}C chemical shifts for the 2 and 4 positions are observed to be much higher than those for the charged 1, 3, and 5 positions[113, 114)]. Thus, the data for these charged species are in line with the expectations based on the resonance hybrids of a pentadienyl group (vide supra). For the three iron complexes included in Table 1[115–117)], the ^{13}C chemical shift values for the 2 and 4 positions tend to be greater than those for the 1, 3, and 5 positions.

NMR spectroscopy has also proven useful in studying rates of C–C bond rotation in allyl and pentadienyl species. The allyl systems are simpler to analyze since only one allyl conformation is possible (ignoring substituent orientations), whereas the pentadienyl

Table 1. ^{13}C NMR parameters for representative pentadienyl species

	C(1,5)	C(2,4)	C(3)	$J_{^{13}C\text{-}H}$ [a]
$C_6H_7^+$	186.6	136.9	178.1	163
$(C_6H_7)Fe(CO)_3^+$	65.4	103.2	89.9	180
$(C_5H_7)Fe(CO)_3^+$	65.4	104.6	98.6	172
$(C_5H_7)_2Fe$	49.8	88.8	90.9	161
C_5H_7Li	66.0	144.0	87.3	142
C_5H_7K	78.7	137.5	79.6	149

[a] Approximate average value

groups can exist in the "W", "U", or "S" conformations. Indeed, allyl cations have been studied in some detail, both experimentally and theoretically[118]. The energies of activation for C–C bond rotation seem to vary from ca. 11.7 kcal/mol for the 1,1,2-trimethylallyl cation to ca. 23.6 kcal/mol for the cis,trans-1,2,3-trimethylallyl cation[119–124]. While electronic factors, particularly stabilization of charge by methyl groups, are generally more important than steric factors, the latter can be important in affecting the relative stabilities and rotational barriers of geometric isomers. In this connection one should note that the barrier for the cis,cis-1,2,3-trimethylallyl cation is 18.1 kcal/mol (cf., cis, trans above). Similarly, rotational barriers for allyl anions have been obtained, and naturally these are quite dependent on the particular cation present[122–124]. For allyllithium, the barrier is only 10.7 ± 0.2 kcal/mol, whereas the potassium and cesium analogs are characterized by values of 16.7 ± 0.2 and 18.0 ± 0.3 kcal/mol. The latter value has been suggested to represent a lower limit for the barrier to C–C bond rotation in an isolated allyl anion.

The situation regarding C–C bond rotation(s) in pentadienyl species is naturally more complex, especially since rotation about an internal C–C bond (i.e., C2–C3 or C3–C4) will lead to a change in conformation ("W", "U", or "S"), whereas external bond rotation (C1–C2 or C4–C5) is more analogous to the case for the allyl groups. Unfortunately, data related to the pentadienyl group is much more limited than that for various allyl groups. In part, this may be due to the fact that pentadienyl cations may undergo facile cyclizations (see Sect. C.VI). However, the anionic species are generally stable (although rearrangements are known[93]), and some data is available concerning them. Early work on pentadienyllithium indicated that rotation about the inner pentadienyl carbon-carbon bonds was much more facile than that around the outer ones ($\Delta G^{\neq}$ ca. 13 vs. 15 kcal/mol)[125–127]. This can essentially be explained by contributions from resonance hybrids such as *IXa* and *IXb*, in which the inner carbon-carbon bonds clearly should be

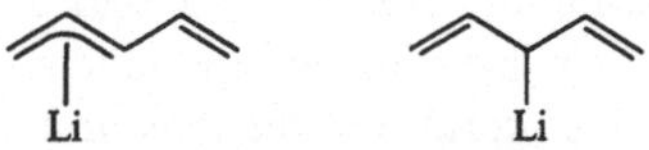

IXa *IXb*

more fluxional. Of course, metal ion participation is again going to be significant, and similar to the allyl anion systems, the rotation barriers of at least the inner bonds are expected to increase from lithium to cesium. In fact, a high barrier has been reported for pentadienylpotassium in solution ("U" conformation)[128], and trans-cis isomerizations for the potassium salts of the 1-methylpentadienyl, 1-methylheptatrienyl, and 1-methyl-1,3-diazapentadienyl anions require energy inputs of 16, 21, and 20 kcal/mol, respectively[126, 127]. Even for a free pentadienyl anion, however, the inner bond rotation should be more free due to the contribution of resonance hybrid *X* with charge localized at the central carbon atom position. Finally, it can be noted that for the highly perturbed 1,3,5-

X

triphenylpentadienyl anion, variable temperature ^{1}H and ^{13}C nmr studies demonstrate an energy of activation for internal carbon-carbon bond rotation of ca. 12–13 kcal/mol[74].

As noted earlier, ^{1}H and ^{13}C nmr studies are also indispensible for determining pentadienyl anion conformations in solution. Generally, these studies involve coordinating solvents such as THF-d_8, and therefore solvated counterions will usually form contact ion pairs or solvent-separated ion pairs with the pentadienyl anions. Most of the conclusions reached from nmr studies rely on the values of various proton-proton coupling constants observed in the ^{1}H nmr spectra. Many of the proton-proton coupling constants appropriate to various pentadienyl anions were first estimated in a nice modelling study involving in part fixed anions *XI* to *XIII*[125]:

XI

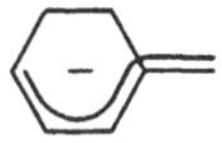
XII

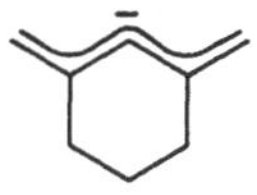
XIII

To illustrate one application, one can observe that *XIV*, the "S" conformation below, is characterized by six different J values – four across the two external C–C bonds (two cis

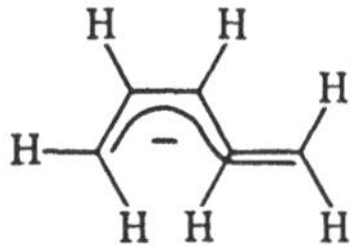
XIV

couplings of ca. 7.5–9.5 Hz and two trans couplings of ca. 15–16 Hz) and two across the two internal C–C bonds (one cis coupling of ca. 6.5–7.5 Hz and one trans coupling of ca. 10–12 Hz)[125, 129, 130]. Note that the proton-proton coupling pattern involving the external C–C bonds will not differ between the conformations, although only in the "U" or "W" conformations can the two external C–C bonds be equivalent. The distinguishing characteristics therefore involve especially the internal C–C bonds, and the "W" conformation will possess only trans proton-proton couplings across these bonds, while the "U" conformation will contain only cis couplings, and the "S" conformation will have both. Naturally, however, the presence of substituent groups in the 2, 3, or 4 positions will mean that other methods will be required to determine the particular conformation. In such cases, chemical trapping experiments (alkylation, silylation, etc.) may be indispensible as indeed is the case for many of the systems described herein. Also, however, variable temperature nmr studies may be quite useful, particularly for symmetrically substituted pentadienyl species. Should these species possess a "S" ground state, at a low enough temperature one end of the pentadienyl moiety should become nonequivalent to the other end.

In various pentadienyllithium compounds, it has been found that the "W" conformation often dominates, although in a few cases the "S" conformer is either the most stable form or has been postulated to be present as a minor component in equilibrium. The "U" conformation is quite uncommon for lithium (excluding cyclic species forced to adopt that conformation), presumably since the small lithium atom can not efficiently bond to all three charged carbon atoms, which would generally be necessary to justify bringing the charged terminal carbon atoms into close proximity of one another. Thus, pentadienyllithium itself has been well established as existing in the "W" form[72, 78, 125, 128], although a minor ($< 10\%$) fraction of a second conformer may also be present[129]. Both 3-methylpentadienyllithium and 1-methylpentadienyllithium exist predominately (ca. 80% or more) in a "W" form, although again small proportions of "S" conformations appear to be present. For the 1-methyl anion, the "W" conformation can exist as either cis or trans forms, and the cis seems to dominate[129]. For 2-methylpentadienyllithium, two nearly equally populated conformers ("W" and one "S") have also been proposed to exist[125], while 1,1-dimethylpentadienyllithium is composed of a major "W" isomer (ca. 80%) and a minor "S" isomer[129]. Interestingly, 2,4-dimethylpentadienyllithium adopts a "U" conformation, apparently since the methyl groups sterically destabilize the "W" and "S" conformations sufficiently so that the "U" form is adopted. Of course, that does not necessitate a η^5 bonding pattern with lithium[72].

Reasonably similar conclusions have also been reached regarding phenyl or trimethylsilyl substituted pentadienyl anions. Thus, 1,3,5-triphenylpentadienyllithium, *XV*, has been observed to exist in two conformations at room temperature, the more abundant of which demonstrated collapse of its ^{1}H nmr resonances on cooling, and therefore was assigned to be a mixture of two comparably stable "S" forms. The other conformer was formulated as "W". The S : W ratio was determined to be ca. 2.1 : 1, and both S ⇄ S′ and S ⇄ W interconversions were observed, the latter requiring more energy[74].

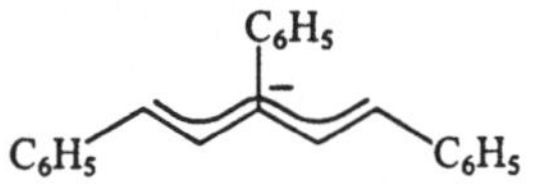

W-XV

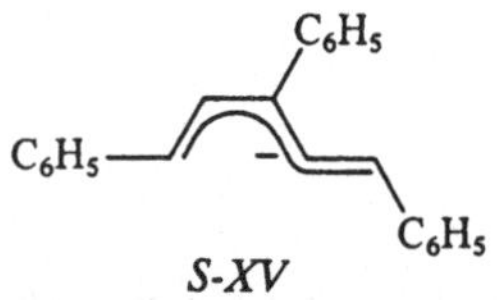

S-XV

1,5-diphenylpentadienyllithium has been similarly examined and a "W" conformation was favored by the authors, although they could not eliminate the possibility of the existence of rapidly equilibrating "S" conformers. In view of the results on the 1,3,5-triphenyl substituted anion, this must be considered to be a real possibility. The phenyl groups were observed to adopt trans configurations with respect to the pentadienyl skeletons[69]. More recently a series of trimethylsilylated pentadienyl anions has been investigated. Both 1-trimethylsilylpentadienyllithium and 1,5-bis(trimethylsilyl)pentadienyllithium exist in "W" conformations, with the trimethylsilyl groups positioned trans to the pentadienyl skeletons. Particularly interesting, however, is the observation that 1,3,5-tris(trimethylsilyl)pentadienyllithium exists entirely in a (presumably η^3) "S" conformation[76]. A lithium dienolate ($LiOC_4H_5$) also exists in the "S" form[131].

With the larger alkali metals, one naturally observes very different behavior compared to lithium. Pentadienylpotassium itself adopts the "W" conformation in liquid ammonia[73], as is the case with the cis- and trans-1-methylpentadienylpotassium compounds[132, 133]. Trans-1-methylpentadienylpotassium actually is less stable than the cis isomer, to which it rapidly isomerizes even at low temperatures ($\tau_{1/2} = 6 \pm 2$ min at $-50°$). Virtually identical results have been found for the 1-methyl-2,3-diazapentadienyl and 1-methylheptatrienyl anions[134, 135]. While the above species behave similarly in NH_3 to the corresponding lithium compounds, much different results are found in THF. Thus, the potassium, rubidium, and cesium salts of the pentadienyl anion and the potassium salts of the 2-methylpentadienyl and 2,4-dimethylpentadienyl anions all adopt the "U" conformation in THF[72, 78, 128, 136], although chemical trapping studies indicate that in the solid phase pentadienylpotassium exists in the "W" conformation, presumably due to the polymeric association of ions[136]. In contrast to the above results, 1-methylpentadienylpotassium appears to exist as a mixture of two isomers, suggested to be trans-"U" and trans-"S" in a 4 : 1 ratio, while 1-trimethylsilylpentadienylpotassium has been reported to exist as a 70 : 30 mixture of trans-"W" and trans-"S" conformers[76, 128]. However, both

3-methylpentadienylpotassium and 1,5-bis(trimethylsilyl)pentadienylpotassium exist in the "W" conformation, the latter having its substituents oriented trans with respect to the pentadienyl skeleton. 3-methylpentadienylpotassium is also known to be "W" in the solid state[128]. In general, it should be clear that there is no assuring that solution and solid state conformations will be identical. Interestingly, a potassium dienolate (KOC_4H_5) is known in both the "W" and "S" conformations, which do *not* readily interconvert[73, 137]. Some sulfur analogs have also been reported[137].

A series of more covalent, fluxional bis(pentadienyl)magnesium compounds has been prepared which are crystalline if isolated as TMEDA adducts[138]. When the pentadienyl ligands are acyclic (C_5H_7, 3-C_6H_9, 4-C_6H_9, 5-C_6H_9, and 2,4-C_7H_{11}), the principal species seems to be terminally σ-bonded, with the ligand generally in a "W" conformation. However, the 2,4-dimethylpentadienyl complex is characterized by a "U" shaped ligand, in order to avoid repulsive nonbonded contacts involving the large methyl groups. Even down to $-100°$, however, the ^1H nmr spectra evidence facile fluxional rearrangements through 1,3-shifts. Only the addition of crown ethers allowed a limiting σ structure to be observed. An x-ray structure of one "U" complex has been reported (see Sect. C.V). In notable contrast to the acyclic pentadienyl ligands, the analogous cycloheptadienyl- and cyclooctadienyl-magnesium compounds involve centrally σ-bonded ligands. As might be expected, the ^{13}C nmr spectra of these cyclic compounds are characterized by C1 and C5 resonances substantially downfield of the C3 resonance, while exactly the opposite behavior was observed for the acyclic compounds. Also reported in this study were $KAl(C_5H_7)_4$ and $KAl(C_5H_7)_2(CH_3)_2$, both of which also possessed terminally σ-bonded structures by nmr, and $Mg(C_5H_7)Br(L)_2$ (L = THF; L_2 = TMEDA).

Other related complexes of beryllium and zinc have been similarly studied[139]. $M(C_5H_7)_2 \cdot 2$ THF (M = Be, Mg, Zn) and $Zn(C_5H_7)Cl \cdot 2$ THF were all observed to be fluxional, i.e., one end of the ligand appeared equivalent to the other via facile 1,3 shifts. However, TMEDA adducts of various bis(pentadienyl)beryllium compounds exhibited nmr spectra in which such motion was frozen out, and the generally observed "W" conformations were terminally σ-bonded. Thus, $Be(C_5H_7)_2$(TMEDA) prepared from KC_5H_7 existed completely in the "W" form, although if a magnesium reagent were used in place of KC_5H_7, it was reported that a "S"/"W" ratio of ca. 8:1 resulted. Similarly, Be(3-$C_6H_9)_2$(TMEDA) also was found to exist solely in the terminally σ-bonded "W" conformation, while the compound prepared from 2-methylpentadienylpotassium is similar, with the methyl group located in the 4 position (away from beryllium) 95% of the time. Essentially the same observation was made for the hexadienyl compound, except that 80% of the methyl groups had a trans orientation with respect to the pentadienyl backbone, while the rest was cis. Curiously, Be(2,4-$C_7H_{11})_2$(TMEDA) appeared to adopt "S" and "W" conformations in a 7:5 ratio. The reason that "U" is avoided while "W" (with apparently strong CH_3–CH_3 repulsions) is adopted is not clear. Unlike the magnesium analogs, the TMEDA adducts of bis(cycloheptadienyl)beryllium and bis(cyclooctadienyl)beryllium are terminally σ-bonded.

Various thermally less stable $Zn(C_5H_7)_2L_2$ complexes (L = $(CH_3)_3NO$, THF) were also quite fluxional to well below room temperature, but again incorporation of a chelating ligand such as TMEDA led to freezing out of the terminally σ-bonded spectra. Even so, however, at ca. 90° some of the spectra demonstrated fluxionality. While $Zn(C_5H_7)$(Cl)(TMEDA) exists entirely in the "W" conformation, $Zn(C_5H_7)_2$(TMEDA) and Zn(3-$C_6H_9)_2$(TMEDA) each contained substantial (20%) quantities of the "S" conformer in addition to the more abundant "W" forms. The compound prepared from 2-methylpentadienylpotassium possesses ligands in nearly a 1 : 1 mixture of 4-methyl "W" and 2-methyl "W" isomers. The hexadienyl analog also existed primarily (90%) in a "W" conformation, with the trans-methyl orientation being slightly favored over the cis (50% : 40%).

IV. EPR Spectroscopy

Whereas NMR spectroscopy is generally applied to the diamagnetic allyl, pentadienyl, or cyclopentadienyl cations or anions, this technique is generally not applicable to the analogous radical species, although CIDNP (chemically induced dynamic nuclear polarization) has been observed for allyl[140)] and pentadienyl radical species[139)]. More useful for the radicals, naturally, is EPR spectroscopy.

EPR spectra of allyl radicals have been intensively studied, and naturally reflect the localization of radical character at the terminal carbon atoms[30, 141–153)]. An important parameter derived from line shape and kinetic analyses is the barrier to C–C bond rotation in these radicals. For the allyl radical itself this value appears to be 15.7 ± 1.0 kcal/mol[30)], and for the syn-methylallyl radical the value is 14.3 kcal/mol[145)]. For various polyfluorinated allyl radicals, the energy of activation drops to 4.5 (2-C_3F_4Cl), 6.1 (C_3F_5), and 7.2 (2-C_3F_4H) kcal/mol[149)], while some cyanoallyls have energies of activation of ca. 10 kcal/mol[147, 152)].

Pentadienyl radicals have also been intensively studied[36, 38, 153–155)]. When generated at low temperatures (below ca. − 70°), two conformations are observed, one being the "W", and the other the less symmetric "S". Radical character was naturally concentrated at the 1, 3, and 5 positions. Generation at warmer temperatures only allows observation of the "W" conformer[36, 38)], due to a facile conversion of the "S" form to the more stable "W" form. Attempts to selectively prepare the "S" form were unsuccessful[155)]. The difference in stabilities for these two forms was estimated as being ca. 1–2 kcal/mol[36, 38)], while the unobserved "U" conformer appears to be much higher in energy[36)]. Of course, the "U" form is in a sense accessible when cyclohexadienyl radicals are generated[156–159)]. A number of these cyclohexadienyl radical species have in fact been investigated, and one major observation made was that the carbon atom of the methylene group lies in the plane of the pentadienyl fragment, although little energy is apparently required for deformation and therefore the radicals vibrate significantly between bent structures[156, 158)]. Further, there is a relatively large hyperfine splitting induced by the methylene protons (ca. 48.0 G), compared to the values for the 1,5 (9.0 G), 2,4 (2.7 G), and 3 (13.1 G) positioned protons. These can be compared to a value of ca. 6.0 G for $\bullet C_5H_5$[160, 161)].

V. X-Ray Diffraction Studies

In all of the pentadienyl structures determined thus far, some outside perturbations are always present, i.e., there are no structural data thus far pertaining to free, yet electronically simple, pentadienyl anions. There are several polycyano-substituted anions known and despite the strong electronic perturbations brought about by the cyano groups, these species are still likely to be reasonably representative, in a conformational sense, of other

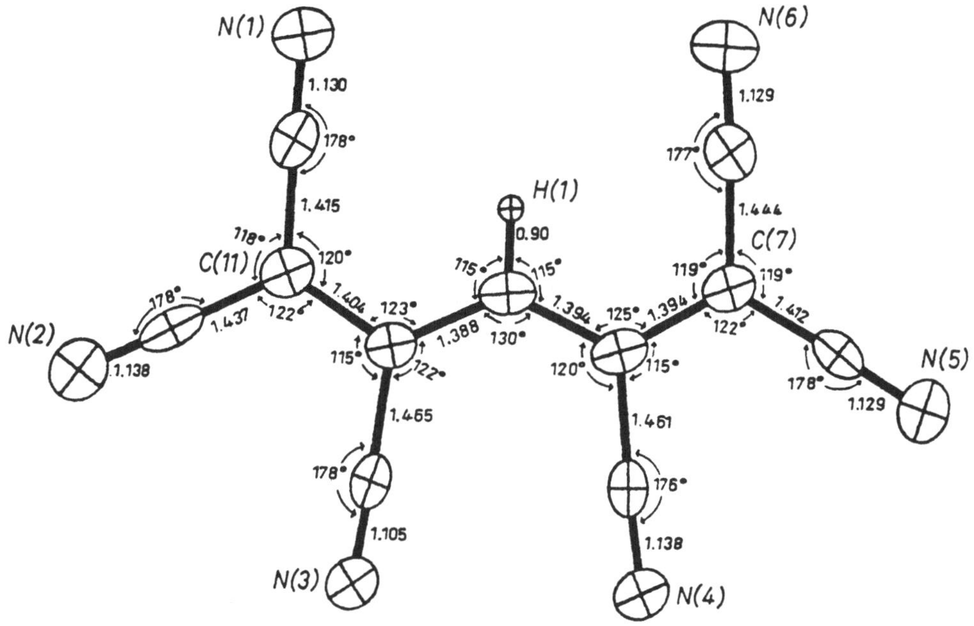

Fig. 3. The solid state structure of the 1,1,2,4,5,5-hexacyanopentadienyl anion (Ref. 162)

pentadienyl anions. For example, tetramethylammonium 1,1,2,4,5,5-hexacyanopentadienide exists in the solid state in the (crystallographically planar) "W" conformation (Fig. 3)[162]. The average pentadienyl C–C bond distance is 1.395(4) Å, while the average C–CN bond distance, being not much longer at 1.439(3) Å, is indicative of the great deal of delocalization present in this anion. Other bonding parameters may be noted from the Figure. It is quite evident from Fig. 3 that substantial repulsion is taking place between the cyano groups attached to the formally uncharged 2 and 4 positions (designated C(8) and C(10) by the authors). It is easy to visualize, then, how larger methyl groups in those positions can destabilize the "W" conformation. Also worthy of note is the fact that the presence of CN groups at the 2 and 4 positions leads to smaller C–C–C bond angles at those positions (123.8(4)°) compared to the 3 position (130.9(5)°).

An entirely analogous conformation has been found in the related 1,1,2,6,7,7-hexacyanoheptatrienide anion (Fig. 4)[163]. The molecule deviates significantly from planarity, the major distortion seeming to be a twist around the centeral carbon atom, which leaves each half of the molecule nearly planar. Here the average pentadienyl C–C bond distance of 1.387 Å may be compared with the value of 1.430 Å for the C–CN bond distances involving the 1 and 7 positions, and the value of 1.468 Å involving the 2 and 6 positions. Interestingly, there is again no evidence of the cyano groups exerting a contraction influence on the adjoining C–C–C bond angles.

A much more unusual structure results for the 3-isopropyl-1,1,2,4,5,5,-hexacyanopentadienide ion (Fig. 5), which is related to the molecule in Fig. 3 by the replacement of hydrogen at the central carbon atom by an isopropyl group[164]. The conformation is now substantially nonplanar as can be seen from the C(1)–C(2)–C(3)–C(4) and C(2)–C(3)–C(4)–C(5) torsion angles of 42.3 and 23.1 degrees, respectively. The confor-

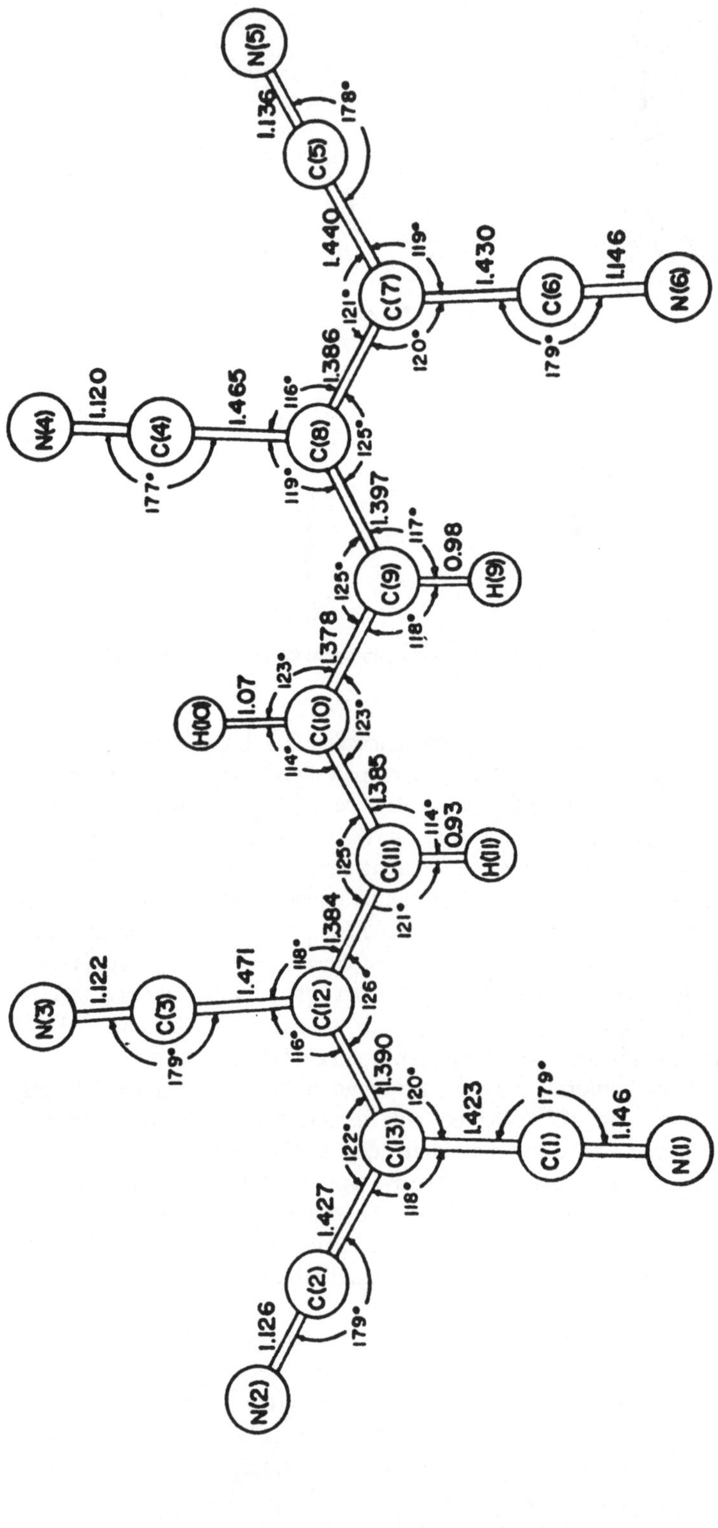

Fig. 4. The solid state structure of the 1,1,2,6,7,7-hexacyanoheptatrienyl anion (Ref. 163)

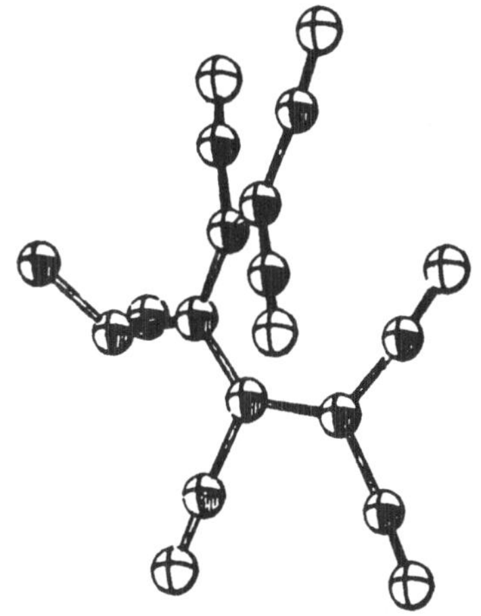

Fig. 5. A perspective view of the nonplanar 3-isopropyl-1,1,2,4,5,5-hexacyanopentadienyl anion. Reprinted with permission from Ref. 164. Copyright 1976 American Chemical Society

mation may be described as a deformed "U", which seems to be twisting toward the "S" conformation. It is perhaps not surprising that such a highly resonance-stabilized anion could be distorted so much from planarity, especially when such high steric demands are present. The average pentadienyl C–C bond distance is 1.41(1) Å, with an average C–CN bond distance of 1.48(1) Å. Interestingly, the C(1)–C(2)–C(3) and C(3)–C(4)–C(5) bond angles average 133.9°, while the C(2)–C(3)–C(4) bond angle is only 115.5°.

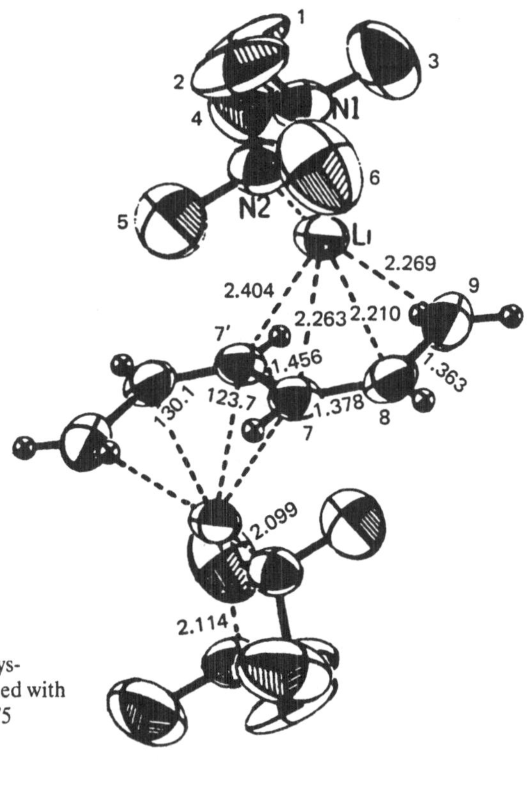

Fig. 6. A perspective view of $[Li(TMEDA)]_2[C_6H_8]$, which lies on a crystallographic center of symmetry. Reprinted with permission from Ref. 166. Copyright 1975 American Chemical Society

While the hexatriene dianion, $C_6H_8^{2-}$, is not strictly a pentadienyl anion, it is certainly a close relative and in fact results from the deprotonation of the hexadienyl anion. In agreement with theoretical arguments[165], the dianion was found to exist in a double-"S" conformation (Z,Z) which is presented in Fig. 6[166]. While the theoretical prediction was based on the expected presence of bonding interactions between the 1,4 and 3,6 positions, the observed geometry may also be due in part to the ability of a sickle-like conformation to provide η^4 coordination to the lithium ion. While the carbon atom portion is nearly planar, the hydrogen atoms in close proximity to the lithium ion (on C(7) and the endo proton on C(9)) tended to be repelled substantially out of the ligand plane. A long central C–C bond distance, 1.456(5) Å vs. 1.370(4) Å for the other two C–C bond types, was interpreted as indicating the importance of a bis(allylic) resonance hybrid, *XVI*:

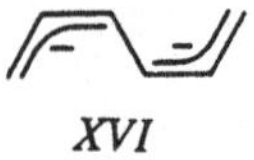

XVI

The Li–C(9),–C(8),–C(7) and –C(7′) bond distances are 2.269(7), 2.210(7), 2.263(7), and 2.404(7) Å, respectively.

The compound $Nd(2,4\text{-}C_7H_{11})_3$, prepared from $NdCl_3$ and three equivalents of $K(2,4\text{-}C_7H_{11})$, is also an example of an ionic pentadienyl species[167]. The structure of this complex (Fig. 7) contains three 2,4-dimethylpentadienyl anions in "U" conformations coordinated to the Nd(III) ion. The average delocalized C–C bond distances in the pentadienyl ligands essentially fall in two sets, an internal set involving the C2–C3 and C3–C4 bonds and an external set involving the C1–C2 and C4–C5 bonds. The internal

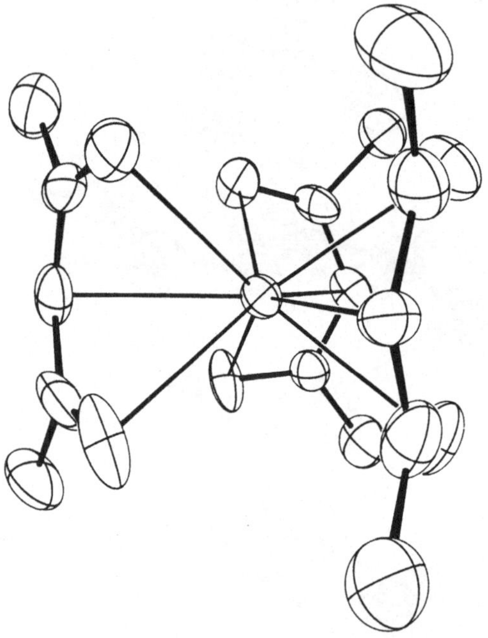

Fig. 7. The structure of tris(2,4-dimethylpentadienyl)neodymium (Ref. 167)

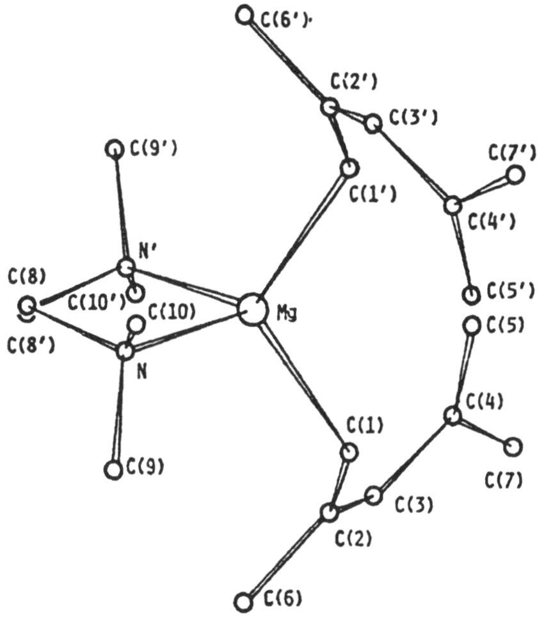

Fig. 8. The solid state structure of the TMEDA adduct of bis(2,4-dimethylpentadienyl)magnesium (Ref. 138)

set was observed to be slightly longer than the external set (1.421(12) vs. 1.373(12) Å), indicative of the contribution of a resonance hybrid having negative charge localized at the 3 position as in *X*. The ligands are somewhat distorted from planarity – the formally uncharged C 2 and C 4 atoms are located ca. 0.07 Å out of the respective planes formed by the formally charged C 1, C 3, and C 5 atoms, in a direction away from the Nd(III) ion. Hence, the Nd–C 2 and Nd–C 4 interactions appear to be repulsive. This is also illustrated by a comparison of the various relative Nd–C bond distances, which average 2.749(10) Å to the C 3 atoms, 2.801(9) Å to the C 1 and C 5 atoms, and 2.855(8) Å to the C 2 and C 4 atoms. The short distance to C 3 suggests that this position is particularly electron rich, as had been suggested by earlier MO studies[165]. An alternative description of the ligand is that it contains two overlapping allylic segments (C 1–C 2–C 3 and C 3–C 4–C 5), whose planes are ca. 9.5° apart. Interestingly, evidence for related pleating of allyl and pentadienyl groups bound to alkali metals has been obtained from nmr spectroscopic studies[113]. The analogous $U(2,4\text{-}C_7H_{11})_3$ compound has also been reported[168], but has not yet been structurally characterized.

Utilization of smaller Mg(II) and Zn(II) species naturally leads to more localized bonding. Thus, $Mg(2,4\text{-}C_7H_{11})_2$(TMEDA) (Fig. 8) contains two σ-bound pentadienyl ligands, each adopting "U" conformations which were slightly nonplanar[138]. The Mg–C(1) bond distance of 2.179(15) Å is reasonable for a localized Mg–C single bond. The localized nature of the Mg–C bonding naturally leads to localized C–C multiple bonding, as in *XVII*, and the C–C distances around the pentadienyl backbone (from Cl)

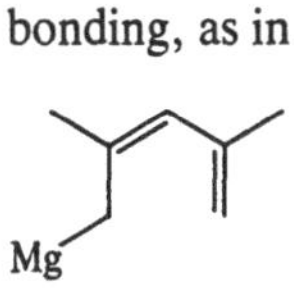

XVII

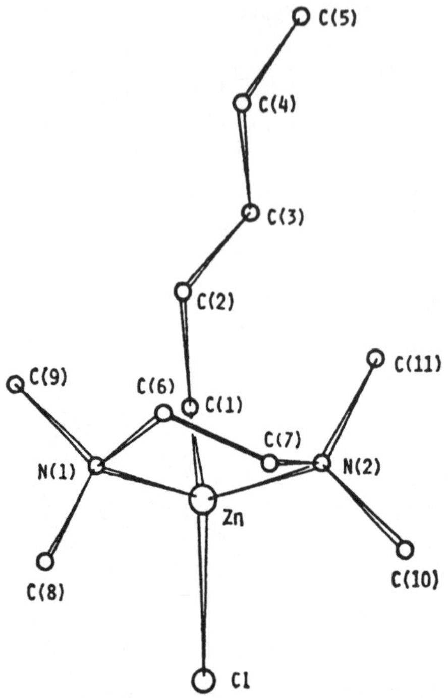

Fig. 9. The molecular structure of the TMEDA adduct of pentadienylzinc chloride (Ref. 139)

are 1.467(20), 1.402(19), 1.466(20), and 1.356(22) Å. A reasonably similar situation exists for $Zn(C_5H_7)Cl(TMEDA)$, except that the ligand now adopts the nearly planar "W" conformation (Fig. 9)[139]. The Zn–C bond distance of 2.031(12) Å is comparable to those of other four-coordinate alkylzinc complexes. As in the previous magnesium compound, the C–C bond distances in the pentadienyl ligand are observed to alternate (1.460(17), 1.293(15), 1.399(13), and 1.273(15) Å).

A number of related cyclohexadienyl species have also been reported. One such class of compounds, the Meisenheimer complexes[169, 170], involves highly delocalized molecules such as the 1,1-dimethoxy-2,4,6-trinitrocyclohexadienyl anion[171]. The entire six-membered ring systems in such complexes are essentially planar, and despite the highly delocalized nature of the molecule, there still appear to be two distinct sets of pentadienyl bond distances with the internal set being slightly longer than the external set (ca. 1.41 vs. 1.37 Å; cf., $Nd(2,4\text{-}C_7H_{11})_3$). The C–C(methylene) bond distances are naturally significantly longer at 1.51 Å. In addition, a structural determination has also been carried out on the heptamethylcyclohexadienyl cation[172]. In many respects the structure is quite similar to those of the Meisenheimer compounds. Thus, the six-membered ring portion is again essentially planar, and the internal delocalized C–C bond distances average 1.407(7) Å while the external delocalized C–C bond distances average 1.365(7) Å. The C–C(methylene) bond distances averaged 1.490(7) Å.

VI. Molecular Orbital Calculations

While a number of theoretical studies have been reported which attempt to model various spectroscopic features of pentadienyl species[173–185], of major interest here will be the general electronic nature of pentadienyl anions, and the electronic effects brought about by the presence of various substituents on the pentadienyl fragment. It must be kept in mind that in many instances the calculational results are greatly affected by the nature of counterions present, if any, and their solvation. In a number of cases, trends observed for free ions are greatly reduced when the solvated counterions are included.

The first conformational study of pentadienyl radicals and anions suggested that since the formally nonbonding pentadienyl π-molecular orbital contained terminal carbon atoms with the same phase (see Fig. 1), a net bonding interaction could result between them if the "U" conformation were adopted[165]. The "U" conformation was therefore considered the likely ground state, although it is not entirely clear that the two terminal carbon atoms would be in close enough proximity to bring about such interactions. In fact, one subsequent study concluded that the π orbital energies were almost independent of conformation, although the σ orbital energies were very much affected[186]. However, it is still possible that in-phase C1–C5 interactions may play a role in other chemical facets of these molecules.

One SCF study on pentadienyllithium concluded that the "W" conformation was more stable than "S" by 5 kcal/mol, and more stable than planar "U" by an incredible 80 kcal/mol, although a nonplanar "U" form was "only" 30 kcal/mol disfavored[187]. An unusual nonplanar η^3 form (through the 1, 4, and 5 positions) was found to be the absolute minimum by 12 kcal/mol. It appears that the calculations on these systems generally involved unrealistic σ (η^1) Li–C_5H_7 interactions, but even so it is surprising that such large differences would be found. Further, for the isolated anion, the "S" and "U" destabilizations compared to "W" are 5 and 160 kcal/mol, respectively, although again the value for a nonplanar "U" form drops to 30 kcal/mol.

A later study on pentadienyllithium made use of the CNDO II method to determine the most likely pentadienyl locations to which the lithium atom would bond for the "W", "S", and "U" conformations[186]. For comparative purposes, the free anions were also calculated but the conformational energies were essentially identical. Introduction of the lithium atom tended to reduce the net charges on the 1, 3, and 5 positions, but negative charge in all species tended to be slightly more localized at the central carbon atom. The η^5-"U" (planar) conformation of LiC_5H_7 was determined to be favored, in contrast to the previous study. The next lowest energy configuration was basically a η^4-"S", followed by nearly comparable η^3-"S" (utilizing the other, E, end of the sickle) and η^3-"W" configurations. An alternative η^3-"W" configuration in which the lithium atom is bound to the 2, 3, and 4 positions, was found to be the least stable of the local minima. It should be recalled that the η^4-"S" geometry has been observed in the dilithium salt of the hexatriene dianion.

A subsequent higher quality MNDO study on the pentadienyl anion and an extensive series of substituted (generally methylated) derivatives has also been reported[188]. The general results of this study seem to best model experimental data, as the "W" form of $C_5H_7^-$ was found to be the absolute minimum, with the "S" and "U" forms ca. 1.0 and 3.7 kcal/mol less stable, respectively. For 3-$C_6H_9^-$ the relative ordering was the same, but the energy differences were 0.6 and 2.7 kcal/mol, respectively, while for the pentadienyl cation the relative values (vs. "W") were 1.4 and 4.1 kcal/mol, respectively. As in the case of most other studies, the pentadienyl external C–C bonds were calculated to be stronger than the C–C internal bonds. For comparative purposes, calculations were also carried out on pentadienylberyllium hydrides. Here the most stable conformation appeared to be a nonplanar "U" form. One end of the pentadienyl ligand was σ (η^1) bound to the beryllium atom, while an olefinic bond at the other end also interacted with the metal (Fig. 10). This nonplanar "U" configuration is actually reasonably similar to that of the previously mentioned SCF study[187].

While one might consider that such a nonplanar structure could be adopted for similar metals such as lithium, zinc, or magnesium, in general coordinating solvents will be present which will bond even more effectively to the metal. It seems possible, however, that unsolvated species such as $M(C_5H_7)_2$ might be isolated under some conditions, and such configurations could be adopted, providing that the more usual η^3 or η^5 configurations are in fact less stable.

Theoretical studies have also provided significant insight into the influences brought about by the placement of various substituents onto the pentadienyl backbone. Naturally one of the most versatile substituents is the methyl group, which is generally considered an electron donor. However, there is substantial evidence that when attached to a very electron rich atom (e.g., an anionic carbon atom), the methyl group will instead behave as an electron acceptor[189–191]. To some extent, then, methylation at such a site might actually be expected to bring about stabilization of the anion, although this is not the generally accepted view. It is well known that OH^- in solution is a weaker base than OR^-, and therefore the alkyl group does not stabilize the negative charge as much as an attached hydrogen atom. However, in the gas phase the order is entirely reversed, and therefore it may well indeed be that alkyl groups do stabilize anions electronically more than protons do, although in the solution phase these same groups may hinder solvation processes which could be even more important for stabilization of the given anion[188, 192–194]. There is, of course, no doubt that methylation stabilizes both radical and cationic species[195].

Nonetheless, some useful information regarding methylation has been obtained from molecular orbital calculations. One CNDO/2 study on the allyl and pentadienyl anions indicated that in all cases the methyl group would serve to withdraw net electron density from the anions[196]. For the allyl anion, a 1-methyl substituent would withdraw a substantial 0.102 electrons, while at the formally uncharged 2 position the effect was consider-

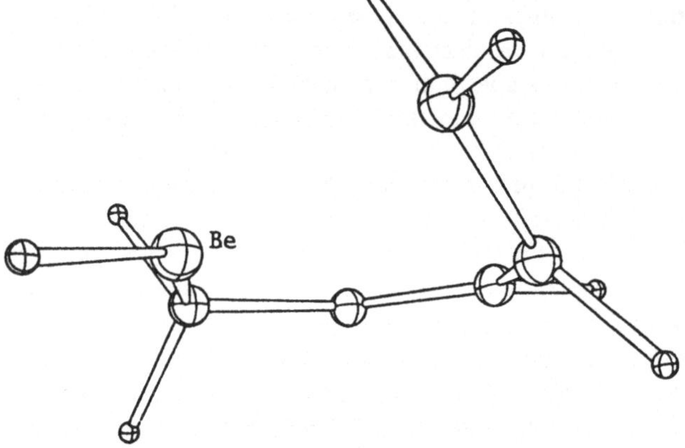

Fig. 10. Illustration of the most stable conformation of unsolvated pentadienylberyllium hydride according to a MNDO calculation (Ref. 188)

ably reduced at 0.030 electrons. For the pentadienyl anion, 1, 2, and 3 substitution led respectively to the withdrawal of 0.068, 0.028, and 0.089 electrons. Of course it must be again stated that the influence a substituent may have on an isolated anion might be altogether different from that produced when a counterion is present. Indeed, a follow-up study on bis(dimethylether)lithium pentadienide did suggest that many of the originally observed trends for the free anions were greatly reduced for a η^3 associated species[197]. Thus, with a methyl group in the terminal position away from the solvated lithium atom, only 0.036 electrons were withdrawn, while methylation at the terminal position near the lithium atom resulted in a similar withdrawal of 0.039 electrons for trans substitution, but a net *donation* of 0.009 electrons for cis substitution.

The previously mentioned MNDO study also addressed the question of the electronic effects of methylation[188]. In agreement with the two CNDO/2 studies[196, 197], methyl substituents did withdraw electron density from the pentadienyl anion, whether the methyl group was located at the 1 (0.035 e), 2 (0.020 e), or 3 (0.014 e) positions. In each case the methyl carbon atom actually bore net positive charge (0.150, 0.128, 0.179, respectively), while the hydrogen atoms bore net negative charge. In contrast, for the pentadienyl cation, the methyl groups served as electron donors (0.158, 0.121, and 0.122 e donated for the 1, 2, and 3 positions respectively).

It is in fact attractive to regard methyl groups as electron acceptors from pentadienyl anions given the fact that pentadienyl backbone C_a–C_b–C_c bond angles generally contract substantially when a methyl group is attached to C_b[164]. Calculations on allyl cations[198–200], radicals[198, 199], and anions[201] in fact predict backbone C–C–C bond angles of ca. 121, 127 and 132.5°, respectively. Thus, the methyl group seems to be serving to reduce the contribution of an anionic resonance hybrid to the preference of one which may be more radical-like or (less likely) even cation-like in character. It is perhaps best to conclude that the methyl group is quite polarizable, and may serve as either a donor or acceptor depending on its environment within the pentadienyl group[202–206].

However, another aspect of these methyl groups pertains to whether or not they serve to stabilize or destabilize a pentadienyl anion. Of course, it is clearly recognized that a stabilizing influence would ensue for pentadienyl cations and radicals. Results of the MNDO study indicate that methylation at an active (i.e., 1, 5, or especially 3) position of a pentadienyl anion is accompanied by a stabilizing effect of ca. 1 kcal/mol, while methylation at a 2 or 4 site is destabilizing, by ca. 2. kcal/mol[188]. For the pentadienyl cation, methylation of the 1, 2, or 3 positions always leads to stabilization, by ca. 6, 1, and 4.5 kcal/mol, respectively.

Two ab-initio studies have also appeared which deal to a large extent with a wide range of substituent effects on cyclohexadienyl radicals[207] and anions[208]. In the case of the radical it was found that both π-acceptor (e.g., CN, NO_2) and π-donor (e.g., F, OCH_3, NH_2) substituents brought about stabilization, particularly the former, when located on the 1, 3, or 5 positions. Location at the 2 or 4 positions brought about minimal (even destabilizing) effects. As representative examples of the stabilizations brought about by substitution at the 1, 2, or 3 positions, the following numbers are cited: CN (1.3, − 0.4, 3.0 kcal/mol), OCH_3 (2.0, − 0.6, 0.6), and CH_3 (1.0, 0.3, 0.7). For the cyclohexadienyl anions, π-donors generally brought about stabilization when located in the 2 position, but destabilization generally resulted for the 1 and 3 positions. On the other hand, π-acceptors brought about stabilization for all positions in the order $3 > 1 > 2$. A general order of stabilization energies, independent of position, was given as $NO_2 >$ CN

> CO_2H > F > OH > NH_2 > CO_2^-. The methyl group influence was more dependent on the position of substitution in line with its more polarizable nature. However, a similar order of anion stabilities was given as CO_2H > OCH_3 > H > R.

A series of calculations on pentadienyl-like anions and polyanions has also been carried out in an attempt to correlate anion stability (or ease of preparation) with the resonance energy per atom (REPA) in the various anions[86]. In general, fairly good correlations were observed and the highest REPA values were found when a net charge per carbon atom of ca. 0.17 was present. When the net charge per carbon atom reached ca. 0.35, the REPA value become zero, and rapidly became negative for higher charge densities.

D. Experimental and Theoretical Studies of Transition Metal-Pentadienyl Compounds

In many regards the most interesting aspects of pentadienyl groups must be the variety of interactions possible with the d orbitals present in transition metals. This is particularly evident given the rich transition metal chemistry involving the allyl and cyclopentadienyl ligands. To date x-ray structural studies have provided the greatest information regarding the nature of the transition metal-pentadienyl bond, but various complementary spectroscopic and theoretical contributions are now also yielding key insight into the bonding. As it is the primary purpose of this treatment to examine the nature of bonding in transition metal-pentadienyl compounds, a comprehensive review of all transition metal-pentadienyl compounds to have been synthesized is not appropriate here. Rather, a detailed coverage of pertinent physical studies will be presented.

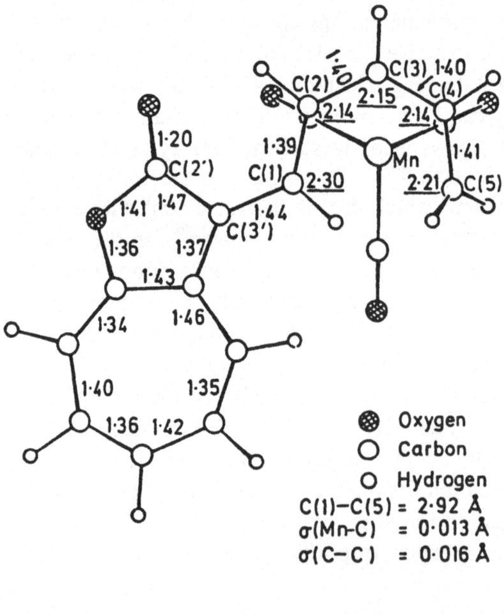

Fig. 11. The crystal structure of a substituted pentadienylmanganese tricarbonyl, $Mn(C_{14}H_{11}O_2)(CO)_3$ (Ref. 209)

I. X-Ray Diffraction Studies

1. Mono(pentadienyl)metal Complexes

One unusual, substituted pentadienyl metal compound has been reported to result from the interaction of tropone with dimanganese decacarbonyl[209]. The compound's formula is $(C_{14}H_{11}O_2)Mn(CO)_3$, and as can be seen from Fig. 11, it is a derivative of (pentadienyl)manganese tricarbonyl. Two molecules of tropone have combined to give the substituted pentadienyl ligand. The average C–C (pentadienyl) bond distance is 1.40(1) Å, while the Mn–C bond distances varied substantially from 2.14–2.30 Å, and therefore are longer than those in $(C_5H_5)Mn(CO)_3$ (2.13–2.18 Å). A noteworthy feature of this structure was the twisting of the terminal CH_2 group out of the pentadienyl ligand plane by 48°. The relative locations of the carbonyl ligands are virtually identical to those observed in related cyclic (e.g., cyclohexadienyl) carbonyl species[210–217]. More recently the parent (pentadienyl)manganese tricarbonyl compound and a rhenium analog have been reported, but have not yet been structurally examined[218–220].

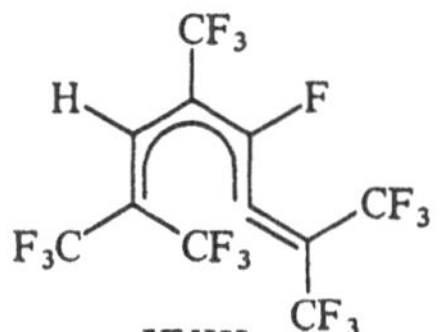

XVIII

In another ligand condensation reaction, it has been observed that tetrakis(trifluoromethyl)allene reacts with $(C_5H_5)Fe(CO)_2^-$ to yield a mixture of products, including $(C_5H_5)Fe(C_{11}HF_{16})$, which contains an unusual polyfluorinated pentadienylidene ligand *XVIII*[221]. The structure of this product is presented in Fig. 12. As can be seen, the

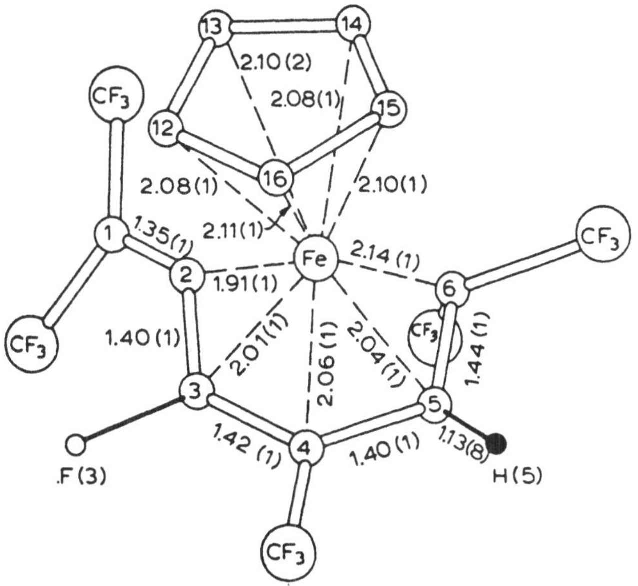

Fig. 12. The solid state structure of $Fe(C_5H_5)(C_{11}HF_{16})$ (Ref. 221)

Fe–C(pentadienyl) bonding is somewhat asymmetric. This data provides a good case for comparing the relative bonding between cyclopentadienyl and pentadienyl ligands since any ligand-ligand repulsion effects will be felt by both ligands. It is interesting, therefore, that the average Fe–C(pentadienyl) bond distance of 2.03 Å is noticeably shorter than the average Fe–C_5H_5 distance of 2.09 Å. Of course, it is not entirely clear how much of this shortening, if any, may be due to the presence of the electronegative fluorine-containing pentadienyl substituents. In fact, in other related complexes containing both the cyclopentadienyl ligand and a cyclohexadienyl-like ligand, the two relative bonding capabilities appear more comparable[222–224], so that it does appear that electron withdrawing substituents preferentially enhance the metal-pentadienyl bonding in this example.

In a third complex reaction, diphenylacetylene and acetonitrile have been observed to condense into the η^6-HC(C_6H_5)–(C(C_6H_5))$_3$C(NH)(CH_3) ligand, bonded to niobium (Fig. 13)[225]. Among a number of resonance hybrids can be included the simplified form *XIX*, which contains a pentadienyl-metal interaction. Of course, the ligand could also be considered to be isoelectronic with hexatriene or the hexatriene dianion, depending on the metal oxidation state (cf., Fig. 6). A great deal of buckling is present in the con-

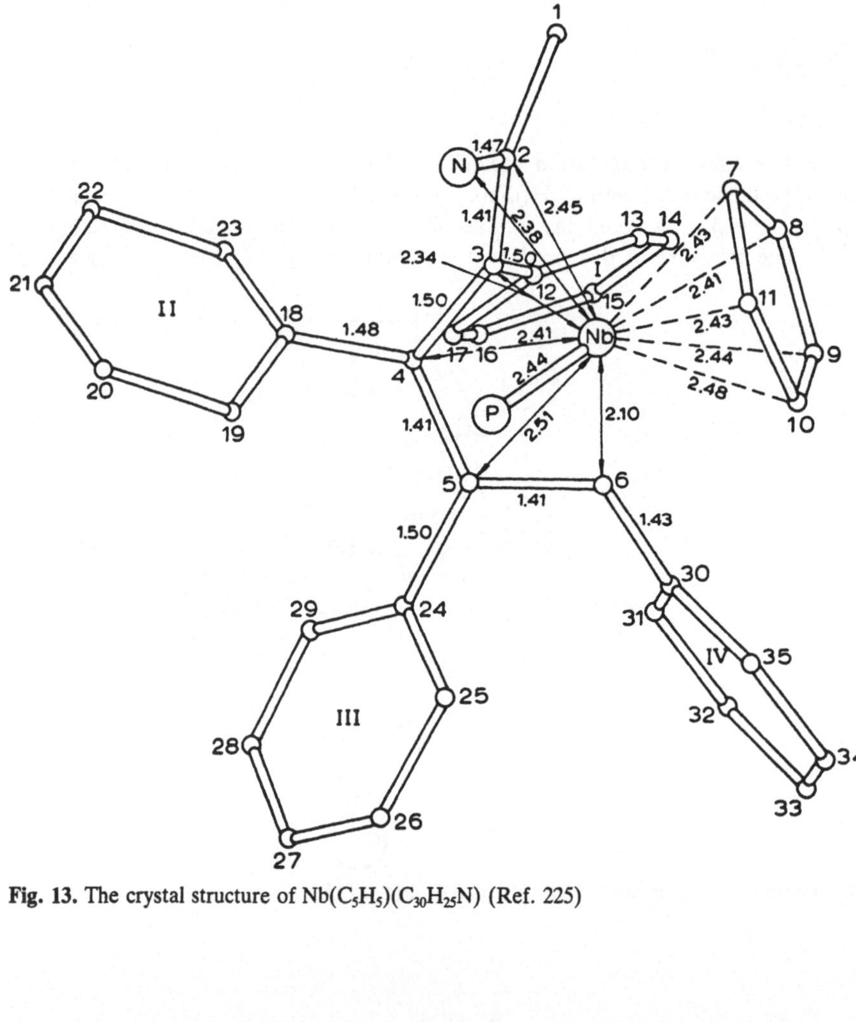

Fig. 13. The crystal structure of $Nb(C_5H_5)(C_{30}H_{25}N)$ (Ref. 225)

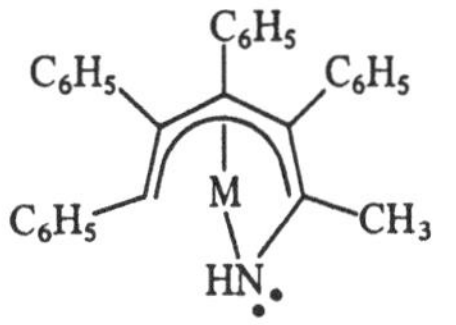

XIX

densed ligand, and is clearly necessary for efficient η^6 interaction with the niobium atom. The pentadienyl portion itself exists in a conformation intermediate between the "S" and "U" forms, being closer to the former. This is somewhat reminiscent of the structure observed for the 3-isopropyl-1,1,2,4,5,5-hexacyanopentadienyl anion (Fig. 5). Interestingly, a number of ("S"-pentadienyl)metal complexes have been reported, especially transient species related to $(C_5H_7)Fe(CO)_3^+$, but none appear to have been structurally characterized to date[226–229].

In a perhaps more straightforward fashion, the reaction of nickel(II) chloride with 1,4-pentadiene and triethylaluminum leads to formation of $[Ni(C_5H_7)]_2$ whose structure is shown in Fig. 14[230, 231]. Here, the pentadienyl ligands adopt the "W" conformation and thus manage to bond to both nickel atoms. The structure can in some sense be considered related to the 16 electron bis(allyl)nickel complex, except that replacement of two allyl groups (3 electron donors) by a pentadienyl ligand (a 5 electron donor) necessitates the gain of a further electron by each nickel atom. This extra electron is obtained by the formation of a formal nickel-nickel single bond (observed Ni–Ni distance = 2.590(1) Å). *XXa* and *XXb* may be considered to be reasonable resonance hybrids:

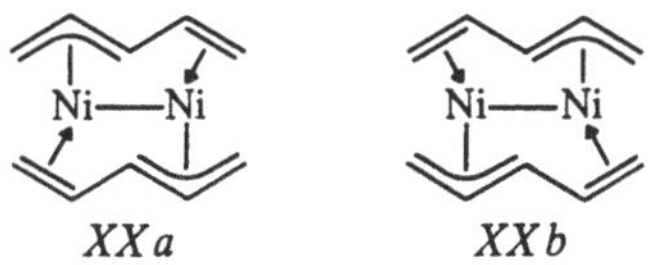

XXa *XXb*

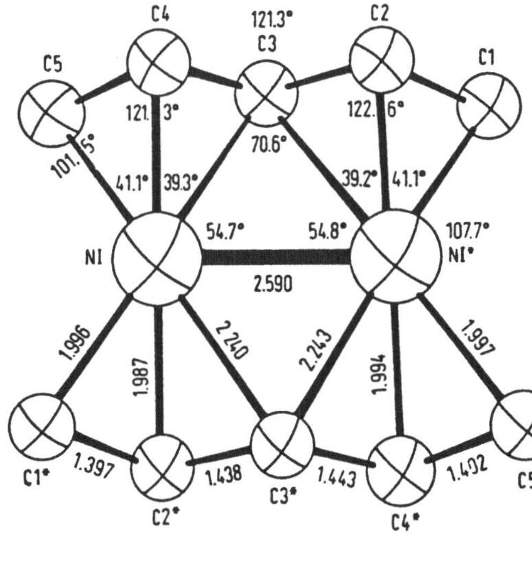

Fig. 14. The solid state structure of bis-(pentadienyl)dinickel which contains a crystallographically imposed center of inversion (Ref. 231)

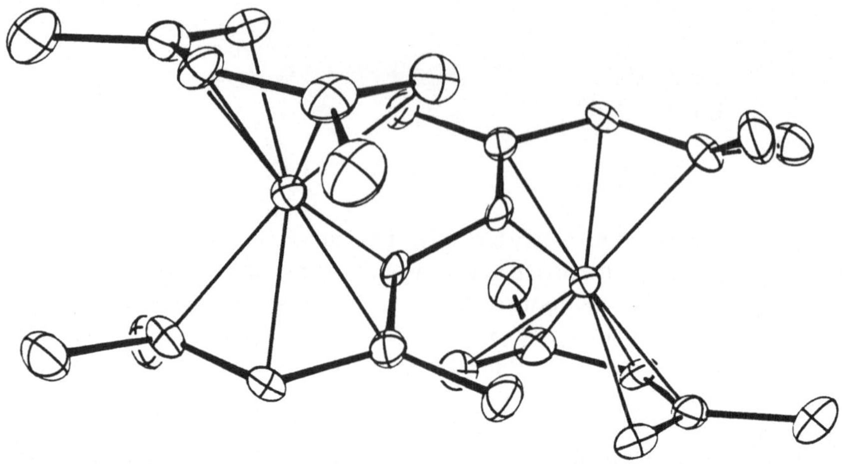

Fig. 15. The solid state structure of the bis(2,4-dimethylpentadienyl)cobalt dimer, which lies on a crystallographic center of inversion (Ref. 236)

In contrast to the three previous structures, the accuracy of this determination was sufficient that a distinction could be made between the C–C (internal) and the C–C (external) bond distances, which averaged 1.440(4) Å and 1.400(4) Å, respectively. The differences of course come about from the usual resonance hybrids in which negative charge (or sigma bond character) is localized at the C3 position. It remains an open question whether or not such a structure could be adopted by a ligand which does not favor the "W" conformation (e.g., 2,4-dimethylpentadienyl). In a general sense, it is probably accurate to expect that the pentadienyl ligands will be quite superb for bonding to two or more transition metals simultaneously[232, 233)] (cf., palladium allyl, cyclopentadienyl, and arene complexes[234)]). In fact, a cyclohexadienyl complex, $HOs_3(CO)_9(C_6H_7)$, is known where the ligand engages in an η^1 (through C3) and two η^2 interactions[235)].

The reaction of cobaltous chloride with the 2,4-dimethylpentadienyl anion has been reported to lead initially to a pentane soluble product which on standing forms a semicrystalline, nearly pentane insoluble product. The final compound turned out to be dimeric, as indicated by mass spectroscopy and ultimately by x-ray diffraction (Fig. 15)[236)]. In this structure a pentadienyl ligand from one cobalt atom has coupled to a pentadienyl ligand on the second to yield two mutually conjugated butadiene ligands, and as a result each cobalt atom achieves the noble gas configuration. The dimer lies on a center of inversion. A remarkable facet of this structure is the fact that the two η^4-butadiene linkages have become mutually conjugated, whereas one would have expected otherwise, as has been the situation in related (pentadienyl)metal dimers[237–239)], e.g., $(C_{10}H_{14})Fe_2(CO)_6$. Presumably the initial product contains nonconjugated η^4-butadiene ligands (*XXIa*) which, during an η^5-η^3 isomerization by the remaining pentadienyl ligand, may undergo either metal-mediated sigmatropic shifts or else reversible metal insertion into C–H bonds to ultimately provide the final isolated product, incorporating two conjugated butadiene units (*XXIb*). Apparently the driving force for this reaction is the greater conjugation in the product, and the higher degree of methylation resulting on the olefinic bonds.

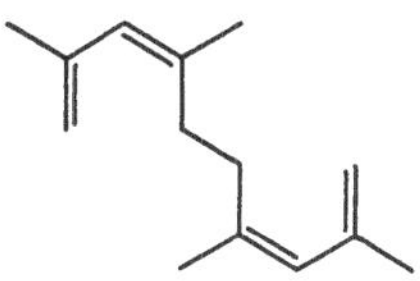
XXIa

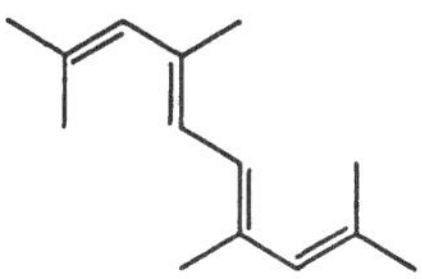
XXIb

The Co-C(1,5) bond distances average 2.136(9) Å, while the corresponding Co–C(2,4) and Co–C(3) distances are 2.069(9) and 2.072(14) Å. The overall average Co–C(pentadienyl) distance of 2.097(6) Å is reasonably close to that observed for $Fe(2,4\text{-}C_7H_{11})_2$ (vide infra). The C(1)–C(2)–C(3) and C(3)–C(4)–C(5) bond angles average 123.5(9)°, with the C(2)–C(3)–C(4) angle appearing to be larger at 126.7(12)°. The internal C–C bond distance of 1.418(14) Å is not clearly longer than the external C–C bond distance, 1.403(13) Å. The bonding to the butadiene ligand is usual in that the internal Co–C bonds seem to be shorter than the external ones, 2.018(9) vs. 2.116(8) Å[240] The external butadiene C–C bond distances average 1.418(13) Å compared to the single internal distance of 1.445(17) Å.

2. *Homoleptic Bis(pentadienyl)metal Compounds*

a. Metal-Ligand Bonding

Certainly most useful at the present time for understanding the bonding in transition metal pentadienyl compounds is the series of structural results for the $M(2,4\text{-}C_7H_{11})_2$ (M = V, Cr, Fe) and $Ru(2,3,4\text{-}C_8H_{13})_2$ compounds[241–243]. While $Ti(2,4\text{-}C_7H_{11})_2$ is known as a liquid[244], its structure has not yet been determined, but it appears that it should be readily inferrable from the other structural data (vide infra).

Pertinent bonding parameters for these compounds are set out in Table 2, and the structures are illustrated in Figs. 16–19. For purposes of comparison, some parameters related to analogous cyclopentadienyl compounds are also included[245–247]. While the conformational differences are the most obvious feature apparent from the figures, of primary interest is the relationship of the M–C bond distances in the open systems to those in the closed. For the present discussions, the pentadienyl ligands have been assumed to have one end equivalent to the other, even though occassionally there were slight distortions present. However, when a distinction is made between ends, the lower numbered end is designated as the one which lies closer to the open face of the other pentadienyl ligand, as in *XXIIa* and *XXIIb*. Further, in each case it is assumed that the

C(1), C(1′)
XXIIa

C(1), C(1′)
XXIIb

two pentadienyl ligands are equivalent, even though only for the chromium complex is there imposed symmetry (C_2) to make this rigorously true. Beginning with the iron compound, it can be seen that the average Fe–C bond distance of 2.089(1) Å for $Fe(2,4\text{-}C_7H_{11})_2$ is clearly longer than the distance of 2.064(3) Å in ferrocene. However, it must be kept in mind that the former is a very crowded molecule, existing in an eclipsed

Table 2. Selected bonding parameters[a]

A. For Various Metal-Pentadienyl Compounds

Parameter	$V(2,4\text{-}C_7H_{11})_2$	$Cr(2,4\text{-}C_7H_{11})_2$	$Fe(2,4\text{-}C_7H_{11})_2$	$Ru(2,3,4\text{-}C_8H_{13})_2$	$Mn_3(3\text{-}C_6H_9)_4$
χ	89.8	82.9	59.7	52.5	35.6
M–C(1,5)	2.179(4)	2.166(6)	2.108(5)[b]	2.162(6)	2.143(5)[c]
M–C(2,4)	2.231(4)	2.162(5)	2.073(4)	2.181(5)	2.066(3)
M–C(3)	2.236(5)	2.167(7)	2.084(3)	2.258(7)	2.148(4)
M–C (avg.)	2.211(2)	2.165(4)	2.089(3)	2.188(3)	2.114(2)
M-center of mass	1.632(2)	1.594(4)	1.508(2)	1.645(4)	1.537(2)
C–C(internal)	1.415(4)	1.414(8)	1.416(4)	1.430(7)	1.408(4)
C–C(external)	1.389(4)	1.378(8)	1.406(4)	1.426(7)	1.427(4)[d]
C(1)–C(2)–C(3)	124.7(4)	125.2(8)	124.1(3)	122.5(7)	127.5(3)
C(2)–C(3)–C(4)	130.2(3)	127.4(8)	125.5(3)	122.8(6)	121.1(4)
C(3)–C(4)–C(5)	125.0(3)	123.5(8)	120.7(3)	122.1(6)	125.2(4)
C(1)–C(5)	3.05(1)	2.93(1)	2.785(5)	2.715(9)	2.818(5)
Ligand Tilt[e]	13.9	18.2	15.0	18.2	6.3
Fold Angle[f]	78.8	72.9	68.1	71.4	68.2

B. For $M(C_5H_5)_2$ Compounds

Parameter	$V(C_5H_5)_2$	$Cr(C_5H_5)_2$	$Fe(C_5H_5)_2$	$Ru(C_5H_5)_2$
M–C	2.280(5)	2.169(4)	2.064(3)	2.196(3)
C–C	1.434(3)	1.431(2)	1.440(2)	1.439(2)
M-center of mass	1.928(6)	1.798(4)	1.660(10)	1.840(7)

[a] Angles are given in degrees, distances are given in Å
[b] Composed of two nonequivalent values of 2.095(4) and 2.122(3) Å
[c] Composed of two nonequivalent values of 2.157(4) and 2.129(4) Å
[d] Composed of two nonequivalent values of 1.441(6) and 1.413(6) Å
[e] Defined as the angle between the two ligand planes
[f] Defined as the angle between a ligand plane and the M–C1–C5 plane

conformation and therefore possessing substantial interligand nonbonded contacts (four C–C separations in the range 2.98(1)–3.22(1) Å were present). Were it not for such contacts, the Fe–C bond distances in the open compound might well be shorter than those in ferrocene.

Reasonably similar observations can be made for the complexes bis(2,3,4-trimethylpentadienyl)ruthenium and bis(2,4-dimethylpentadienyl)chromium, both of which contain metals larger than iron. Thus, in ruthenocene and chromocene the average M–C bond distances have been found to be 2.196(3) and 2.169(4) Å, respectively, whereas for their open analogs, the distances are quite comparable, if not actually shorter, at 2.188(3) and 2.165(4) Å, respectively. Thus it does seem that the Fe–C bond distances in Fe(2,4-

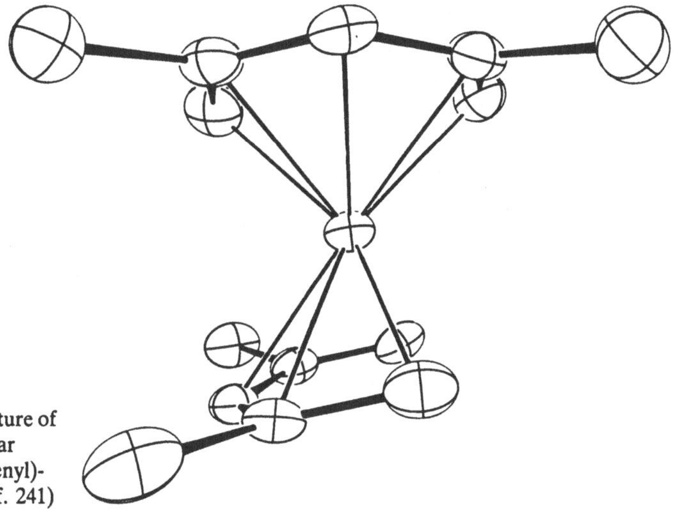

Fig. 16. The crystal structure of the virtually perpendicular bis(2,4-dimethylpentadienyl)-vanadium molecule (Ref. 241)

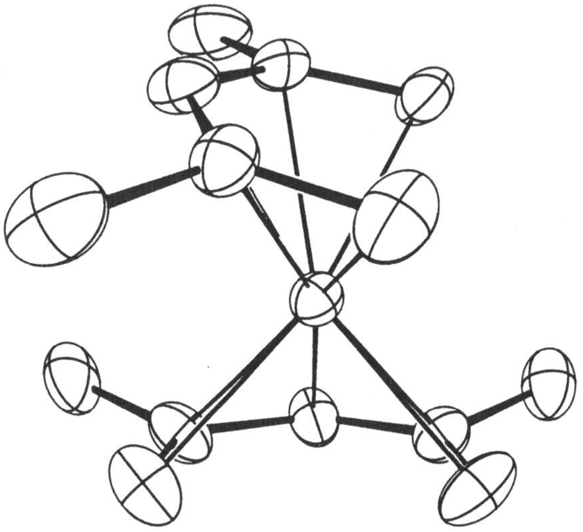

Fig. 17. Perspective view of bis(2,4-dimethylpentadienyl)-chromium. The molecule lies on a crystallographic twofold rotation axis (Ref. 241)

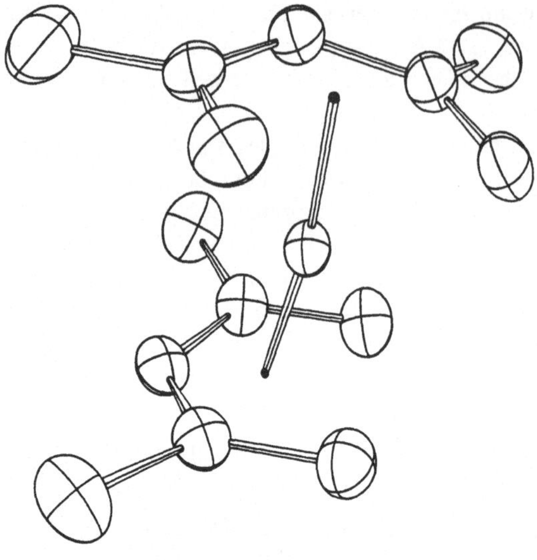

Fig. 18. The solid state structure of the virtually eclipsed bis(2,4-dimethylpentadienyl)iron molecule. Reprinted with permission from Ref. 242. Copyright 1983 American Chemical Society

$C_7H_{11})_2$ are actually lengthened somewhat due to the particularly severe intramolecular ligand-ligand repulsions present. However, it must be emphasized that the ruthenium and chromium structures are themselves greatly crowded, which strongly indicates that their metal-carbon bond distances would also be further shortened were it not for this crowding. Since all of the iron and ruthenium compounds are diamagnetic, and both chromium compounds possess two unpaired electrons[248)], a very fundamental relationship becomes apparent: in comparable situations where similar electronic configurations exist, there is a clear correspondence in metal-carbon bond distances between the metallocenes and the "open metallocenes". This correspondence in fact provides substantial justification for the designation of these complexes as "open metallocenes".

A very interesting situation exists, however, for the bis(2,4-dimethylpentadienyl)-vanadium compound, which possesses only one unpaired electron compared to vanadocenes's three. Thus, while vanadocene is characterized by an average V–C bond distance of 2.280(5) Å, the low spin open analog was found to have an average V–C bond distance which is significantly shorter at 2.211(2) Å. Using bond distance as a criterion, therefore, the "open vanadocene" actually seems more stable than vanadocene itself! Of course, the low spin nature of this compound is primarily responsible for this shortening, yet evidence for true stabilization can also be observed for the diamagnetic "open titanocene" $Ti(2,4\text{-}C_7H_{11})_2$, which is quite stable thermally at and even above room temperature, and does not form nitrogen adducts, in marked contrast to $Ti(C_5H_5)_2$ and even $Ti[C_5(CH_3)_5]_2$[249)], each of which possess two unpaired electrons[250)]. While structural data for $Ti(2,4\text{-}C_7H_{11})_2$ is not yet available, a reasonable estimate of the average Ti–C bond distance can still be made, given the size difference of Ti(II) vs. V(II) as judged by average metal-carbon bond distances in $Ti(2,4\text{-}C_7H_{11})_2(PF_3)$ and $V(2,4\text{-}C_7H_{11})_2(PF_3)$ (2.326 Å vs. 2.275 Å, respectively, vide infra). Addition of this difference to the average V–C bond distance of 2.211(2) Å in $V(2,4\text{-}C_7H_{11})_2$ leads to a predicted Ti–C bond distance of ca. 2.26–2.27 Å in $Ti(2,4\text{-}C_7H_{11})_2$. It can be noted, then, that this distance is comparable to, or even shorter than, the V–C bond distance in vanadocene, and therefore readily accounts for the higher stability of the "open titanocene". For purposes of comparison, the metal-carbon bond distances in the metallocenes and "open metallocenes" may be seen in Fig. 20.

Of course, in the metal-pentadienyl compounds the individual metal-carbon bonds need not necessarily be identical in length, as is clearly apparent from Table 2. Perhaps

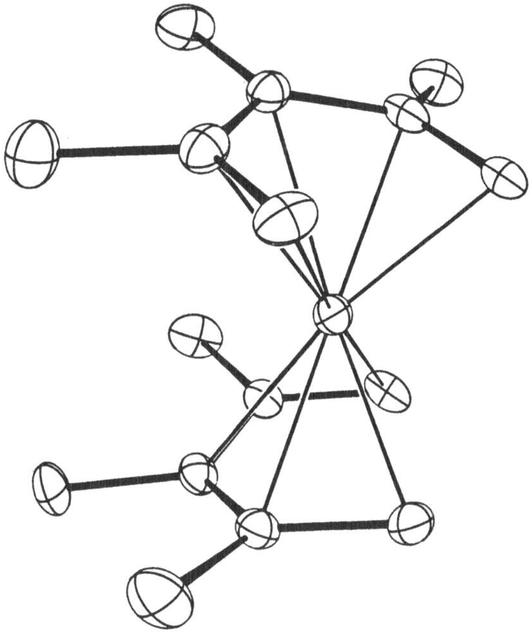

Fig. 19. The solid state structure of bis(2,3,4-trimethylpentadienyl)-ruthenium. Reprinted with permission from Ref. 243. Copyright 1983 American Chemical Society

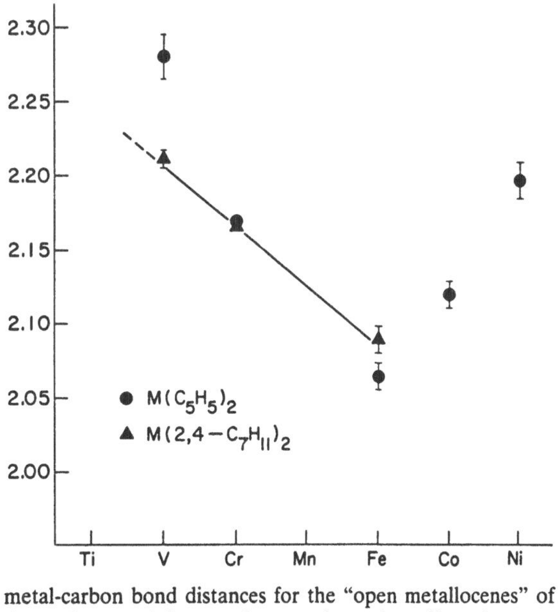

Fig. 20. Comparison of the average metal-carbon bond distances for the "open metallocenes" of vanadium, chromium, and iron with those of some of the metallocenes themselves. Error bars are drawn at the $\pm 3\sigma$ confidence limits

the most dramatic difference observable involves the terminal carbon atoms at positions 1 and 5. In the case of the iron compound, these carbon atoms are clearly furthest from the metal atom, while for chromium all metal-carbon distances are virtually equivalent, and for vanadium the terminal carbon atoms become closest to the metal. Of course, either electronic or size arguments could be offered to explain this trend. However, the fact that in $Ru(2,3,4-C_8H_{13})_2$ the terminal carbon atoms are closest to the metal atom, a situation like that for vanadium rather than iron, suggests that these trends are probably due to the relative sizes of the metal atoms such that metal-carbon bonding is progressively enhanced to the terminal positions when a larger metal is involved. Whether this is a result of steric, overlap, or some other phenomenon is not clear.

The relative ligand orientations adopted by these "open metallocenes" are also of fundamental interest, and can be seen from Figs. 16–19 to be quite variable. While one in general must be cautious about assuming that solid state conformations are the actual ground state, note here that the adoption of three different space groups for the $M(2,4-C_7H_{11})_2$ (M = V, Cr, Fe) compounds suggests that the conformations are determining the solid state packing more than the packing affects the conformations. Further, the observed trends appear quite rational, and there is compelling evidence (see Sects. D.II and D.IV) that the observed conformations must indeed be at least close to the ground state configurations. A conformation angle χ can be defined between two ligand-related planes in order to allow convenient discussion of the relative ligand orientations. Each plane will be composed of the metal atom, the carbon atom in the central (3) position, and the midpoint between the terminal (1 and 5) positions. The cis-eclipsed orientation (*XXIII a*) will be defined as having a conformation angle of 0°, while the anti-eclipsed (*XIII b*) will be 180°, i.e.,

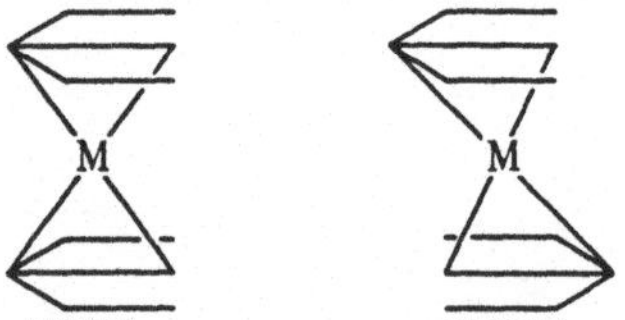

XXIII a *XXIII b*

The conformation angle for the iron compound has been found to be 59.7°, virtually equivalent to the ideal value of 60° expected for the gauche-eclipsed orientation *XXIV*. It is particularly important that this eclipsed conformation was adopted despite the fact that substantial inter-ligand repulsions are present. There clearly must be a powerful elec-

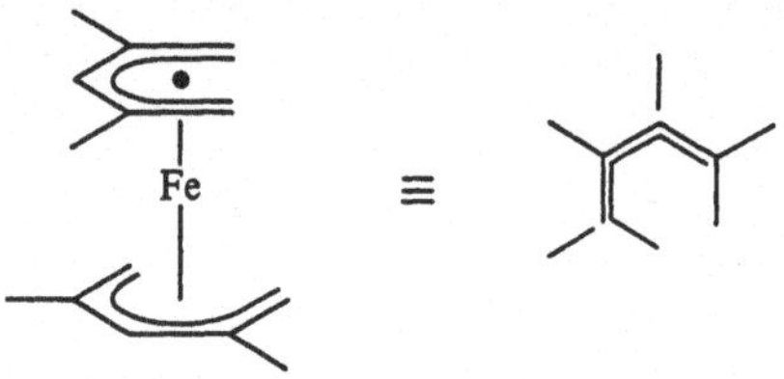

XXIV a *XXIV b*

tronic driving force which outweighs the steric contribution. It can be noted in this regard that the gauche-eclipsed conformation (and the anti conformation as well) is characterized by a near-octahedral disposition of the six formally charged carbon atoms of the pentadienyl ligands.

The situation for $Ru(2,3,4\text{-}C_8H_{13})_2$ is quite analogous. Here, however, there is a slight but significant twist away from the gauche-eclipsed orientation, with the conformation angle being 52.5°. The rather clear origin of this twist is the fact that two substantial intramolecular CH_3–CH_3 nonbonded contacts are present, which are relieved somewhat by the twisting. In fact, the choice of ligand in that study was made specifically in order to see if such destabilizing influences could lead to the adoption of the anti conformation *XXVa*. This, however, did not take place, which clearly emphasizes the favorability of the gauche-eclipsed conformation *XXVb*.

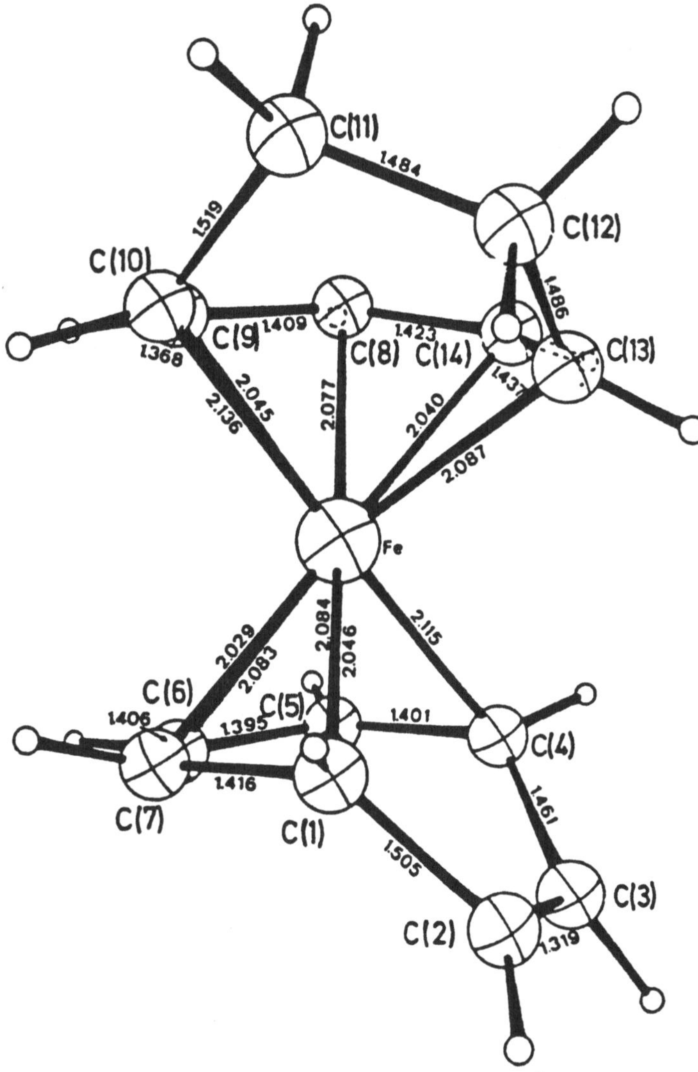

Fig. 21. The crystal structure of $Fe(C_7H_7)(C_7H_9)$, which contains two dienyl ligands in the same relative conformation as observed for $Fe(2,4\text{-}C_7H_{11})_2$ (Ref. 252)

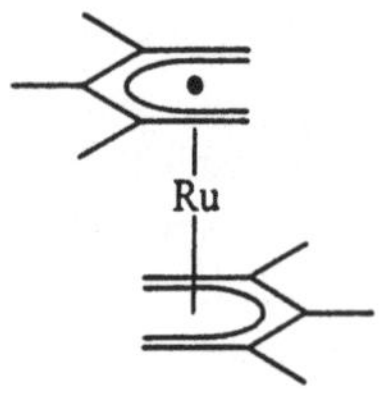

XXV a

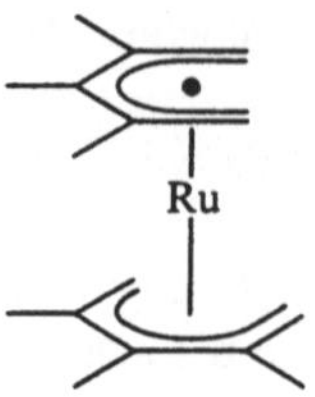

XXV b

There is further data to establish the generality of the gauche-eclipsed ground state for iron and ruthenium. A number of structural studies have been carried out on related compounds such as $Fe[\eta^5\text{-}6\text{-}t\text{-}C_4H_9\text{-}1,3,5\text{-}(CH_3)_3C_6H_3]_2$[251], $Fe(\eta^5\text{-}C_7H_7)(\eta^5\text{-}C_7H_9)$[252], and $Ru(\eta^5\text{-}C_7H_7)(\eta^5\text{-}C_7H_9)$[253], and in each case thus far the same gauche-eclipsed configuration has been found. This is illustrated in Fig. 21 for $Fe(\eta^5\text{-}C_7H_7)(\eta^5\text{-}C_7H_9)$. Various nmr studies have also demonstrated the unsymmetric nature of the ground state in a number of these compounds (Sect. D.II).

In contrast to the above, the conformation of the $V(2,4\text{-}C_7H_{11})_2$ and $Cr(2,4\text{-}C_7H_{11})_2$ compounds are markedly different. The former possesses a nearly perfectly perpendicular conformation ($\chi = 89.8°$), while the chromium compound is intermediate ($\chi = 82.9°$)[241]. A crude molecular orbital scheme has recently been devised to explain these and other trends. Starting from the metallocene d orbital pattern induced by ligand field influences[4], one can consider the differences that should result on replacement of the cyclopentadienyl ligand by a pentadienyl ligand. With z being oriented along the high symmetry axis of the metallocenes, it can be seen that the wider pentadienyl ligand will be closer the xy plane but farther from the z axis. As a result, d orbitals with z components will experience net stabilization compared to the metallocenes, while orbitals with xy components will experience destabilization. It therefore appears feasible that the approximate relative orbital splitting patterns shown in Fig. 22 would result. It must be kept in mind that this model is only very approximate, and that no degeneracies are actually required. Nonetheless this model provides a very effective way to understand the conformational natures of these compounds, as well as their magnetic natures. Thus, if the formally d_{z^2} orbital is substantially lower than the formally d_{xy} and $d_{x^2-y^2}$ orbitals, one would expect the d^2 "open titanocenes" and d^3 "open vanadocenes" to have low spin configurations (0 and 1 unpaired electrons, respectively), relative to the corresponding metallocenes. It might well be questioned, however, whether the above described ligand

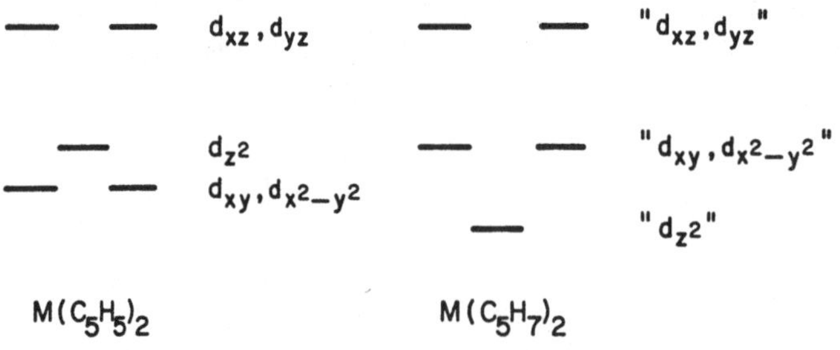

Fig. 22. Possible approximate relative d orbital splitting patterns for the metallocenes (*left*) and "open metallocenes" (*right*)

field influences would be capable of bringing about enough splitting to allow adoption of the low spin configurations. In fact, it seems likely that some other factor must be even more important. In this regard, MO studies on the "open ferrocenes" (Sect. D.IV) indicate that very substantial mixing of metal d orbitals takes place, both with ligand orbitals and with other metal d orbitals, whereas such interactions are much less important for the higher symmetry metallocenes[254]. Hence while it again must be emphasized that the attached orbital labels and the relative energies in Fig. 22 are only at best approximate, it is also quite clear that the orbital mixing influences could easily bring about greater splitting of the (formally metal d) orbitals than might otherwise be expected. It is useful to note in this regard that the calculated net d orbital populations for the "open ferrocenes" are in reasonable accord with this simple model with d_{z^2} (1.99 e) > $d_{x^2-y^2}$ (1.66 e) ≃ d_{xy} (1.66 e) > d_{xz} (1.20 e) > d_{yz} (0.81 e)[254].

The above splitting scheme can further be effectively utilized to account for the conformational trends observed for the "open metallocenes". Particularly relevant here is the fact that the formal d_{xy} and $d_{x^2-y^2}$ orbital populations are sequentially increased from 1 to 4 as one changes the metal atom from vanadium to iron. The orbitals being populated are those which can engage in δ bonding between the metal and ligand orbitals (M → L), and while δ bonding in the metallocenes themselves appears quite weak, the extent of δ bonding in the "open ferrocenes" has been indicated by MO studies (Sect. D.IV) to be quite strong. Very possibly, then, addition of three electrons to partially δ bonding orbitals takes place in going from vanadium to iron, and brings about the change in conformation from perpendicular (V) to intermediate (Cr) to eclipsed (Fe). Because of the complex mixing of orbital character in the "open metallocenes", the actual bonding interactions taking place are rather intricate, and therefore somewhat difficult to present pictorially. Nonetheless, even though the above model is somewhat approximate, it is clearly effective in explaining the low spin natures of the $M(2,4\text{-}C_7H_{11})_2$ (M = Ti, V) compounds and the relative conformations adopted by these same compounds (M = V, Cr, Fe). It is clear that some very fundamental aspects and relationships involving the "open metallocenes" are now being uncovered and our understanding of these systems will naturally improve dramatically as time progresses.

b. Pentadienyl Ligand Bonding Parameters

The pertinent bonding parameters for the pentadienyl ligands themselves are also summarized in Table 2. For the most part the parameters (excluding conformations) are not greatly different from those observed for the anionic species in Sect. C.V. Nevertheless, some important further observations can be made, including some comparisons with cyclopentadienyl analogs. For each compound the average external C–C bond distance is shorter than the internal one, although for iron and especially ruthenium the difference is not statistically conclusive. If one assumes that the C–C bond distances in one first row transition metal structure are essentially equivalent to those in another – and there is no clear difference from the data – then one obtains overall average internal and external C–C distances for these three structures of 1.415(3) and 1.391(3) Å, respectively. The difference between these numbers, 0.024(4) Å, is clearly statistically significant. Thus, even including the added perturbation of covalent interaction with a transition metal, the pentadienyl groups still display shortening of the external C–C bonds due to resonance hybrid *IIIb*. It is also interesting to note that both the internal and the external pentadienyl C–C bond distances are clearly shorter than the average cyclopentadienyl C–C distance of 1.435(2) Å. Clearly the pentadienyl framework bonding is stronger than that in cyclopentadienyl, which can again be accounted for rather easily – while a resonance hybrid for the cyclopentadienyl anion would indicate a formal C–C bond order of 1.4, that for the pentadienyl anion would be 1.5. To some extent cyclopentadienyl ring strain might also contribute to this difference as well.

Also similar to some of the observations made for pentadienyl anions, a noted contraction of C–C–C bond angle occurs around a pentadienyl skeletal carbon atom on methylation. Thus, for the 3-methyl substituted manganese compound, a difference of 5.2(5)° is noted between the C2–C3–C4 angle and the average of the C1–C2–C3 and C3–C4–C5 angles (Table 2). However, in the 2,4-dimethyl substituted compounds, the reverse trend is observed with the differences averaging 5.3(4)°, 3.0(10)°, and 3.1(4)°, respectively, for vanadium, chromium, and iron, respectively. Finally, for Ru(2,3,4-$C_8H_{13})_2$, all angles are equivalent within experimental error. Naturally the ligand contractions experienced on methylation markedly affect the relative girths of the pentadienyl ligands. Thus, the 2,3,4-trimethylpentadienyl ligand in the ruthenium compound is clearly the smallest one to be encountered thus far, having an average C1–C5 separation of only 2.715(9) Å, and one can expect, therefore, that the simple C_5H_7 ligand should be widest of all. However, it is at the same time also clear that the pentadienyl ligands are capable of flexing somewhat to better accommodate a given system. Thus, of the M(2,4-$C_7H_{11})_2$ structures, the vanadium one has the widest ligand (C1–C5 = 3.05(1) Å), while chromium is intermediate (2.93(1) Å), and iron is smallest (2.785(5) Å). The ligand size for the vanadium compound is somewhat smaller than that observed for Nd(2,4-$C_7H_{11})_3$, 3.14(2) Å (Sect. C.V). Thus it seems that as the ligand comes closer to the metal it undergoes contraction, perhaps to optimize metal-ligand overlap, although other suggestions could also be made. A similar observation has been made for the allyl ligand, however, and attributed to just such overlap considerations. This seems to further emphasize that many features found in one of these ligand systems might have counterparts in the others.

A final feature of interest for these pentadienyl ligands is the fact that their attached substituents are generally tilted substantially down toward the metal atom, as can be seen in Table 3. In general, methyl groups in the 2 and 4 positions tend to bend down by an average of ca. 9°, although the value seems dependent somewhat on conformation. The bending of methyl groups in the 3-position seems somewhat smaller, however, at ca. 0–4°. In ferrocene a similar substituent bending was observed, although the magnitude is essentially smaller at 3.7(9)°. In this case the bending has been attributed to an attempt to increase metal-ligand overlap by pointing the ligand p orbitals more toward the metal atom, as indicated in *XXVI*[245]. The generally much greater bending for the larger pentadienyl ligands is quite consistent with this view.

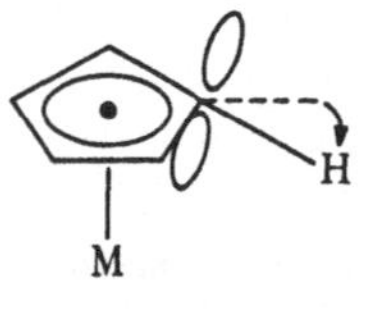

XXVI

Table 3. Bending (degrees) of refined substituents toward the central metal atom in various metal pentadienyl compounds

Position	$Mn_3(3\text{-}C_6H_9)_4$	$V(2,4\text{-}C_7H_{11})_2$	$Cr(2,4\text{-}C_7H_{11})_2$	$Fe(2,4\text{-}C_7H_{11})_2$	$Ru(2,3,4\text{-}C_8H_{13})_2$
2		11.9	9.9	6.0	8.7
3	3.7	5.8			0.3
4		6.3	9.3	11.2	10.3

A somewhat more complicated distortion occurs for the endo and exo substituents on the terminal pentadienyl carbon atoms. In the $V(2,4\text{-}C_7H_{11})_2$ structure, the hydrogen atom positions were readily refined[241)], and it was found that the H_{exo} atoms were bent down toward the vanadium atom by an average of 11.5°, while the H_{endo} atoms were bent away from the metal by an average of 34.7°. In the $Ru(2,3,4\text{-}C_8H_{13})_2$ structural report, the hydrogen atom positions were also subjected to limited refinement, and the corresponding angles were ca. 17 and 42°, respectively[243)]. One can again invoke electronic arguments for these bending distortions, although steric factors, especially H_{endo}–H_{endo} repulsions, could also be playing some role. In fact, in the more ionic $Nd(2,4\text{-}C_7H_{11})_3$, which contains a wider pentadienyl ligand, it did *appear* that the H_{endo} and H_{exo} atoms were located more nearly in the idealized pentadienyl plane[167)]. Thus, it is clear that electronic considerations are again important in bringing about the terminal carbon atom substituent tilts in at least the transition metal compounds, but steric influences may also be involved.

3. Bis(pentadienyl)metal Complexes with Additional Ligands

Two metals, titanium and vanadium, readily form mono(ligand) adducts on reaction of their respective $M(2,4\text{-}C_7H_{11})_2$ compounds with CO or PF_3[255)]. Both $M(2,4\text{-}C_7H_{11})_2(PF_3)$

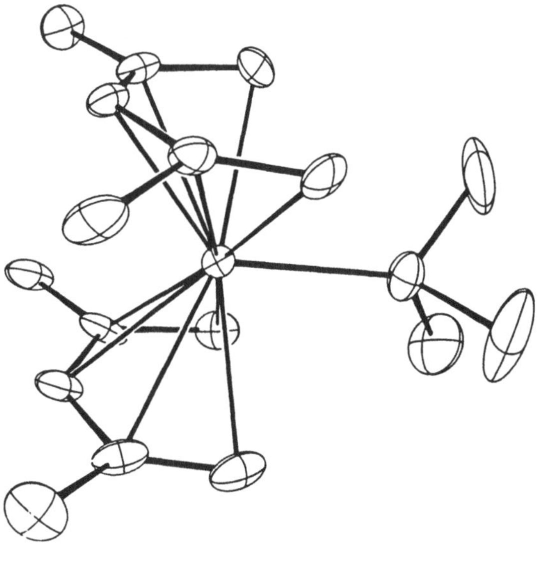

Fig. 23. The solid state structure of the trifluorophosphine adduct of bis(2,4-dimethylpentadienyl)-titanium. The corresponding vanadium compound is isomorphous (Ref. 256)

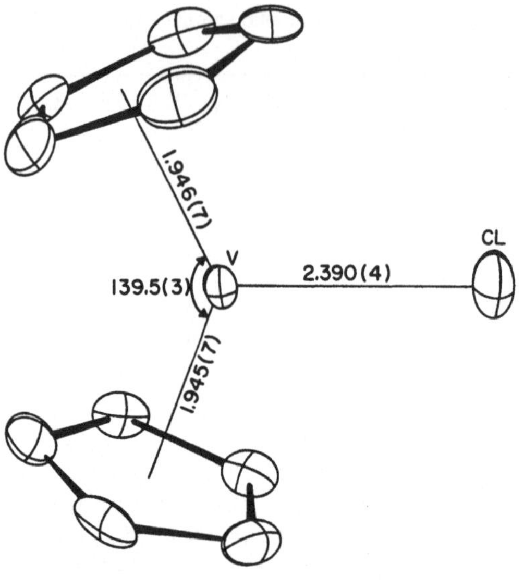

Fig. 24. The solid state structure of bis-(cyclopentadienyl)vanadium(III) chloride, which possesses crystallographically imposed mirror plane symmetry (Ref. 257)

compounds have already been structurally characterized, although some disorder has resulted in each case due to the presence of pseudosymmetry[256]. A perspective view of the titanium structure can be seen in Fig. 23. The vanadium compound is isomorphous. As can be seen, the pentadienyl ligands have now adopted the syn-eclipsed conformation, with the additional ligand located by the open pentadienyl edges. Quite clearly, pentadienyl ligands are much larger than either the cyclopentadienyl or the pentamethylcyclopentadienyl ligands, and the pentadienyl complex geometries bear no relationship to the classic "bent metallocene" configurations, such as adopted by $V(C_5H_5)_2Cl$ (Fig. 24)[4, 257–261]. This is due to the greater girth of the open ligand, which typically has C1–C5 separations of ca. 3 Å (see Table 2), whereas in C_5H_5 or $C_5(CH_3)_5$, the C1 to C5 separation actually corresponds to a bonded interaction (ca. 1.4 Å). As can be seen in Table 2, because of the larger pentadienyl size, the open ligands must be closer to the metal atom than a C_5H_5 or $C_5(CH_3)_5$ ligand in order to attain comparable M–C bond distances. Taken together, these two geometric factors make the pentadienyl ligands far more sterically demanding than either C_5H_5 or even $C_5(CH_3)_5$. This is also readily apparent when one considers the existence of the 16 electron $Ti(2,4\text{-}C_7H_{11})_2(CO)$, in contrast to the 18 electon $Ti(C_5(CH_3)_5)_2(CO)_2$[249]. However, it is possible that the syn-eclipsed conformation is here stabilized also by electronic factors, although it seems clear that steric constraints would be sufficiently important by themselves. In this regard it can be noted that analogous allyl and butadiene complexes adopt similar syn-eclipsed geometries, even though their steric constraints are considerably reduced. This can be seen in Fig. 25 for $Ni(C_3H_5)_2(P(CH_3)_3)$[262] and $Mn(C_4H_6)_2(CO)$[263–265]. Once again it appears that trends in one open system also have their analogs in another. While it is clear that these systems are not "bent metallocene" analogs, $Nd(2,4\text{-}C_7H_{11})_3$ (Sect. C.V) may be considered in such a fashion. Thus it remains possible, at least sterically if not electronically, that true "bent metallocene" analogs will be achievable for larger transition metals.

Because the structural data is plagued substantially from pseudo-symmetry, only a few general comparisons may be made. The average metal carbon bond distances appear

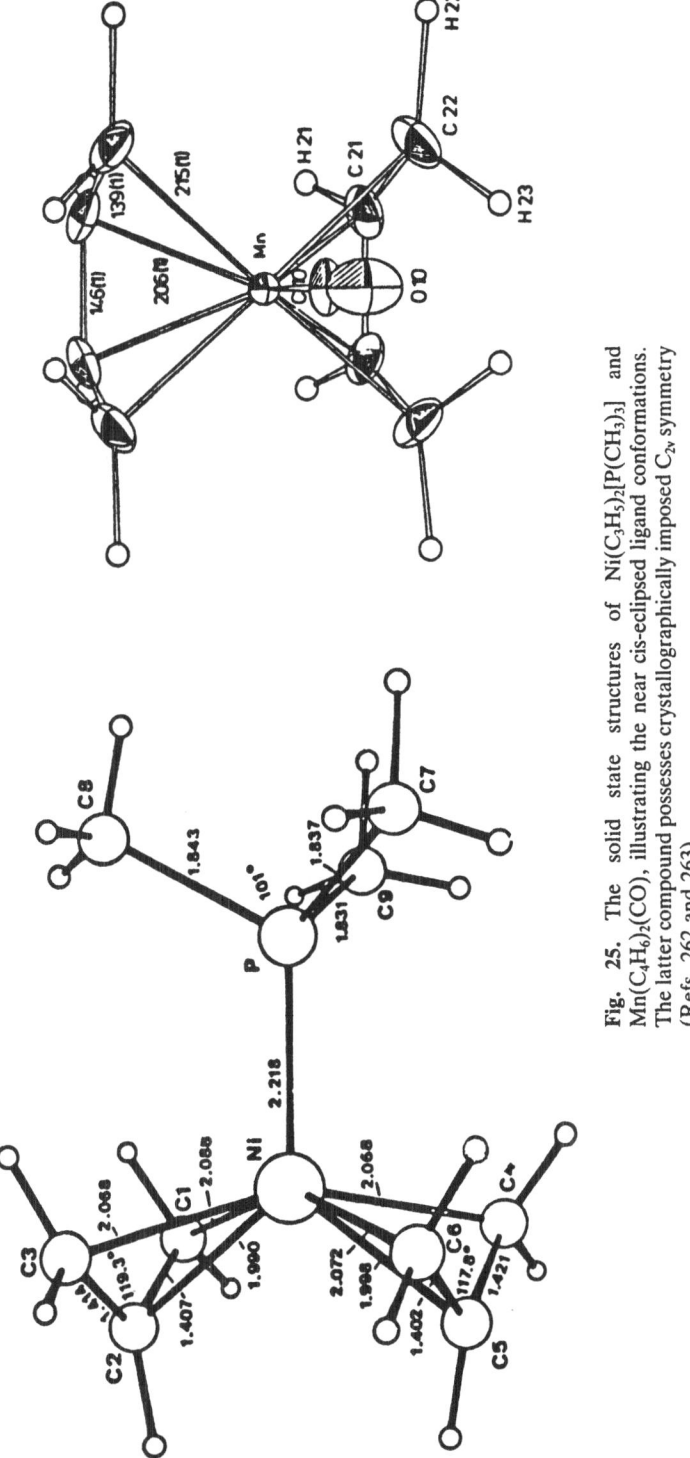

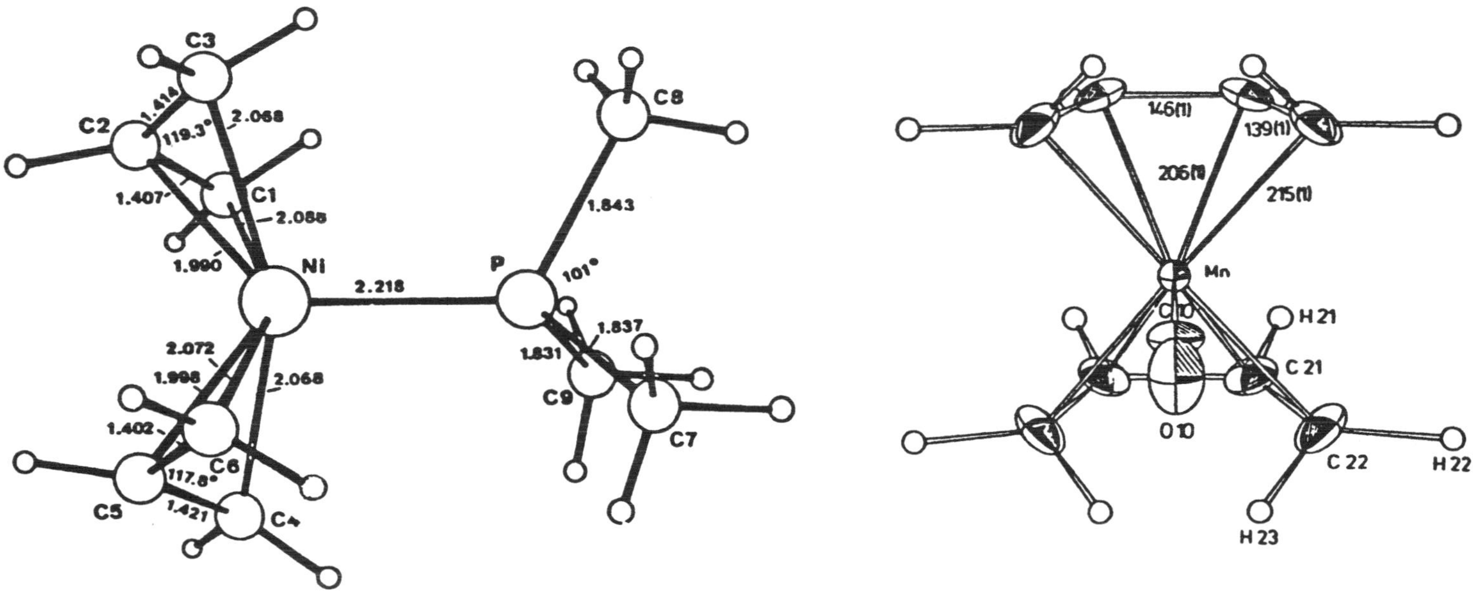

Fig. 25. The solid state structures of $Ni(C_3H_5)_2[P(CH_3)_3]$ and $Mn(C_4H_6)_2(CO)$, illustrating the near cis-eclipsed ligand conformations. The latter compound possesses crystallographically imposed C_{2v} symmetry (Refs. 262 and 263)

to be ca. 2.326 Å (Ti) and 2.275 Å (V). The metal-phosphorus bond distances are not affected by the disorder, and hence have been much more accurately determined as 2.325(1) Å (Ti) and 2.250(1) Å (V). No distinction could be made between the internal and external carbon-carbon bonds, but for the two structures the average overall carbon-carbon bond distance was 1.408 Å. For the titanium compound, the angle between the two ligand least-squares planes (defined by five carbon atoms each) was 1.7°, while the conformation angle χ was 3.4°. For vanadium these angles were 1.6° and 3.0°, respectively.

Reference to the structure of $Mn_3(3\text{-}C_6H_9)_4$ in Fig. 26 will indicate the justification for considering this compound as at least a structural relative to the $M(2,4\text{-}C_7H_{11})_2(PF_3)$ compounds. Essentially two $Mn(3\text{-}C_6H_9)_2$ units can be seen to be approaching syn-eclipsed conformations (χ = 35.6°) with the third manganese atom being situated near the open edges of the pentadienyl ligands[266]. Of course, the central manganese atom is actually interacting with a single terminal carbon atom from each pentadienyl ligand (Mn–C = 2.334(4) Å), which is a direct result of the fact that the conformation is not exactly syn-eclipsed. In fact, it is actually closest to a gauche-staggered conformation, for which χ = 30.0°. Pertinent bonding parameters have been included in Table 2. For the most part the parameters are not surprisingly different from those of the other compounds in the table, and in particular the similarity with $Fe(2,4\text{-}C_7H_{11})_2$, as well as the presence of five unpaired electrons, led to the compound's initial formulation as $Mn^{2+}[Mn(3\text{-}C_6H_9)_2^-]_2$, although further discussion of the electronic nature is presented in Sect. D.IV. Unlike the other first-row metal complexes included in Table 2, the external C–C bond distances are seen to be longer than the internal ones. This is actually a result of the preferential lengthening of the external pentadienyl C–C bond (to 1.441(6) Å) which has the extra interaction with the central manganese atom. The most unusual feature of this structure, however, is the coordination geometry of the central manganese atom. The range of C–Mn(2)–C′ bond angles is fairly narrow, 103.6(3)–118.4(3)°, and clearly these four carbon atoms form a nearly perfect tetrahedron around Mn(2). The two additional manganese atoms are capping tetrahedral edges at an average distance of 2.516(1) Å, with the Mn(1)–Mn(2)–Mn(3) bond angle being 177.51(6)°.

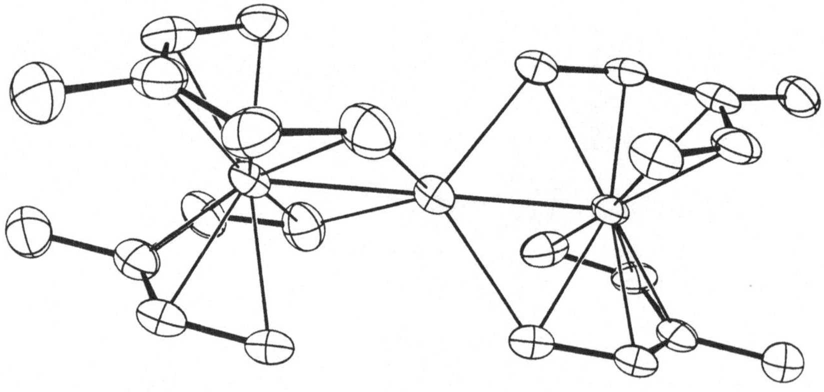

Fig. 26. The solid state structure of tetrakis(3-methylpentadienyl)trimanganese. Reprinted with permission from Ref. 266. Copyright 1984 American Chemical Society

II. NMR Spectroscopy

NMR studies have naturally focussed on the diamagnetic pentadienyl complexes of titanium[244)], iron[242)], and ruthenium[243)], although for the paramagnetic $Nd(2,4\text{-}C_7H_{11})_3$ and $U(2,4\text{-}C_7H_{11})_3$ compounds mentioned in Sect. C.V, the relaxation times were rapid enough to allow nmr spectra to be obtained[167, 168)]. Other than straightforward compound characterization, nmr data has been most useful for establishing the presence of an unsymmetric ground state conformation, and for determining the rotational energy barrier to ligand oscillation, for the various bis(pentadienyl)metal complexes. Thus, while four-line ^1H nmr spectra are observed for both $Fe(2,4\text{-}C_7H_{11})_2$ and $Ru(2,4\text{-}C_7H_{11})_2$ at room temperature, on decreasing the temperature line broadening and collapse occur, until ultimately at a low enough temperature a seven line pattern is evident, indicative of an unsymmetric ground state conformation. Clearly this data precludes the existence of either the anti-eclipsed or syn-eclipsed ground states, and in conjunction with the x-ray structural studies all but proves that the observed solid-state gauche-eclipsed conformer is indeed the ground state. Analogous observations have been made for other symmetric ligand complexes and the ligand oscillational barriers are summarized in Table 4. For $Ti(2,4\text{-}C_7H_{11})_2$, a seven line pattern is observable even at room temperature and coalescence is reached only with substantial heating. Hence a much higher barrier to ligand oscillation is present. It is interesting to note that it is the titanium compound, which possesses the expected longest average metal-carbon bond distance, which has the highest barrier, and the iron compounds, with the shortest average metal-carbon bond distances, which have the lowest barriers. Clearly electronic, rather than steric, factors dominate, although it can be observed that sequential methylation in either the iron or ruthenium series does bring about an increase in the barrier by ca. 0.4 kcal/mol for each group added.

Interesting results have also been obtained for complexes containing unsymmetrically substituted ligands. Thus, $Fe(2\text{-}C_6H_9)_2$ and $Fe(2,3\text{-}C_7H_{11})_2$ display respectively 12 and 14 resonances in the room temperature proton decoupled ^{13}C nmr spectra, leading to the conclusion that two isomers of each are present. For $Fe(2\text{-}C_6H_9)_2$, these isomers may be most conveniently depicted as *XXVII a* and *XXVII b*. However, it must be kept in mind

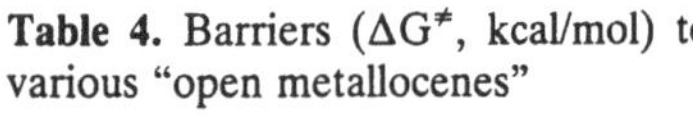

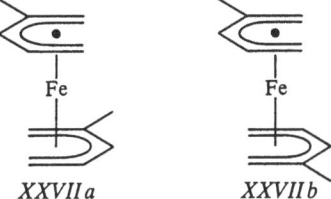

XXVII a *XXVII b*

Table 4. Barriers ($\Delta G^{\neq}$, kcal/mol) to pentadienyl ligand oscillation in various "open metallocenes"

Ligand	Fe	Ru	Ti
C_5H_7	8.4		
$3\text{-}C_6H_9$	8.7		
$2,4\text{-}C_7H_{11}$	9.1	9.7	15.3
$2,3,4\text{-}C_8H_{13}$		10.2	

that the ground states should be gauche-eclipsed. Interestingly, half of the resonances are observed on cooling to collapse and ultimately reappear as doublets, while the other half remain sharp at low temperatures. This has been attributed to the fact that the first isomer depicted above should freeze out as two equivalent gauche-eclipsed enantiomers, while the second isomer should freeze out as two nonequivalent diastereomers. Hence, resonance collapse is only necessitated for the first isomer.

III. EPR Spectroscopy

At present, very little has been reported in terms of electron paramagnetic resonance spectroscopy. Some interesting data has been obtained, however, on some of the vanadium compounds, especially as a comparison to the analogous cyclopentadienyl compounds[255]. Noteworthy in particular are the values of the vanadium hyperfine (I = 7/2) coupling constants. Thus, $V(2,4\text{-}C_7H_{11})_2$ has been reported to exhibit an eight line pattern with the vanadium hyperfine splitting being 77.2 G (Gauss) in THF. Unusually, the monocarbonyl adduct demonstrated a nearly identical hyperfine splitting of 79.4 G, while the mono(trifluorophosphine) adduct has a vanadium hyperfine splitting of 77.1 G, with further splitting (53.6 G) occurring due to the presence of the phosphorus atom. For comparison, vanadocene monocarbonyl has a vanadium hyperfine splitting of only 27.8 G. It has been suggested that the higher value for $V(2,4\text{-}C_7H_{11})_2(CO)$ might be due to greater metal charge restricting unpaired electron density delocalization, which is supported by a higher C–O stretching frequency in the infrared spectrum (1942 vs. 1881 cm^{-1} for $V(C_5H_5)_2(CO)$), although s orbital contributions could not be ruled out. It should be noted here that the greater charge withdrawal by pentadienyl ligands runs counter to the expected trend, given that the pentadienyl electron affinity and ionization potential are both lower than those for cyclopentadienyl (Sect. B.I).

IV. Molecular Orbital Calculations

While a number of reports have appeared which deal with molecular orbital calculations on metal-cyclohexadienyl compounds[267, 268], one in particular has important general implications for metal-pentadienyl chemistry[269]. It has, in fact, been shown that introduction of a CH_2 bridge does not bring about any large changes in pentadienyl π-molecular orbitals, so results obtained here should indeed be relevant to metal-pentadienyl chemistry. Extended Hückel calculations on $[Pt(C_6H_7)(PH_3)_2]^+$ were employed in an attempt to understand why the compound possesses a 16 electron ground state configuration with η^3-C_6H_7 coordination instead of adopting an 18 electron configuration with η^5-C_6H_7 coordination. The calculations did actually find that the potential minimum was the 16 electron complex with η^3-C_6H_7 coordination, favored by 12.9 kcal/mol over the most stable η^5 configuration *XXVIII a*. The reason for this was found to be an enhanced overlap for a net bonding interaction between the ligand and the metal d_{xz} orbital, and net decreased overlap for an antibonding interaction involving the d_{yz} orbital, for the η^3 bonding mode. Of two η^5 configurations considered, the first appeared more stable by 6.4 kcal/mol, although the presence of electron withdrawing substituents was predicted to decrease this difference.

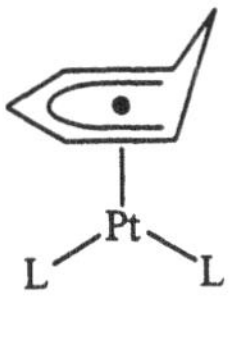

XXVIIIa

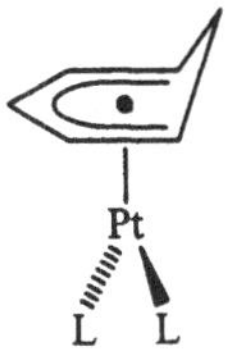

XXVIIIb

A combination of semiempirical LCAO calculations of both the INDO and charge-iterative extended Hückel types has been reported for the bis(pentadienyl)iron ("open ferrocene") compounds[254]. Two potential minima were found, the lower being the anti-eclipsed, *XXIXa,* with the higher being gauche-eclipsed, *XXIXb,* the observed conformer in the solid state structure of bis(2,4-dimethylpentadienyl)iron. The two high

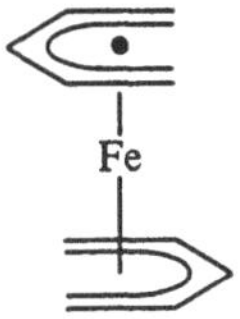

XXIXa

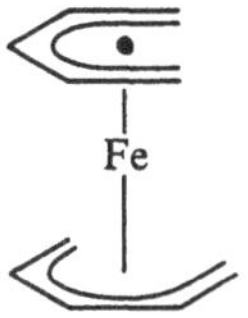

XXIXb

energy conformers also are eclipsed, the higher one being syn-eclipsed, *XXIXc* while the other is a second gauche-eclipsed form, *XXIXd.* It was noted that the two stable conformers have essentially octahedral orientations of the two sets of formally charged pentadienyl carbon atoms, while the two unstable conformers have trigonal prismatic orientations. Compared to ferrocene, much higher rotational barriers were predicted from calculations, in accord with later variable temperature nmr studies (see Sect. D.II). The calculations evidenced several other very dramatic differences relative to ferrocene.

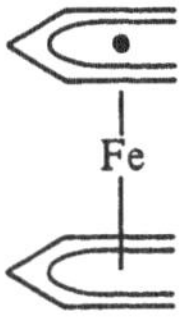

XXIXc

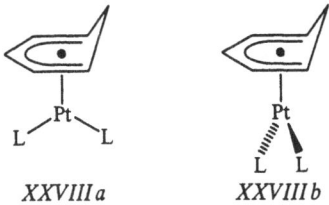

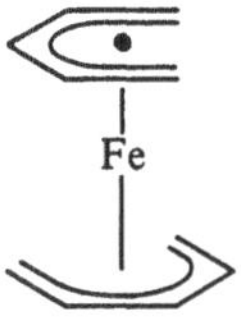

XXIXd

First, while the molecular orbitals in ferrocene are all relatively strongly metal or ligand localized, three molecular orbitals in the "open ferrocenes" were found to possess comparable ligand and metal character. The main reason for this difference was the much lower symmetry present in the "open ferrocenes", which allowed for the greater orbital flexibility. In fact, a great deal of mixing was also observed between various metal d orbitals as well. One of several examples of this was substantial $d_{z^2}/d_{x^2-y^2}$ mixing, which ultimately resulted in a molecular orbital possessing both σ and δ bonding character. Also apparent in this study was the fact that δ bonding in the "open ferrocenes" was much stronger than that in ferrocene itself, which is in qualitative accord with various conclusions reached regarding charge and conformations (Sect. D.II). Because of this enhanced δ coupling and the mixing of metal d orbitals with one another, the "open ferrocenes" possessed much more nearly equal d orbital populations than ferrocene itself. The sums of the individual calculated Fe–C Wiberg bond indices for the "open ferrocenes" were similar to the sum of those for ferrocene. Approximate individual values were Fe–C(1,5) = 0.35, Fe–C(2,4) = 0.26, Fe–C(3) = 0.30, C–C (internal) = 1.27, and C–C (external) = 1.39. Finally, methyl substitution on the pentadienyl ligands appeared to have a noticeable electronically destabilizing influence, and this observation was consistent with reported photoelectron spectroscopic data.

It is of considerable importance to note that some of the general theoretical conclusions are supported experimentally by a Mössbauer spectroscopic study[270]. Selected data are presented in Table 5. The most dramatic difference to be observed involves the quadrupole splitting parameter, with the "open ferrocene" values being much lower than that for ferrocene. As the quadrupole splitting parameter of a given system is heavily dependent on the relative populations of the metal d orbitals, it is clear that the "open ferrocenes" possess much different d orbital populations than ferrocene itself. The theoretical study in fact suggested that the $d_{x^2-y^2}$ and d_{xy} populations for the open system were substantially greater than those in ferrocene, while the opposite observations were made for the d_{xz} and d_{yz} orbitals. Such differences readily explain the dramatic changes observed for the quadrupole splitting parameter. Thus, the Mössbauer data provides important experimental support of the conclusions reached in the theoretical work. Also important to note are the significantly lower values of the isomer shift parameters for the "open ferrocenes". These values demonstrate that the open system iron compounds possess greater s electron density at the iron nucleus, which could come about either as a result of the pentadienyl ligand being a better σ donor, or as a result of a higher positive charge on iron, which would serve to contract the s electron density toward the nucleus. In fact, infrared studies on both $V(2,4\text{-}C_7H_{11})_2(CO)$[255] (Sect. D.III) and $Mn(C_5H_7)(CO)_3$[219] have demonstrated that the pentadienyl ligands withdraw more electron density from their respective metals than do the cyclopentadienyl ligands in their counterparts, $V(C_5H_5)_2CO$ and $Mn(C_5H_5)(CO)_3$, which tends to support the charge explanation. However, σ donor effects could also play a role.

Two molecular orbital studies have thus far been reported for pentadienyl complexes possessing metal-metal interactions. The first of these was a semiempirical INDO calculation on $[Ni(C_5H_7)]_2$ (Sect. D.I) combined with a photoelectron spectroscopic study of the compound[271]. As in the calculations of the "open ferrocenes", the nickel compound demonstrated a great deal of mixing between metal and ligand orbitals – five orbitals possessing comparable (30–70%) character of each. The Wiberg bond indices were as follows: Ni–C(1) = 0.248, Ni–C(2) = 0.110, Ni–C(3) = 0.146, Ni–Ni = 0.089, C(1)–C(2) = 1.582, and C(2)–C(3) = 1.229. These may be compared to the following for bis(allyl) nickel: Ni–C(1) = 0.284, Ni–C(2) = 0.071, C–C = 1.460. In both cases, as in the "open ferrocenes", the strongest metal-carbon interactions are seen to involve the formally charged carbon atoms. Also clear is the fact that the covalent coupling between nickel atoms is rather weak. As would be expected, less charge is localized on nickel in the

Table 5. Selected Mössbauer parameters for ferrocene and various "open ferrocenes" (mm/s)

Parameter	$Fe(C_5H_5)_2$	$Fe(C_5H_7)_2$	$Fe(2\text{-}C_6H_9)_2$	$Fe(3\text{-}C_6H_9)_2$	$Fe(2,3\text{-}C_7H_{11})_2$	$Fe(2,4\text{-}C_7H_{11})_2$
Isomer Shift	0.542(9)	0.479(10)	0.462(5)	0.482(5)	0.461(3)	0.498(7)
Quadrupole Splitting	2.453(17)	1.402(15)	1.206(8)	1.255(41)	1.261(6)	1.516(10)

formally univalent pentadienyl compound (+0.545) compared to bis(allyl)nickel (+0.783).

The second polymetallic pentadienyl compound to be studied by theoretical calculations was $Mn_3(3\text{-}C_6H_9)_4$[266]. As for the previous nickel compound, a semiempirical INDO model was employed for the study. Three electronic configurations were considered – a doublet with spin +3/2 on the central manganese atom and −1/2 on each of the terminal atoms, a quartet with spin +1/2 on each manganese atom, and a sextet with spin +3/2 on the central atom and +1/2 on each of the terminal ones. The doublet and sextet were found to be essentially identical in energy, but the quartet was much higher – 29.8 kcal/mol. Magnetic studies from 2–300 °K have in fact provided support for a sextet ground state. Interestingly, in contrast to the initial reasonable formulation of the compound as an associated salt, $Mn^{2+}[Mn(3\text{-}C_6H_9)_2^-]_2$, the calculations indicate substantial net negative charge ($-0.16|e|$) on the central manganese atom, and net positive charge on the terminal ones ($0.25|e|$) each). The relative d orbital splitting patterns and Wiberg bond indices for the terminal manganese centers are reasonably similar to those observed for the "open ferrocenes", regardless of spin state. However, those for the central manganese atom differ substantially, and vary noticeably with spin state. The Wiberg bond indices for the central manganese atom to the proximate four terminal carbon atoms average only 0.079 for the sextet state, and to the other four terminal carbon atoms average only 0.021. The Mn–Mn Wiberg bond indices were also small, being only 0.027 for the sextet state. This value is substantially less than that for the previous nickel compound, and indicates that the manganese-manganese interactions are dominated by electrostatic forces as a result of substantial but opposite charges existing on adjacent manganese centers. The calculations would therefore almost seem to describe the central manganese atom as a nearly "naked" manganese atom. However, if it is possible to bring about Lewis base coordination to this atom, it seems quite likely that net positive charge could be induced onto it, more akin to the original formulation as Mn^{2+} $[Mn(3\text{-}C_6H_9)_2^-]_2$[248].

E. Concluding Remarks

It is now quite clear that pentadienyl ligands possess many unique and important characteristics, and these have already been manifested in the large and very significant series of reports dealing with both main group and transition metal pentadienyl chemistry. Yet metal-pentadienyl chemistry is still only in a very early stage, and much more remains to be discovered, not only chemically but also with regard to electronic, theoretical, structural, conformational, and other spectroscopic aspects. Many more interesting series of compounds will undoubtedly be prepared, and the metal-pentadienyl field might ultimately rival even the fields of its better known counterparts, the allyl and cyclopentadienyl ligands. In any case it is already evident that many structural and bonding trends in the metal-pentadienyl compounds are often analogous to trends observed for these better known counterparts. Hence it can be expected that the study of the pentadienyl system should also serve to bridge the gap between the other two, thereby providing a more unified understanding of the various open or closed, 3- or 5-membered ligand systems.

F. Abbreviations

THF	tetrahydrofuran
TMEDA	N,N,N',N'-tetramethylethylenediamine
C_6H_9	methylpentadienyl
C_7H_{11}	dimethylpentadienyl
C_8H_{13}	trimethylpentadienyl
C_9H_{15}	tetramethylpentadienyl
C_7H_7	cycloheptatrienyl
C_7H_9	cycloheptadienyl
CNDO	complete neglect of differential overlap
INDO	intermediate neglect of differential overlap
MNDO	moderate neglect of differential overlap
SCF	self-consistent field

G. References

1. Kealy, T. J., Pauson, P. L.: Nature (London) *168,* 1039 (1951)
2. Miller, S. A., Tebboth, J. A., Tremaine, J. F.: J. Chem. Soc. 632 (1952)
3. Wilkinson, G.: J. Organometal. Chem. *100,* 273 (1975)
4. Lauher, J. W., Hoffmann, R.: J. Am. Chem. Soc. *98,* 1729 (1976)
5. Sohn, Y. S., Hendrickson, D. N., Gray, H. B.: ibid. *93,* 3603 (1971)
6. Ammeter, J. H. et al.: ibid. *100,* 3686 (1978)
7. Bagus, P. S., Wahlgren, U. I., Almlof, J.: J. Chem. Phys. *64,* 2324 (1976)
8. Jonassen, H. B. et al.: J. Am. Chem. Soc. *80,* 2586 (1958)
9. Jolly, P. W., Wilke, G.: The Organic Chemistry of Nickel, Vols. I and II, Academic Press, New York (1974, 1975)
10. Kaduk, J. A., Poulos, A. T., Ibers, J. A.: J. Organometal. Chem. *127,* 245 (1977)
11. Putnik, C. F. et al.: J. Am. Chem. Soc. *100,* 4107 (1978)
12. Clarke, H. L.: J. Organometal. Chem. *80,* 155 (1974)
13. Brown, D. A., Owens, A.: Inorg. Chim. Acta. *5,* 675 (1971)
14. Rohmer, M.-M., Demuynck, J., Veillard, A.: Theor. Chim. Acta. *36,* 93 (1974)
15. Rösch, N., Hoffmann, R.: Inorg. Chem. *13,* 2656 (1974)
16. Böhm, M. C., Gleiter, R., Batich, C. D.: Helv. Chim. Acta. *63,* 990 (1980)
17. Perevalova, E. G., Nikitina, T. V.: in Organometallic Reactions, Vol. 4 (Becker, E. I., Tsutsui, M., eds.) Wiley-Interscience, New York 1972, p. 163
18. Wilke, G. et al.: Angew. Chem. Intl. Ed. Engl. *5,* 151 (1966)
19. Bönnemann, H.: ibid. *12,* 964 (1973)
20. Mahler, J. E., Pettit, R.: J. Am. Chem. Soc. *84,* 1511 (1962)
21. Streitwieser, A., Jr.: Molecular Orbital Theory for Organic Chemists, Wiley, New York 1961
22. Egger, K. W., Golden, D. M., Benson, S. W.: J. Am. Chem. Soc. *86,* 5420 (1964)
23. Golden, D. M., Rodgers, A. S., Benson, S. W.: ibid. *88,* 3196 (1966)
24. Field, R. J., Abell, P. I.: ibid. *91,* 7226 (1969)
25. Golden, D. M., Gac, N. A., Benson, S. W.: ibid. *91,* 2136 (1969)
26. Doering, W. von E., Beasley, G. H.: Tetrahedron *29,* 2231 (1973)
27. Egger, K. W., Cocks, A. T.: Helv. Chim. Acta. *56,* 1537 (1973)
28. Zimmerman, A. H., Brauman, J. I.: J. Am. Chem. Soc. *99,* 3565 (1977)
29. Rossi, M., King, K. D., Golden, D. M.: ibid. *101,* 1223 (1979)

30. Korth, H.-G., Trill, H., Sustmann, R.: ibid. *103*, 4483 (1981)
31. Egger, K. W., Benson, S. W.: ibid. *88*, 241 (1966)
32. James, D. G. L., Suart, R. D.: Chem. Commun., 484 (1966)
33. Frey, H. M., Krantz, A.: J. Chem. Soc. A, 1159 (1969)
34. Egger, K. W., Jola, M.: Intl. J. Chem. Kin. *2*, 265 (1970)
35. Alonso, J. H., Dolbier, W. R., Jr., Frey, H. M.: ibid. *6*, 893 (1974)
36. Sustmann, R., Schmidt, H.: Chem. Ber. *112*, 1440 (1979)
37. Trenwith, A. B.: J. Chem. Soc. Faraday I *76*, 266 (1980)
38. Davies, A. G. et al.: J. Chem. Soc. Perkin II, 633 (1981)
39. Lossing, F. P.: Canad. J. Chem. *49*, 357 (1971)
40. Sen Sharma, D. K., Franklin, J. L.: J. Am. Chem. Soc. *95*, 6562 (1973)
41. Egger, K. W., Cocks, A. T.: Helv. Chim. Acta *56*, 1516 (1973)
42. Rossi, M., Golden, D. M.: J. Am. Chem. Soc. *101*, 1230 (1979)
43. Golden, D. M., Benson, S. W.: Chem. Rev. *69*, 125 (1969)
44. Furuyama, S., Golden, D. M., Benson, S. W.: Intl. J. Chem. Kin. *3*, 237 (1971)
45. DeFrees, D. J., McIver, R. T., Jr., Hehre, W. J.: J. Am. Chem. Soc. *102*, 3334 (1980)
46. Tachikawa, E., Tang, Y.-N., Rowland, F. S.: ibid. *90*, 3584 (1968)
47. Mackay, G. I. et al.: Canad. J. Chem. *56*, 131 (1978)
48. Zimmerman, A. H., Gygax, R., Brauman, J. I.: J. Am. Chem. Soc. *100*, 5595 (1978)
49. Richardson, J. H., Stephenson, L. M., Brauman, J. I.: J. Chem. Phys. *59*, 5068 (1973)
50. Engelking, P. C., Lineberger, W. C.: ibid. *67*, 1412 (1977)
51. Houle, F. A., Beauchamp, J. L.: J. Am. Chem. Soc. *100*, 3290 (1978)
52. Pignataro, S., Cassuto, A., Lossing, F. P.: ibid. *89*, 3693 (1967)
53. Streitwieser, A., Jr., Nair, P. M.: Tetrahedron *5*, 149 (1959)
54. Harrison, A. G. et al.: J. Am. Chem. Soc. *82*, 5593 (1960)
55. Pottie, R. F., Lossing, F. P.: ibid. *85*, 269 (1963)
56. Arnett, E. M., Johnston, D. E., Small, L. E.: ibid. *97*, 5598 (1975)
57. McMahon, T. B., Kebarle, P.: ibid. *98*, 3399 (1976)
58. Bartmess, J. E., Scott, J. A., McIver, R. T., Jr.: ibid. *101*, 6046 (1979)
59. Jaun, B., Schwarz, J., Breslow, R.: ibid. *102*, 5741 (1980)
60. Grutzner, J. B., Jorgensen, W. L.: ibid. *103*, 1372 (1981)
61. Breslow, R., Washburn, W.: ibid. *92*, 427 (1970)
62. Bordwell, F. G. et al.: ibid. *97*, 3226 (1975)
63. Streitwieser, Jr., A., Nebenzahl, L. L.: ibid. *98*, 2188 (1976)
64. Breslow, R., Grant, J. L.: ibid. *99*, 7745 (1977)
65. Schlosser, M., Schneider, P.: Helv. Chim. Acta. *63*, 2404 (1980)
66. Bates, R. B., Gosselink, D. W., Kaczynski, J. A.: Tet. Lett. *3*, 199 (1967)
67. Bates, R. B. et al.: J. Am. Chem. Soc. *95*, 926 (1973)
68. Bates, R. B. et al.: ibid. *92*, 6345 (1970)
69. Brenner, S., Klein, J.: Isr. J. Chem. *7*, 735 (1969)
70. Zimmerman, H. E. et al.: J. Am. Chem. Soc. *101*, 6367 (1979)
71. Oppolzer, W., Snowden, R. L., Simmons, D. P.: Helv. Chim. Acta. *64*, 2002 (1981)
72. Schlosser, M., Rauchschwalbe, G.: J. Am. Chem. Soc. *100*, 3258 (1978)
73. Heiszwolf, G. J., Kloosterziel, H.: Recl. Trav. Chim. Pays-Bas *86*, 807 (1967)
74. Hunter, D. H. et al.: Can. J. Chem. *54*, 1464 (1976)
75. Oppolzer, W., Burford, S. C., Marazza, F.: Helv. Chim. Acta. *63*, 555 (1980)
76. Yasuda, H. et al.: Organometallics *2*, 21 (1983)
77. Yasuda, H., Narita, T., Tani, H.: Tet. Lett. *27*, 2443 (1973)
78. Yasuda, H. et al.: Bull. Chem. Soc. Jpn. *52*, 2036 (1979)
79. Newcomb, M., Ford, W. T.: J. Org. Chem. *39*, 232 (1974)
80. Bates, R. B. et al.: J. Am. Chem. Soc. *91*, 4608 (1969)
81. Kloosterziel, H., VanDrunen, J. A. A.: Recl. Trav. Chim. Pays-Bas *88*, 1084 (1969)
82. Bates, R. B., Brenner, S., Cole, C. M.: J. Am. Chem. Soc. *94*, 2130 (1972)
83. Bates, R. B., Brenner, S., Mayall, B. I.: ibid. *94*, 4765 (1972)
84. Bates, R. B. et al.: ibid. *96*, 5640 (1974)
85. Bahl, J. J. et al.: ibid. *99*, 6126 (1977)
86. Bates, R. B. et al.: ibid. *103*, 5052 (1981)

87. Klein, J., Glily, S., Kost, D.: J. Org. Chem. *35,* 1281 (1970)
88. Klein, J., Glily, S.: Tetrahedron *27,* 3477 (1971)
89. Marvell, E. N.: in Thermal Electrocyclic Reactions (Wasserman, H. H., ed.) Academic Press, New York 1980
90. Deno, N. C.: in Isotopes in Organic Chemistry, Vol. I (Buncel, E., Lee, C. C., eds.) Elsevier, Amsterdam 1975
91. Sorensen, T. S., Rauk, A.: in Pericyclic Reactions, Vol. II (Marchand, A. P., Lehr, R. E., eds.) Academic Press, New York 1977
92. Huisgen, R.: Angew. Chem. Intl. Ed. Engl. *19,* 947 (1980)
93. Bates, R. B., McCombs, D. A.: Tet. Lett. *12,* 977 (1969)
94. Hunter, D. H., Sim, S. K.: Canad. J. Chem. *50,* 669, 678 (1972)
95. Hunter, D. H., Steiner, R. P.: ibid. *53,* 355 (1975)
96. Shoppee, C. W., Henderson, G. N.: J. Chem. Soc., Perkin I, 1028 (1977)
97. Schmidt, R. R.: Angew. Chem. Intl. Ed. Engl. *14,* 581 (1975)
98. Perkins, M. J., Ward, P.: J. Chem. Soc., Perkin I, 667 (1974)
99. Deno, N. C., Pittman, C. U., Jr.: J. Am. Chem. Soc. *86,* 1871 (1964)
100. Sorensen, T. S.: Canad. J. Chem. *42,* 2768 (1964)
101. Sorensen, T. S.: ibid. *43,* 2744 (1965)
102. Deno, N. C., Pittman, C. U., Jr., Turner, J. O.: J. Am. Chem. Soc. *87,* 2153 (1965)
103. Olah, G. A., Pittman, C. U., Jr., Sorensen, T. S.: ibid. *88,* 2331 (1966)
104. Campbell, P. H. et al.: ibid. *91,* 6404 (1969)
105. Sorensen, T. S.: ibid. *89,* 3782 (1967)
106. Sorensen, T. S., Rajeswari, K.: ibid. *93,* 4222 (1971)
107. Bladek, R., Sorensen, T. S.: Canad. J. Chem. *50,* 2806 (1972)
108. Brookhart, M., Davis, E. R.: J. Am. Chem. Soc. *92,* 7622 (1970)
109. Hegarty, A. F., Coy, J. H., Scott, F. L.: J. Chem. Soc., Perkin II, 104 (1975)
110. Lehr, R. E. et al.: J. Am. Chem. Soc. *98,* 4867 (1976)
111. Sustmann, R., Lübbe, F.: ibid. *98,* 6037 (1976)
112. Olah, G. A. et al.: ibid. *100,* 6299 (1978)
113. Schlosser, M., Stähle, M.: Angew. Chem. Intl. Ed. (Engl.) *21,* 145 (1982)
114. O'Brien, D. H., Hart, A. J., Russell, C. R.: J. Am. Chem. Soc. *97,* 4410 (1975)
115. Olah, G. A., Yu, S. H., Liang, G.: J. Org. Chem. *41,* 2383 (1976)
116. Birch, A. J., Westerman, P. W., Pearson, A. J.: Aust. J. Chem. *29,* 1671 (1976)
117. Wilson, D. R., Ernst, R. D., Cymbaluk, T. H.: Organometallics *2,* 1220 (1983)
118. Allinger, N. L., Siefert, J. H.: J. Am. Chem. Soc. *97,* 752 (1975)
119. Olah, G. A., Bollinger, J. M.: ibid. *90,* 6082 (1968)
120. Bollinger, J. M., Brinich, J. M., Olah, G. A.: ibid. *92,* 4025 (1970)
121. Deno, N. C., Haddon, R. C., Nowak, E. N.: ibid. *92,* 6691 (1970)
122. Freedman, H. H., Sandel, V. R., Thill, B. P.: ibid. *89,* 1762 (1967)
123. Schleyer, P. v. R. et al.: ibid. *91,* 5174 (1969)
124. Thompson, T. B., Ford, W. T.: ibid. *101,* 5459 (1979)
125. Bates, R. B., Gosselink, D. W., Kaczynski, J. A.: Tet. Lett. *3,* 205 (1967)
126. Staley, S. W.: in Pericyclic Reactions, Vol. I (Marchand, A. P., Lehr, R. E., eds.) Academic Press, New York 1977
127. Hunter, D. H.: in Isotopes in Organic Chemistry, Vol. I (Buncel, E., Lee, C. C., eds.) Elsevier, Amsterdam 1975
128. Yasuda, H. et al.: Bull. Chem. Soc. Jpn. *54,* 1481 (1981)
129. Ford, W. T., Newcomb, M.: J. Am. Chem. Soc. *96,* 309 (1974)
130. Yasuda, H., Tani, H.: J. Macromol. Chem. *9,* 1007 (1975)
131. Bates, R. B., Kroposki, L. M., Potter, D. E.: J. Org. Chem. *37,* 560 (1972)
132. Heiszwolf, G. J., Van Drunen, J. A. A., Kloosterziel, H.: Recl. Trav. Chim. Pays-Bas *88,* 1377 (1969)
133. Kloosterziel, H., Van Drunen, J. A. A.: ibid. *89,* 270 (1970)
134. Kloosterziel, H.: ibid. *92,* 1167 (1973)
135. Kloosterziel, H., Van Drunen, J. A. A.: ibid. *88,* 1471 (1969)
136. Yasuda, H., Yamauchi, M., Nakamura, A.: J. Organometal. Chem. *202,* C1 (1980)
137. Kloosterziel, H., Van Drunen, J. A. A., Galama, P.: Chem. Commun., 885 (1969)

138. Yasuda, H. et al.: Bull. Chem. Soc. Jpn. *53,* 1089 (1980)
139. Yasuda, H. et al.: ibid. *53,* 1101 (1980)
140. Benn, R. et al.: J. Organometal. Chem. *146,* 103 (1978)
141. Kochi, J. K., Krusic, P. J.: J. Am. Chem. Soc. *90,* 7157 (1968)
142. Crawford, R. J., Hamelin, J., Strehlke, B.: ibid. *93,* 3810 (1971)
143. Edge, D. J., Kochi, J. K.: ibid. *94,* 6485 (1972)
144. Kawamura, T., Meakin, P., Kochi, J. K.: ibid. *94,* 8065 (1972)
145. Gorton, P. J., Walsh, R.: J. Chem. Soc., Chem. Commun.: 783 (1972)
146. Hefter, H. J., Wu, C.-H. S., Hammond, G. S.: J. Am. Chem. Soc. *95,* 851 (1973)
147. Sustmann, R., Trill, H.: ibid. *96,* 4343 (1974)
148. Krusic, P. J., Meakin, P., Smart, B. E.: ibid. *96,* 6211 (1974)
149. Smart, B. E. et al.: ibid. *96,* 7382 (1974)
150. Griller, D., Cooper, J. W., Ingold, K. U.: ibid. *97,* 4269 (1975)
151. Sustmann, R., Brandes, D.: Chem. Ber. *109,* 354 (1976)
152. Sustmann, R. et al.: ibid. *110,* 255 (1977)
153. Bascetta, E. et al.: J. Chem. Soc., Chem. Commun.: 110 (1982)
154. Griller, D., Ingold, K. U., Walton, J. C.: J. Am. Chem. Soc. *101,* 758 (1979)
155. Ingold, K. U., Walton, J. C.: J. Chem. Soc., Chem. Commun.: 604 (1980)
156. Yim, M. B., Wood, D. E.: J. Am. Chem. Soc. *97,* 1004 (1975)
157. Griller, D. et al.: ibid. *97,* 5526 (1975)
158. Kira, M., Sakurai, H.: ibid. *99,* 3892 (1977)
159. Griller, D. et al.: ibid. *103,* 7761 (1981)
160. Krusic, P. J., Kochi, J. K.: ibid. *90,* 7155 (1968)
161. Kira, M., Watanabe, M., Sakurai, H.: ibid. *102,* 5202 (1980)
162. Sass, R. L., Nichols, T. D.: Z. Kristallogr. *140,* 1 (1974)
163. Edmonds, J., Herdklotz, J. K., Sass, R. L.: Acta. Cryst. *B 26,* 1355 (1970)
164. Gardner, H. C., Kochi, J. K.: J. Am. Chem. Soc. *98,* 558 (1976)
165. Hoffmann, R., Olofson, R. A.: ibid. *88,* 943 (1966)
166. Arora, S. K. et al.: ibid. *97,* 6271 (1975)
167. Ernst, R. D., Cymbaluk, T. H.: Organometallics *1,* 708 (1982)
168. Cymbaluk, T. H., Liu, J.-Z., Ernst, R. D.: J. Organometal. Chem., *255,* 311 (1983)
169. Strauss, M. J.: Chem. Rev. *70,* 667 (1970)
170. Destro, R., Gramaccioli, C. M., Simonetta, M.: Acta Cryst. *24 B,* 1369 (1968)
171. Ueda, H. et al.: Bull. Chem. Soc. Jpn. *41,* 2866 (1968)
172. Baenziger, N. C., Nelson, A. D.: J. Am. Chem. Soc. *90,* 6602 (1968)
173. Dewar, M. J. S., Hashmall, J. A., Venier, C. G.: ibid. *90,* 1953 (1968)
174. Carsky, P., Zahradnik, R.: J. Phys. Chem. *74,* 1249 (1970)
175. Baird, N. C.: Tetrahedron *28,* 2355 (1972)
176. Kispert, L. D. et al.: J. Am. Chem. Soc. *94,* 5979 (1972)
177. Jorgensen, P., Bellum, J.: Mol. Phys. *26,* 725 (1973)
178. Atkinson, A., Hopkinson, A. C., Lee-Ruff, E.: Tetrahedron *30,* 2023 (1974)
179. Hinchliffe, A., Cobb, J. C.: J. Mol. Struct. *23,* 273 (1974)
180. Kloosterziel, H.: Recl. Trav. Chim. Pays-Bas *93,* 215 (1974)
181. Hinchliffe, A.: J. Mol. Struct. *27,* 329 (1975)
182. Haddon, R. C.: Aust. J. Chem. *30,* 1 (1977)
183. Tezuka, T. et al.: J. Am. Chem. Soc. *103,* 1367 (1981)
184. Herndon, W. C.: J. Org. Chem. *46,* 2119 (1981)
185. Takada, T., Dupuis, M.: J. Am. Chem. Soc. *105,* 1713 (1983)
186. Bushby, R. J., Patterson, A. S.: J. Organometal. Chem. *132,* 163 (1977)
187. Bongini, A. et al.: ibid. *92,* C 1 (1975)
188. Dewar, M. J. S., Fox, M. A., Nelson, D. J.: ibid. *185,* 157 (1980)
189. Brauman, J. I., Blair, L. K.: J. Am. Chem. Soc. *92,* 5986 (1970)
190. Hine, J., Hine, M.: ibid. *74,* 5266 (1952)
191. Schubert, W. M., Murphy, R. B., Robins, J.: J. Org. Chem. *35,* 951 (1970)
192. Bordwell, F. G., Drucker, G. E., McCollum, G. J.: ibid. *41,* 2786 (1976)
193. Charton, M.: J. Am. Chem. Soc. *99,* 5687 (1977)
194. Cumming, J. B., Kebarle, P.: ibid. *99,* 5818 (1977)

195. Mayr, H., Förner, W., Schleyer, P. v. R.: ibid. *101,* 6032 (1979)
196. Grunwell, J. R., Sebastian, J. F.: Tetrahedron *27,* 4387 (1971)
197. Sebastian, J. F., Hsu, B., Grunwell, J. R.: J. Organometal. Chem. *105,* 1 (1976)
198. Peyerimhoff, S. D., Buenker, R. J.: J. Chem. Phys. *51,* 2528 (1969)
199. Buenker, R. J., Peyerimhoff, S. D.: Theor. Chim. Acta *24,* 132 (1972)
200. Radom, L. et al.: J. Am. Chem. Soc. *95,* 6531 (1973)
201. Boerth, D. W., Streitwieser, A., Jr.: ibid. *100,* 750 (1978)
202. Libit, L., Hoffmann, R.: ibid. *96,* 1370 (1974)
203. Brauman, J. I., Blair, L. K.: ibid. *93,* 4315 (1971)
204. DeFrees, D. J. et al.: ibid. *99,* 6451 (1977)
205. Pross, A., Radom, L.: ibid. *100,* 6572 (1978)
206. Bartmess, J. E., Scott, J. A., McIver, Jr., R. T.: ibid. *101,* 6056 (1979)
207. Birch, A. J., Hinde, A. L., Radom, L.: ibid. *102,* 4074 (1980)
208. Birch, A. J., Hinde, A. L., Radom, L.: ibid. *102,* 6430 (1980)
209. Barrow, M. J. et al.: J. Chem. Soc., Chem. Commun., 1239 (1971)
210. Semmelhack, M. F. et al.: J. Am. Chem. Soc. *101,* 3535 (1979)
211. Bird, P. H., Churchill, M. R.: Chem. Commun., 145 (1968)
212. Mawby, A., Walker, P. J. C., Mawby, R. J.: J. Organometal. Chem. *55,* C39 (1973)
213. Bird, P. H., Churchill, M. R.: Chem. Commun., 777 (1967)
214. Churchill, M. R., Julis, S. A.: Inorg. Chem. *17,* 2951 (1978)
215. Van Vuuren, P. J. et al.: J. Am. Chem. Soc. *93,* 4394 (1971)
216. Howard, J. A. K. et al.: J. Chem. Soc., Chem. Commun., 673 (1974)
217. Churchill, M. R., Scholer, F. R.: Inorg. Chem. *8,* 1950 (1969)
218. Abel, E. W., Moorhouse, S.: J. Chem. Soc., Dalton Trans., 1706 (1973)
219. Seyferth, D., Goldman, E. W., Pornet, J.: J. Organometal. Chem. *208,* 189 (1981)
220. Paz-Sandoval, M. d. l. A., Powell, P.: ibid. *219,* 81 (1981)
221. Nesmeyanov, A. N. et al.: ibid. *111,* C9 (1976)
222. Bottrill, M. et al.: J. Chem. Soc., Dalton Trans., 292 (1980)
223. Bailey, N. A. et al.: ibid. 829 (1980)
224. Espinet, P. et al.: ibid. 1048 (1980)
225. Kirillova, N. I. et al.: J. Organometal. Chem. *63,* 311 (1973)
226. Clinton, N. A., Lillya, C. P.: J. Am. Chem. Soc. *92,* 3065 (1970)
227. Sorensen, T. S., Jablonski, C. R.: J. Organometal. Chem. *25,* C62 (1970)
228. Lillya, C. P., Sahatjian, R. A.: ibid. *25,* C67 (1970)
229. Brookhart, M., Harris, D. L.: ibid. *42,* 441 (1972)
230. Rienäcker, R., Yoshiura, H.: Angew. Chem. Intl. Ed. Engl. *8,* 677 (1969)
231. Krüger, C.: ibid. *8,* 678 (1969)
232. Sappa, E., Milone, L., Tiripicchio, A.: J. Chem. Soc., Dalton Trans., 1843 (1976)
233. King, J. A., Jr., Vollhardt, K. P. C.: J. Am. Chem. Soc. *105,* 4846 (1983)
234. Werner, H.: in Advances in Organometallic Chemistry, Vol. 19 (Stone, F. G. A., West, R., eds.) Academic Press, New York 1981, p. 155
235. Bryan, E. G. et al.: J. Chem. Soc., Chem. Commun., 254 (1976)
236. Wilson, D. R., Ernst, R. D.: submitted for publication
237. Jotham, R. W. et al.: J. Organometal. Chem. *118,* 59 (1976)
238. Sapienza, R. S. et al.: ibid. *121,* C35 (1976)
239. Connelly, N. G. et al.: J. Chem. Soc., Dalton Trans., 1317 (1981)
240. Mingos, D. M. P.: in Comprehensive Organometallic Chemistry, Vol. 3 (Wilkinson, G., Stone, F. G. A., Abel, E. W., eds.) Pergamon Press, Oxford 1982, Chapt. 19
241. Campana, C. F. et al.: Inorg. Chem. *23* (1984)
242. Wilson, D. R., Ernst, R. D., Cymbaluk, T. H.: Organometallics *2,* 1220 (1983)
243. Stahl, L., Ernst, R. D.: ibid. *2,* 1229 (1983)
244. Liu, J.-Z., Ernst, R. D.: J. Am. Chem. Soc. *104,* 3737 (1982)
245. Haaland, A.: Acc. Chem. Res. *12,* 415 (1979)
246. Haaland, A., Nilsson, J. E.: Acta. Chem. Scand. *22,* 2653 (1968)
247. Hardgrove, G. L., Templeton, D. H.: Acta. Cryst. *12,* 28 (1959)
248. Wilson, D. R., Liu, J.-Z., Ernst, R. D.: J. Am. Chem. Soc. *104,* 1120 (1982)

249. Pez, G. P., Armor, J. N.: in Advances in Organometallic Chemistry, Vol. 19 (Stone, F. G. A., West, R., eds.) Academic Press, New York 1981, p. 1
250. Bercaw, J. E. et al.: J. Am. Chem. Soc. *94,* 1219 (1972)
251. Mathew, M., Palenik, G. J.: Inorg. Chem. *11,* 2809 (1972)
252. Blackborow, J. R. et al.: J. Chem. Soc., Dalton Trans., 2205 (1977)
253. Schmid, H., Ziegler, M. L.: Chem. Ber. *109,* 125 (1976)
254. Böhm, M. C. et al.: J. Am. Chem. Soc. *104,* 2699 (1982)
255. Ernst, R. D., Liu, J.-Z., Wilson, D. R.: J. Organometal. Chem. *250,* 257 (1983)
256. Wilson, D. R., Ernst, R. D.: submitted for publication
257. Fieselmann, B. F., Stucky, G. D.: J. Organometal. Chem. *137,* 43 (1977)
258. Cetinkaya, B. et al.: ibid. *188,* C31 (1980)
259. Petersen, J. L., Dahl, L. F.: J. Am. Chem. Soc. *97,* 6416 (1975)
260. Petersen, J. L., Dahl, L. F.: ibid. *97,* 6422 (1975)
261. Petersen, J. L. et al.: ibid. *97,* 6433 (1975)
262. Henc, B. et al.: J. Organometal. Chem. *191,* 449 (1980)
263. Huttner, G., Neugebauer, D., Razavi, A.: Ang. Chem. Intl. Ed. *14,* 352 (1975)
264. Krüger, C., Tsay, Y.-H.: ibid. *10,* 261 (1971)
265. Koerner von Gustorf, E., Jaenicke, O., Polansky, O. E.: ibid. *11,* 532 (1972)
266. Böhm, M. C. et al.: Inorg. Chem., *22,* 3815 (1983)
267. Whitesides, T. H., Lichtenberger, D. L., Budnik, R. A.: ibid. *14,* 68 (1975)
268. Hoffmann, R., Hofmann, P.: J. Am. Chem. Soc. *98,* 598 (1976)
269. Mingos, D. M. P., Nurse, C. R.: J. Organometal. Chem. *184,* 281 (1980)
270. Ernst, R. D., Wilson, D. R., Herber, R. H.: J. Am. Chem. Soc. *106,* 1646 (1984)
271. Böhm, M. C., Gleiter, R.: Chem. Phys. *64,* 183 (1982)

A New Look at the Stereochemistry and Electronic Properties of Complexes of the Copper(II) Ion*

Brian J. Hathaway

University College, Cork, Ireland

As a consequence of the non-spherical symmetry of the copper(II) ion, d^9 configuration, and of the influence of the Jahn-Teller, and pseudo Jahn-Teller effect on six-coordinate geometries, the stereochemistries of the copper(II) ion are characterized by non-rigid geometries (fluxional behaviour), and ranges of distorted geometries (Plasticity Effect). The latter may be connected by a series of Structural Pathways, which may be characterised by an 'Electronic Criterion of Stereochemistry' for a related series of complexes, e.g. the $[Cu(bipy)_2X][Y]$ complexes.

* This Article is dedicated to the late Professor Ing. JAN GAZO, Dr. Sc., who passed away suddenly on 15 February, 1983, in appreciation of his contribution to the chemistry of copper(II) over a period of thirty years

Structure and Bonding 57

1 Introduction

Like all transition metal ions the copper(II) ion is characterised by its ability to form coordination complexes[1-3] with a wide range of ligands, and illustrating a variety of stereochemistries[4-7]. It is less typical in that these stereochemistries are dominated by the formation of nonregular structures, involving significant bond length and bond angle distortions from the regular octahedral, tetrahedral and square coplanar geometries[4,7,8-10]. While the major responsibility for this atypical behaviour must arise from the d^9 electron configuration[4,11], and the operation of the Jahn-Teller effect per se in six-coordinate geometries[12-14]; there are secondary consequences of the Jahn-Teller effect that must also be considered. These are:
(a) The Plasticity Effect[6,7]
(b) Fluxional Copper(II) Stereochemistries[15,16]
(c) Cooperative Jahn-Teller Effect[17]
(d) Second Order Jahn-Teller effects[18].

If more classical chelate ligand effect and bulk ligand effect are also recognised[5], the stereochemical variety available in copper(II) structures is prodigous[4,9,11]. The ready availability of the technique of single-crystal X-ray crystallography over the past decade has produced a wealth of structure determination of copper(II) complexes (a factor of *five* times more than that for nickel(II). But while the structural data is now prolific, the correlation of this structural data with the electronic properties, such as electronic and e.s.r. spectra is much less clearly established[4,5,9,11,16]. An earlier attempt[4] to establish such a relationship, was limited to the more regular stereochemistries of the copper(II) ion, and is now sufficiently out of date to justify a new look at the question of the electronic properties and stereochemistry of the copper(II) ion. The present article is an attempt to do this. As the literature on the copper(II) ion is now very extensive[19-21], this reappraisal is carried out with particular emphasis on the authors own fields of interest, in order to keep the article to a reasonable length. In order to avoid duplication, material adequately covered in the earlier literature[4,11] will not be repeated. This particularly applies to the survey of copper(II) stereochemistries[4], and the establishment of their one-electron orbital energy levels[11].

2 The Jahn-Teller Effect

In order[5] to understand the occurrence of the regular octahedral stereochemistry[22,23] in $K_2Pb[Cu(NO_2)_6]$ [1], and the regular trigonal octahedral stereochemistry[24,25] of $[Cu(en)_3][SO_4]$, [2], it is necessary to look more closely at the structural consequences of the Jahn-Teller theorem[12,14] in respect to six-coordinate CuL_6 chromophores. Qualitatively[12,16] the electronic properties of the d^9 configuration of the copper(II) ion in an orbitally degenerate electronic ground state can no longer involve separately defined electronic and vibrational energies (the Born-Oppenheimer approximation), but a vibronic potential energy surface, as in Fig. 1 (a) is required[26]. The even mode of vibration of e_g symmetry, made up of two displacement coordinates, S_{2a}, and S_{2b}, Fig. 1 (b), is the only mode that can couple with the electronically degenerate ground state of E_g symmetry, of energy E_0, in a cubic system and remove the orbital degeneracy. The energy surfaces which arise from this coupling, E_- and E_+, take the form shown in Fig. 1 (a),

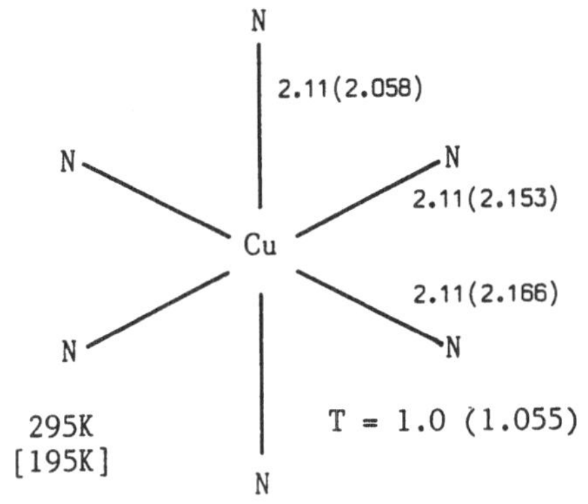

[1] $K_2Pb[Cu(NO_2)_6]$

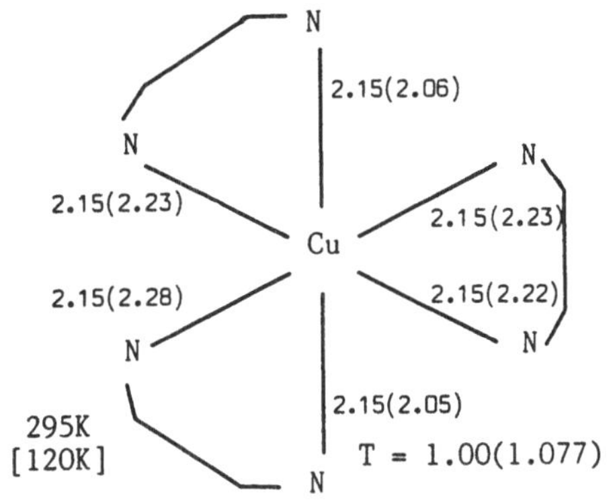

[2] $[Cu(en)_3][SO_4]$

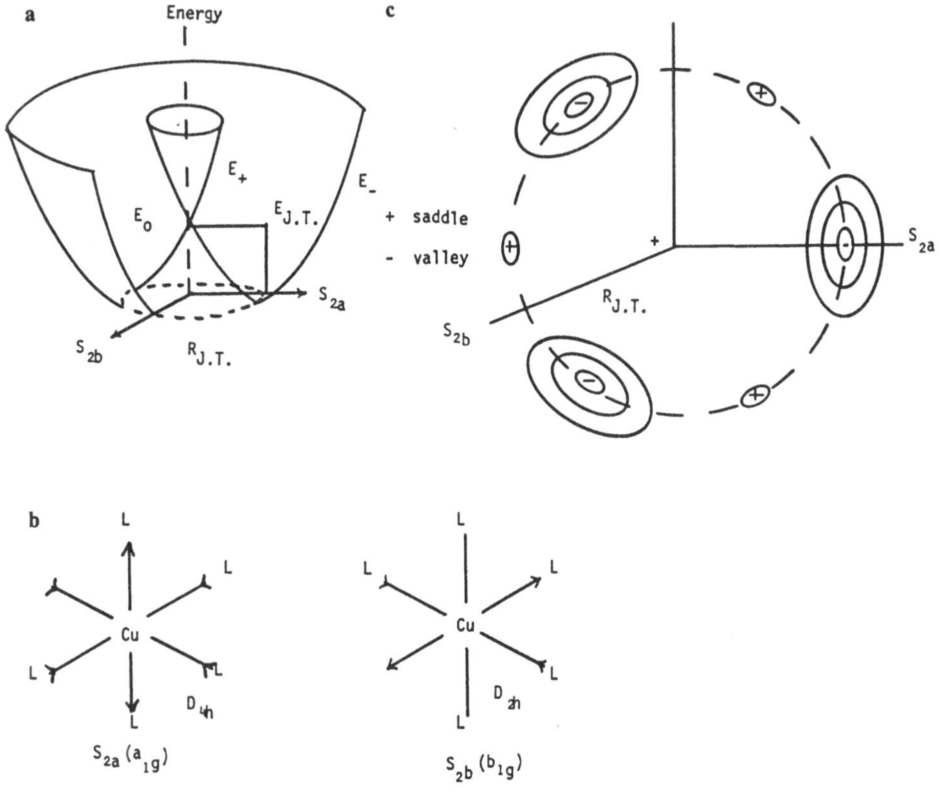

Fig. 1. **a** The adiabatic potential energy surface (Mexican Hat); **b** the normal coordinates S_{2a} and S_{2b}; **c** the projection of the potential energy surface **a** warped by the inclusion of higher order terms viewed down the principal axes of **a**, with $R_{J.T.}$ = radius of the minimum potential

and is known as the "Mexican Hat" model. The surface E_- involves a potential energy minimum. E_{JT}, the Jahn-Teller stabilisation energy, relative to E_o, at a distance R_{JT}, the Jahn-Teller radius, from the origin. If only first order coupling terms are involved the potential energy well has full cylindrical symmetry, but if strong Jahn-Teller coupling and higher order terms are involved the lower energy surface is warped; if the sign of the coupling constant is negative, Fig. 2 (A), minima occur in the potential surface for $\theta = 0$, 120 and 240°, and these values correspond at equilibrium to three equivalent elongated tetragonal octahedral distortions, D_{4h} symmetry, along the three orthogonal z, x and y axes, respectively, Fig. 2 (A), points I, II and III. In crystals, lattice packing effects and cooperative Jahn-Teller effects may further warp the potential energy surface such that the strict D_{4h} symmetry is removed. Fig. 2 illustrates the three different situations that can arise if three *elongated* tetragonal octahedral stereochemistries are involved, type (A), all three wells are of equal energy, type (B), two wells are equal and of low energy, and type (C), one low energy well occurs.

If all three wells are of equal energy, Fig. 2 (A), and if B < thermal energy kT (ca. 200 cm^{-1}), at any *one* copper site a *three* dimensional dynamic interconversion of the elongation axes occurs with equal thermal population of wells I, II and III. Consequently, the crystallographically determined structure will be octahedral, but is better described as a pseudo-octahedral stereochemistry, in view of the *three* dimensional dynamic behaviour involved. Figure 2 (B) depicts a situation that well III is of considerably higher energy than wells I and II, which are of approximately equal energy and hence approximately equally occupied. If, B < thermal (kT) then the two 90° misaligned CuL_6 chromophores are thermally accessible and at any *one* copper site a *two* dimensional dynamic interconversion of the elongation axes occurs. The crystallographically determined CuL_6 stereochemistry will appear compressed octahedral, but is better described as a *pseudo* compressed stereochemistry, in view of the *two* dimensional dynamic behaviour involved. If the potential energies of wells I and II are significantly different, Fig. 2 (C), and if B < thermal energy (kT), then the observed structure will be elongated rhombic octahedral, but the observed tetragonality will be high, T = 0.90, and temperature variable (T = mean inplane Cu-L bond distance/mean out-of-plane Cu-L bond distance). If B is larger than the thermal energy (kT), the elongated octahedral structure of well I, Fig. 1 (C), predominates and the *static* non-temperature variable elongated rhombic octahedral stereochemistry occurs, as found for the majority of elongated rhombic octahedral complexes of the copper(II) ion, with tetragonalities in region T = 0.80–0.85.

Thus, the three diagrams of Fig. 2 (A)–(C) contain all the structural situations required to account for the *six-coordinate* stereochemistries of the copper(II) ion. Figure 2 (A) describes the genuine dynamic Jahn-Teller systems[27)], such as in $K_2Pb[Cu(NO_2)_6]$ [1][22)] and $[Cu(en)_3][SO_4]$ [2][24, 25)], which all involve high symmetry chromophores (O_h or D_3) and involve six *equivalent* ligands. Figure 2 (B) accounts for the occurrence of the compressed octahedral stereochemistry[23, 28, 29)], as in $Cs_2Pb[Cu(NO_2)_6]$ [3] and $Rb_2Pb[Cu(NO_2)_6]$, [4], and Fig. 2 (C), accounts for the existence of

(i) fluxional elongated rhombic octahedral copper(II) complexes[30, 31)] as in $[NH_4]_2[Cu(OH_2)_6][SO_4]_2$, [5], and $[Cu(phen)_3][NO_3]_2 \cdot 2\ H_2O$, [6], and
(ii) of the *static* elongated rhombic octahedral complexes,[32)] as in $Cs_2[Cu(OH_2)_6][SO_4]_2$, [7], which dominates the stereochemistry of the copper(II) ion.

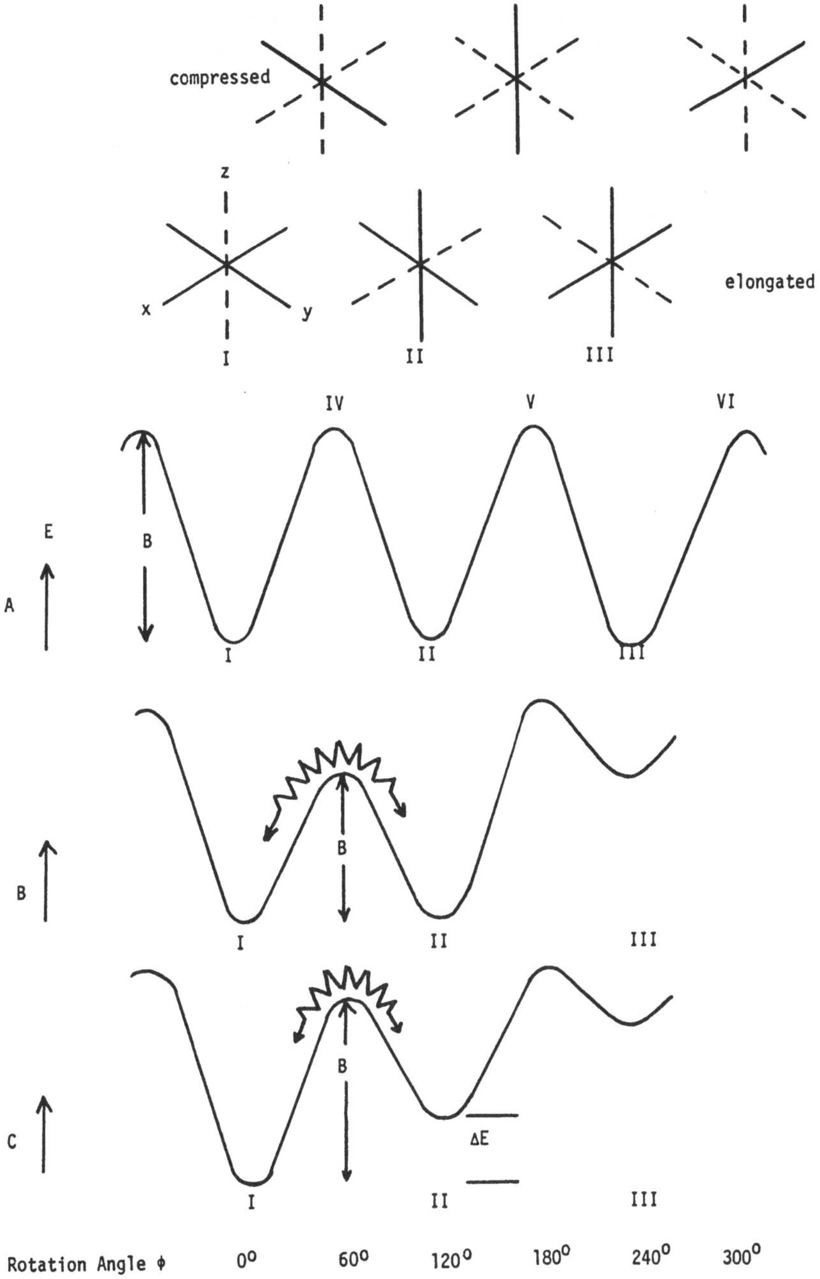

Fig. 2 A–C. The circular cross-section of the warped potential energy surface for **A** three wells of equal energy; **B** two wells of equal energy, and **C** one low energy well

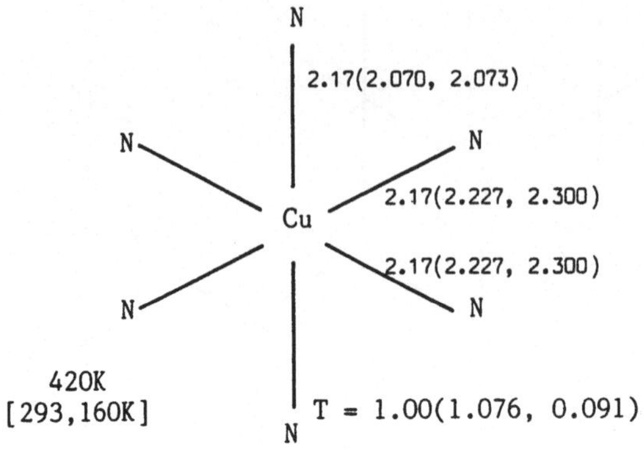

[3] $Cs_2Pb[Cu(NO_2)_6]$

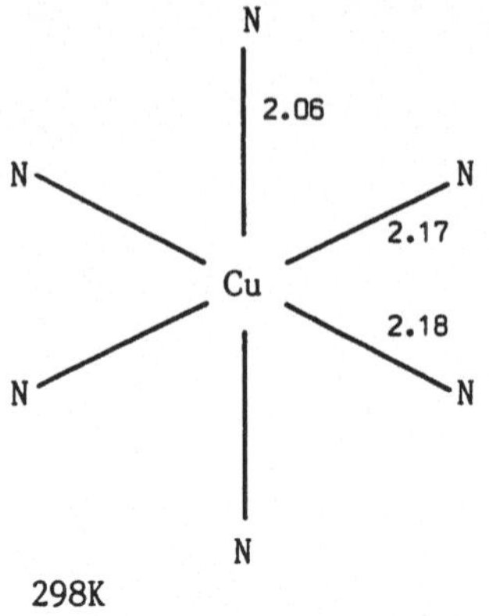

[4] $[Rb_2PbCu(NO_2)_6]$

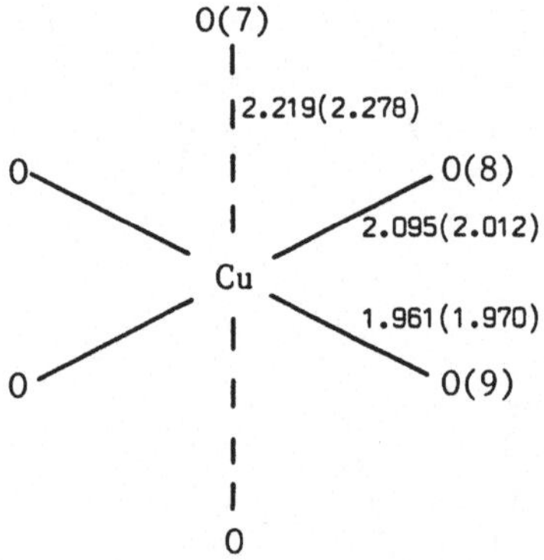

[5] $[NH_4]_2[Cu(OH_2)_6][SO_4]_2$

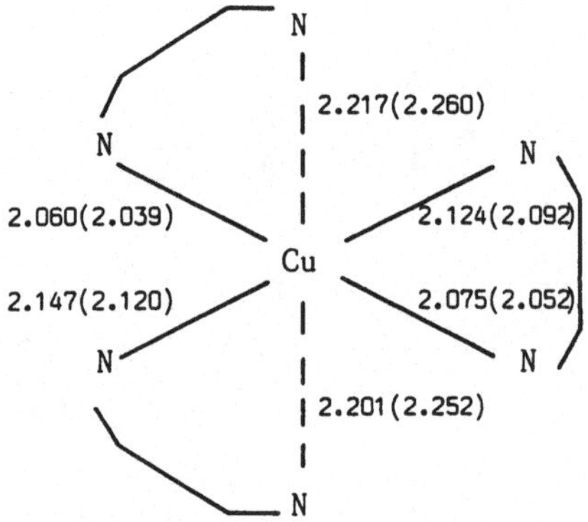

[6] $[Cu(phen)_3][NO_3]_2.2H_2O$

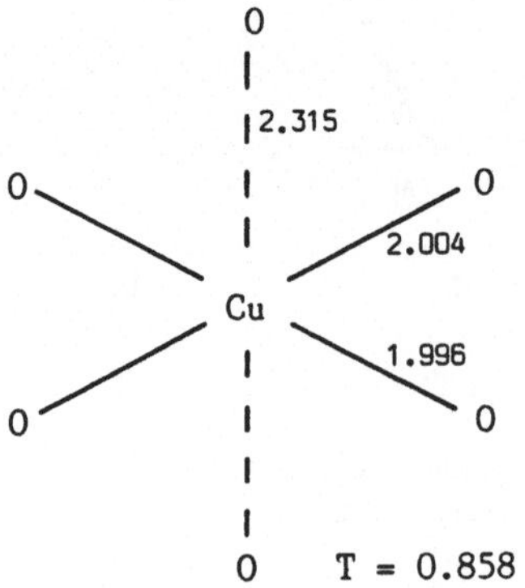

[7] $[Cs]_2[Cu(OH_2)_6][SO_4]_2$

3 The Pseudo Jahn-Teller Effect

The adiabatic potential energy surface of Figs. 1 and 2 only applies strictly to six-coordinate complexes of the copper(II) ion with six *equivalent* ligands. With non-equivalent ligands, as in a *cis* or *trans* CuL_4X_2 chromophore, the general behaviour of Fig. 2 can still be applied, except that the regular six-coordinate geometry of Fig. 2 (A) can never arise. The behaviour inherent in Fig. 2 (B and C) may be applied and is referred to as the pseudo Jahn-Teller effect, Fig. 3 (A), (B) and (C). Two potential energy surfaces are still involved which are split at the origin by $2\Delta'$ due to the small differences in bond lengths

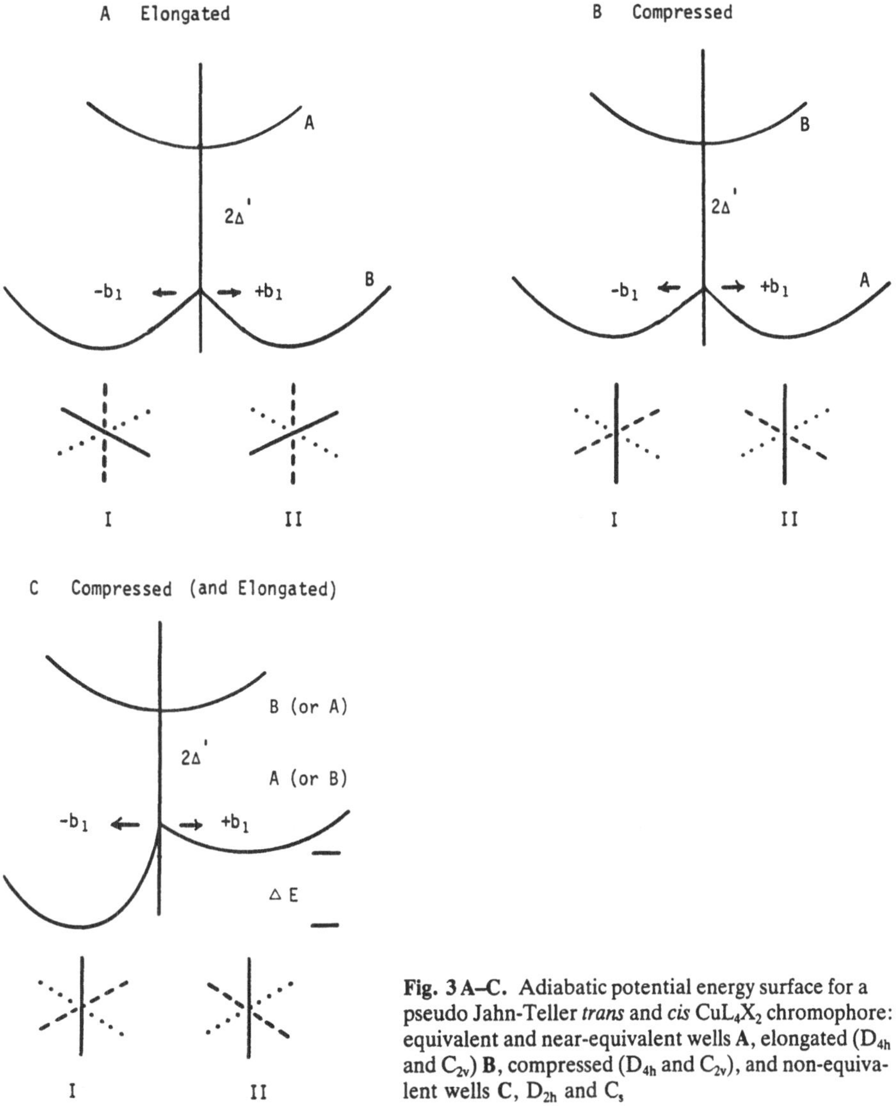

Fig. 3 A–C. Adiabatic potential energy surface for a pseudo Jahn-Teller *trans* and *cis* CuL_4X_2 chromophore: equivalent and near-equivalent wells **A**, elongated (D_{4h} and C_{2v}) **B**, compressed (D_{4h} and C_{2v}), and non-equivalent wells **C**, D_{2h} and C_s

and bond angles of the L and X ligands of a CuL_4X_2 chromophore. The active vibrational[14] mode of b_1 symmetry may create two wells in the lower potential energy surface for both the overall elongated CuL_4X_2 chromophore, Fig. 3 (A), $^2B_1$ ground state and for the overall compressed CuL_4X_2 chromophore, Fig. 3 (B), $^2A_1$ ground state. In both cases two equivalent octahedral chromophores of orthorhomibic symmetry are present which, for thermal equilibration, would yield CuL_4X_2 chromophores of D_{4h} symmetry, with an elongated stereochemistry, Fig. 3 (A), and a *pseudo* compressed stereochemistry, Fig. 3 (B). In host sites of lower than tetragonal symmetry wells I and II become non-equivalent, to produce an energy difference of ΔE, Fig. 3 (C). If ΔE is of the magnitude of thermal energy $kT = 200\ cm^{-1}$, an assymmetric thermal population of wells I and II is produced and the observed structure will be a weighted average of the two wells, namely elongated rhombic octahedral, and compressed rhombic octahedral for situations close to those in Fig. 3 (A) and (B), respectively. If the initial splitting $2\,\Delta'$ is small with respect to the Jahn-Teller coupling the lower potential surface will involve well defined wells I and II, Fig. 3. In the reverse case the wells I and II may be poorly defined.

The pseudo Jahn-Teller Effect as described in Fig. 3 (A), (B) and (C) then accounts for the stereochemistry of the wide range of six-coordinate copper(II) complexes involving non-equivalent ligands. CuL_4X_2, with both *trans* and *cis* octahedral geometries. Figure 3 (A) accounts for the large number of *static* elongated octahedral complexes with a *trans*-CuL_4X_2 chromophore, tetragonal, D_{4h}, as in $[Cu(NH_3)_4(SCN)_2]$ [8][33] and rhombic, D_{2h}, as in $[Cu(en)_2(FBF_3)_2]$[34] [9]. In [9] the inplane rhombic distortion is only small (ca. 0.1 Å) and never overrides the predominant z-axis elongation, and hence a ^2B ground state always applies. Figure 3 (B) accounts for the compressed tetragonal octahedral stereochemistry, with a *trans* CuL_4X_2 chromophore, and near D_{4h} symmetry, of $[Cu(dien)_2][NO_3]$ [10][16, 35] and $[Cu(methoxyacetate)_2(OH_2)_2]$[16, 36], Figs. 4, 5 both with ^2A ground states. It also accounts for the strict C_2 symmetry of the compressed rhombic octahedral stereochemistry, with a *cis*-CuL_4X_2 chromophore, ^2A ground state of $[Cu(phen)_2(O_2CCH_3)][C10_4]\cdot 2\,H_2O$ [11][37] and the near C_2 symmetry of $[Cu(bipy)_2(ONO)][NO_3]$ [12][38] and $[Cu(phen)_2(O_2CCH_3)][ClO_4]$ [13][39] both with an ^2A ground state. Figure 3 (C) then accounts for the very rhombic *in plane* distortion of b_1 symmetry, in the *trans* CuL_4X_2 chromophore of $[Cu(dien)_2]Br_2\cdot H_2O$ [14][40], and in the *cis* CuL_4O_2

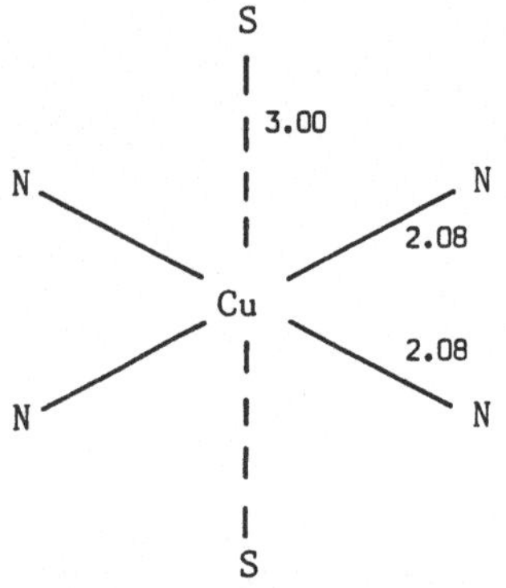

[8] $[Cu(NH_3)_4(SCN)_2]$

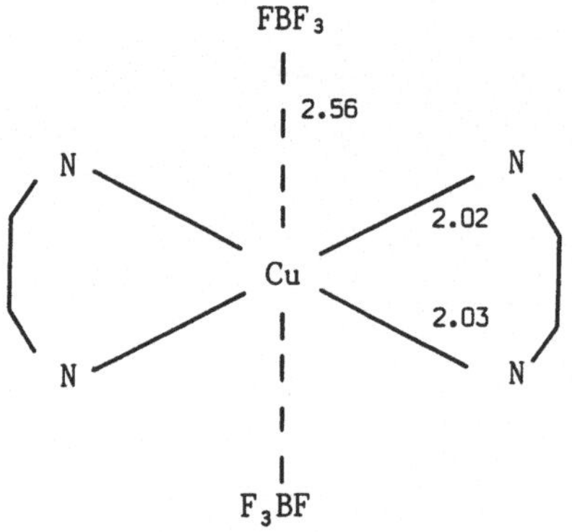

[9] $[Cu(en)_2(FBF_3)_2]$

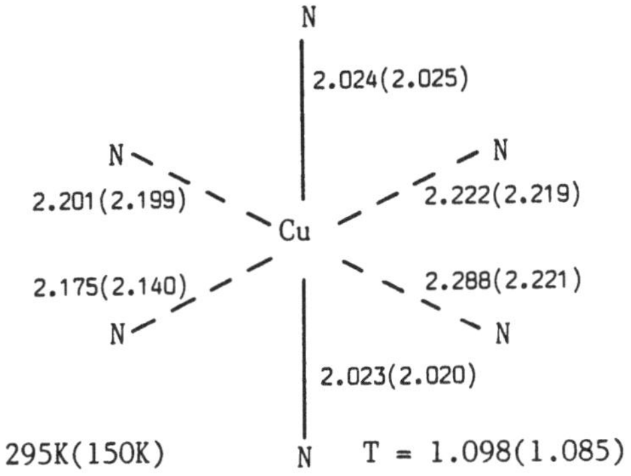

[10] $[Cu(dien)_2][NO_3]_2$

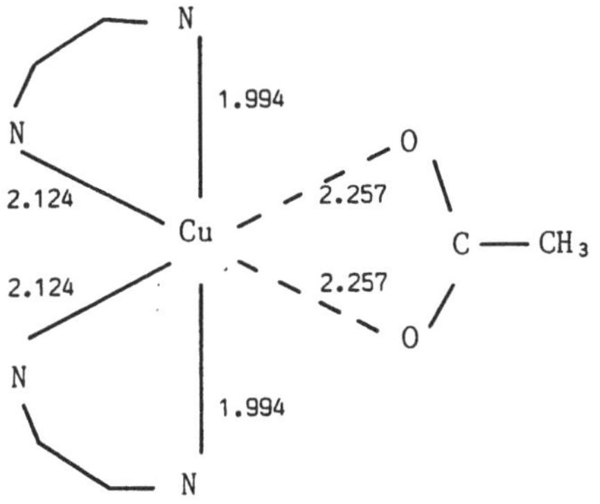

[11] $[Cu(phen)_2(O_2CCH_3)][ClO_4]\cdot 2H_2O$

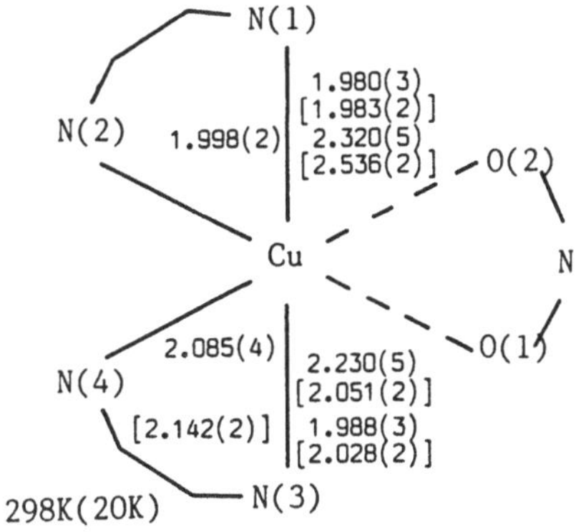

[12] $[Cu(bipy)_2(ONO)][NO_3]$

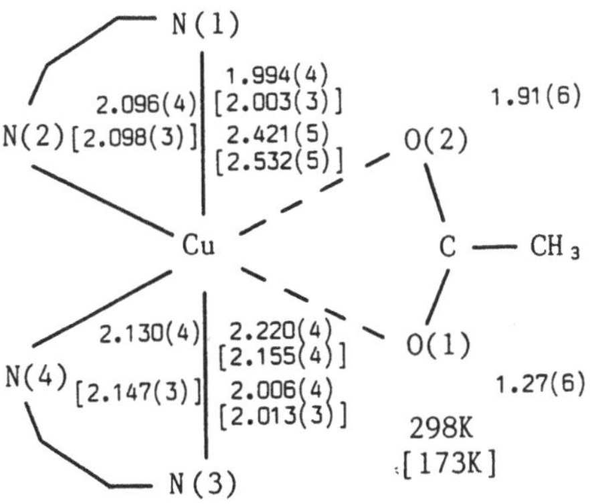

[13] $[Cu(phen)_2(O_2CCH_3)][ClO_4]$

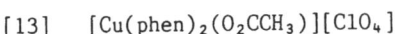

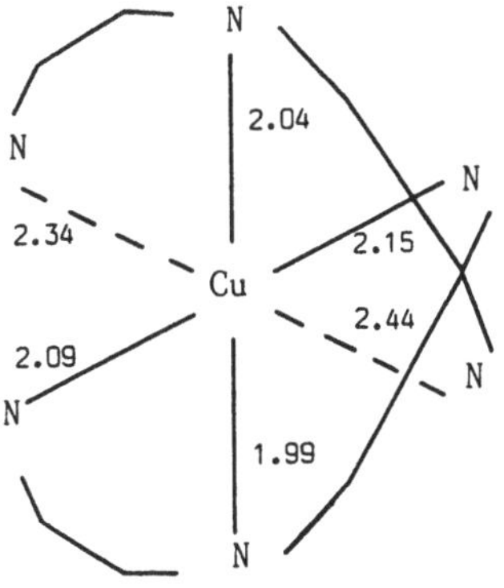

[14] $[Cu(dien)_2]Br_2\cdot H_2O$

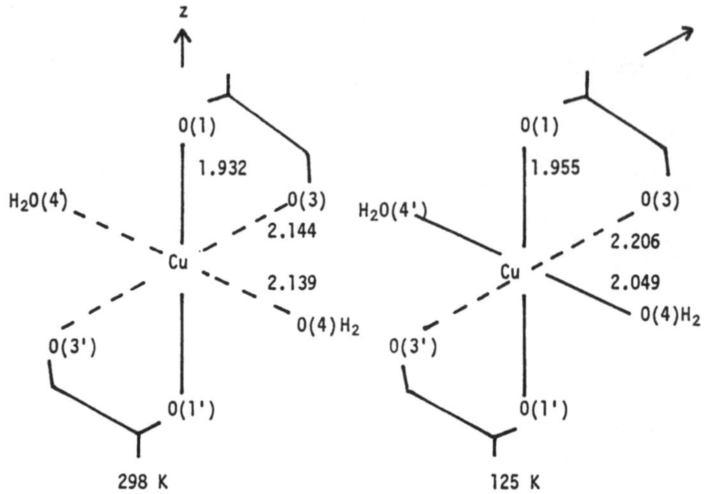

Fig. 4. [Cu(methoxyacetate)$_2$(OH$_2$)$_2$]: Molecular structures at 298 and 125 K; change in principal axis (z) by 90°

chromophore of [Cu(bipy)$_2$(ONO)] [NO$_3$] [12] at 20 K. In the structure of [14] and [12] the inplane distortion of b_1 symmetry is so large that the ground state changes from ^2A to ^2B with the elongation in the x-direction. In the near high symmetry structures above, the axial elongations, along z, and the inplane rhombic distortions, along x or y, are restricted by the "bite" of the chelate ligands present and for $\Delta E < kT$, Fig. 3 (C) predicts that these structures will be temperature variable.

The observation of the pseudo octahedral and pseudo compressed octahedral stereochemistries for the copper(II) ion then arise as an artifact of fluxional behaviour, Fig. 2 (A) for CuL$_6$ chromophores and Fig. 3 (B) for the latter in CuL$_4$X$_2$ complexes. In

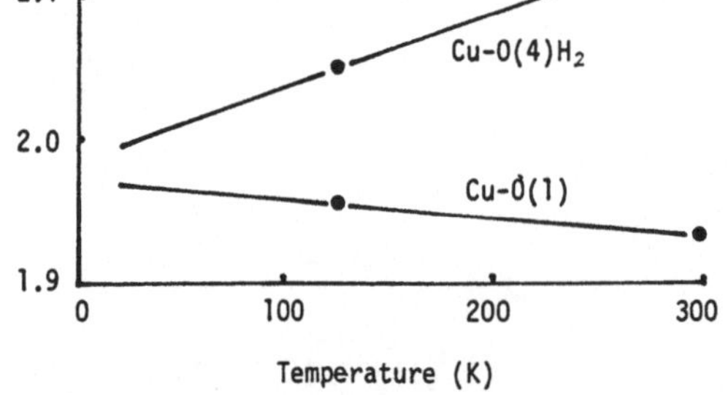

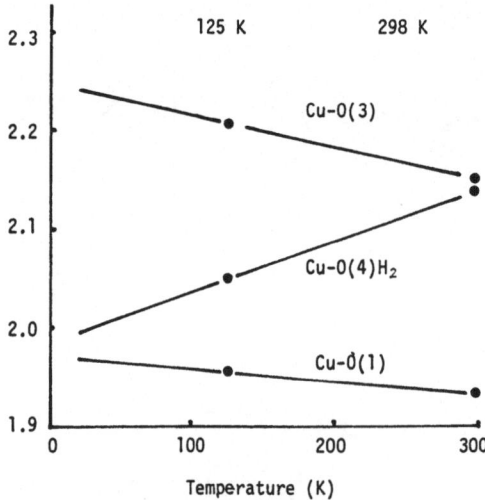

Fig. 5. [Cu(methoxyacetate)$_2$(OH$_2$)$_2$]: Cu-O distance v temperature K

all cases the strictly regular stereochemistries are only possible in crystals involving high crystallographic site symmetry. Lattice effects must then be responsible for holding the copper(II) ion in an environment of high symmetry and high energy, the Cooperative Jahn-Teller Effect[17], relative to the more stable low symmetry structures of the individual wells of Fig. 2, and of Fig. 3, which are stabilised in lattices of low symmetry such as monolinic and triclinic.

The ability of the pseudo Jahn-Teller effect[14] to account for the structure and electronic properties (see later) of six-coordinate copper(II) complexes has been questioned on the basis[41] that the effect is an order of magnitude too small to account for the distortion involved, 0.4–1.0 Å. Nevertheless, an exactly comparable potential energy diagram to that of Fig. 3 has been proposed[42] using the approach of near degenerate non-rigid molecules. As it is impossible, at present, to distinguish these two alternatives the present article retains the more unifying approach of the pseudo Jahn-Teller effect

4 The Observation of the Jahn-Teller Effect

The major value[26] of the "Mexican Hat" model for these six-coordinate copper(II) complexes involving six equivalent or non-equivalent ligands is that it predicts:

1. The structures of the CuL_6 chromophores may be temperature variable with bond-length changes of the order of the Jahn-Teller radius, Fig. 1 (A), $R_{JT} = 0.1–0.4$ Å, accessible by low temperature X-ray crystallography, which if determined at low enough temperature may yield ΔE values.
2. The e.s.r. spectra of CuL_6 chromophores[16] may be temperature variable and yield data on the barrier height, B and ΔE, Fig. 2 (C), which should both be of the magnitude 0–400 cm^{-1}.
3. The Jahn-Teller energy E_{JT}, Fig. 1 (A), should be available from optical spectroscopy[26], as E_{JT} should be in the order of 1500–2500 cm^{-1}, but should *not* be significantly temperature variable[16].

The different information that is available from these three techniques is related to the different time scales[43] that are involved in their measurement. X-ray diffraction has an interaction time of 10^{-16}s, but is averaged over vibrational motion, so, the structure is time averaged (≈ 1 s), e.s.r. spectroscopy has an interaction time of 10^{-9} s, and is also averaged by vibrational motion, while electronic spectroscopy has an interaction time of 10^{-15} s and is *not* time averaged, but reflects the extreme static configuration of the molecular vibration. Consequently, the x-ray and e.s.r. techniques provide averaged structural data in these fluxional systems, while electronic spectroscopy relates to the underlying static stereochemistry. It is anticipated that EXAFS spectroscopy[44 a)] with an interaction time of 10^{-15} s, will also yield bond distances that relate to the underlying static structure of these fluxional chromophores, but the full potential of this new technique is only just being realised[44 b)].

5 Temperature Variable Crystallographic Data

The most convincing evidence for the fluxional model of six-coordinate copper(II) complexes is obtained from temperature variable crystallography. Historically the first such data was obtained for the classic octahedral CuN_6 chromophore of $K_2Pb[Cu(NO_2)_6]$, [1][22], which is cubic at room temperature with a regular octahedral CuN_6 chromophore, but at 276 and 195 K is orthorhombic with a compressed rhombic octahedral CuN_6 chromophore. More recently the structure[23, 28] of $Cs_2Pb[Cu(NO_2)_6]$, [3], has been measured over a temperature range; it is cubic[45] at 420 K with an octahedral CuN_6 chromophore, Fig. 2 (A) behaviour; it is orthorhombic[28] at 298 K with a compressed CuN_6 chromophore, Fig. 2 (B) behaviour, and at 160 K is monoclinic[46] with an elongated rhombic octahedral CuN_6 chromophore, Fig. 2 (C) behaviour. Consequently, the temperature variation of the crystal structure of $Cs_2Pb[Cu(NO_2)_6]$, covers the full range of fluxional behaviour, as depicted in the type Fig. 2 (A)–(C) behaviour of a fluxional system, in which the 160 K data was determined by the interesting technique of profile analysis of neutron diffraction powder data. The room temperature structure[24, 25] of $[Cu(en)_3][SO_4]$, [2], is trigonal with a regular CuN_6 chromophore of D_3 symmetry at 298 K, but at 120 K, the space group is triclinic[47], with a compressed rhombic octahedral CuN_6 chromophore stereochemistry. The compressed rhombic octahedral stereochemistry[29] also occurs in $Rb_2Pb[Cu(NO_2)_6]$, [14], at room temperature. The room temperature structure[30] of $[NH_4]_2[Cu(OH_2)_6][SO_4]_2$, [5], is elongated rhombic octahedral, but with a relatively high tetragonality of 0.914, Table 1 (A), Fig. 6 (a), which drops to 0.891 at 203 K and to 0.874 at 123 K. With decreasing temperature there is a significant increase in the long Cu-O (7) distance, a decrease in the Cu-O (8) distance, and a slight, but hardly significant increase in the short Cu-O (9) distance. The correlation of the tetragonalities of the Tutton Salts, Table 1 (b), with their axial bond distances, Fig. 6 (b), does suggest

Table 1. The Cu-OH_2 bond distances of (a) $[NH_4]_2[Cu(OH_2)_6][SO_4]_2$ and (b) $[M^I]_2[Cu(OH_2)_6]$-$[SO_4]_2$ at room temperature and low temperature

(a) $[NH_4]_2[Cu(OH_2)_6][SO_4]_2$[30]			
	298	203	123
Cu-O (7)	2.219	2.250	2.278
Cu-O (8)	2.095	2.041	2.012
Cu-O (9)	1.961	1.967	1.970
Tetragonality[b]	0.914	0.891	0.874

(b) $[M^I]_2[Cu(OH_2)_6][SO_4]_2$			
	K[48]	Rb[49]	Cs[31]
Cu-O (7)	2.069	2.031 (2.000)[a]	2.004
Cu-O (8)	2.278	2.307 (2.317)[a]	2.315
Cu-O (9)	1.943	1.957 (1.978)[a]	1.966
Tetragonality	0.881	0.864 (0.858)[a]	0.858

[a] low temperature data (77 K)[51]

[b] Tetragonality $T = R_S/R_L$, where R_S = mean inplane copper-ligand distance and R_L = mean out-of-plane copper-ligand distance

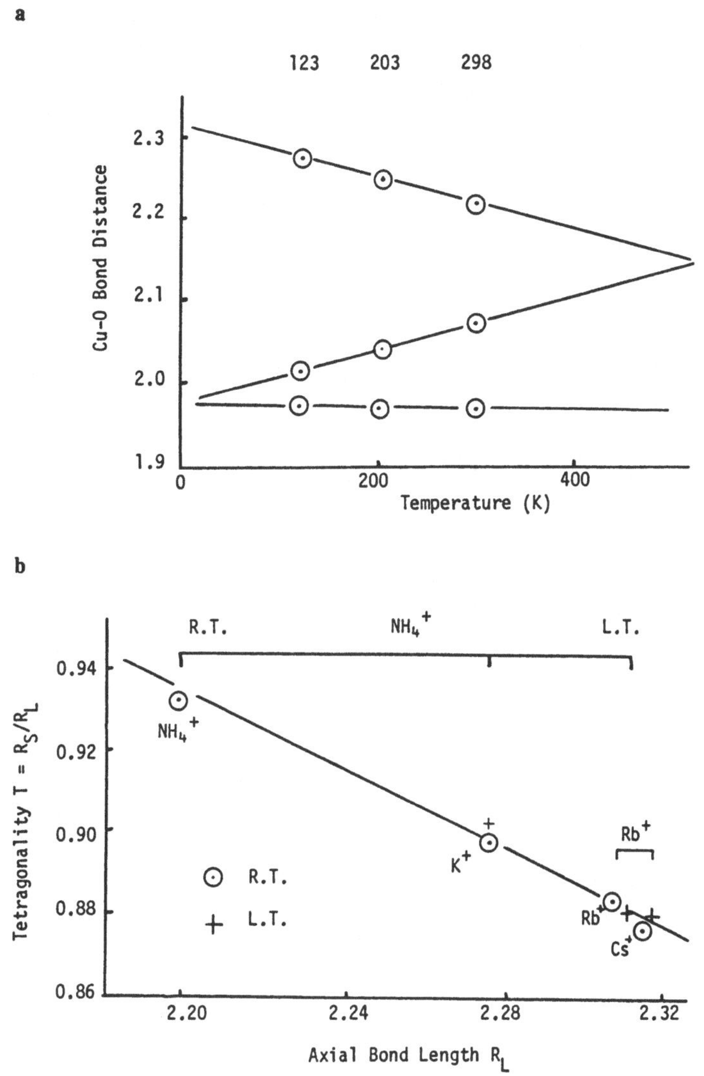

Fig. 6 a, b. Copper Tutton Salts **a** Cu-O v Temperature; **b** Tetragonality v Axial Cu-O bond length, R_L

that the ammonium salt is much more fluxional than the potassium[48] salt, and that the rubidium[49] and caesium[31] salts are virtually static in structure-type, Fig. 2 (C), behaviour. The variation of the Cu-O distance with temperature (K) suggests an almost linear variation with temperature, Fig. 6 (a), which extrapolate to Cu-O distances of 2.32, 1.96 and 1.98 Å equivalent to a nearer axial, elongated stereochemistry at 4 K, a result that has been confirmed[50] by the recent powder neutron diffraction structure determination of $[ND_4]_2[Cu(OD_2)_6][SO_4]_2$ at 5 K, with Cu-O distances of 1.95, 2.01 and 2.30 Å. Using the extrapolated X-ray crystallographic data[30] and a $\ell n\, n_1/n_2 = \Delta E/0.695\,T$ relationship, the estimated ΔE, Fig. 2 (C), for $[NH_4]_2[Cu(OH_2)_6][SO_4]_2$ is

160 ± 20 cm^{-1}. The low temperature structure[51] of $Rb_2[Cu(OH_2)_6]$, $[SO_4]_2$, [6], at 77 K showed so little variation with temperature, Fig. 6 (b), that it was reported non-temperature variable. The room temperature structure[51] of $[Cu(phen)_3][NO_3]_2 \cdot 2H_2O$, [6], has a clear elongated rhombic octahedral CuN_6 chromophore, tetragonality 0.951 at 298 K, which is significantly reduced to 0.920 at 153 K. For six-coordinate copper(II) complexes with non-equivalent ligands, the compressed rhombic octahedral stereochemistry[35] of $[Cu(dien)_2][NO_3]_2$, [10], has been determined at 298 and 150 K, but shows no significant change in structure[53] from the 298 K data and clearly requires an even lower temperature to reveal the underlying static elongated rhombic octahedral CuN_6 chromophore stereochemistry. The most accurate low-temperature crystal structure[36] of a fluxional copper(II) system, where no change of phase is involved, monoclinic $P2_1/n$, is that for $[Cu(methoxyacetate)_2(OH_2)_2]$, Fig. 4, which at 298 K has a compressed tetragonal octahedral CuO_6 chromophore and changes to an elongated rhombic octahedral chromophore at 125 K, with the predicted change in principal axis from the Cu-O (4) direction at 298 K to that of the Cu-O (3) direction at 125 K, Fig. 4, consistent with the sense of the change from Fig. 3 (B) to Fig. 3 (C).

Although the *cis*-distorted octahedral CuN_4O_2[38] chromophore is not generally considered as a potentially fluxional copper(II) system, the recent low temperature crystal structure of the two near-regular *cis*-distorted octahedral complexes, $[Cu(bipy)_2(ONO)][NO_3]$[54, 55] [12], and $[Cu(phen)_2(O_2CCH_3)][ClO_4]$, [13][39], show a significant increase in the Cu-O (2) distance, a decrease in the Cu-O (1) distance, and a small increase in the Cu-N (4) distance, Fig. 7 (a). These changes are of the correct order of magnitude for a fluxional copper(II) system, and are in the right sense, for a change from a regular *cis*-distorted octahedral chromophore to a distorted square pyramidal (4 + 1 + 1*) CuN_4O_2 chromophore[9]. Consequently, the *regular* cis-octahedral CuN_4O_2 structure must be considered to arise from a time average structure of two 90° misaligned distorted square pyramidal (4 + 1 + 1*) structure, and that the *cis*-distorted octahedral structure arises as a consequence of fluxional behaviour, Fig. 3 (B). The low temperature data for [12] down to 20 K, yields from the $\ell n\, n_1/n_2$ plot, Fig. 7 (b), a ΔE value[55] of 75 ± 10 cm^{-1}, and for [13][39], down to 173 K, ΔE = 125 ± 15 cm^{-1}.

It should be pointed out that for all of these fluxional copper(II) systems the temperature variable crystal structure data does not distinguish between a dynamic and static disorder[56] of the CuL_6 chromophores (see Ref. 21, Table 4).

6 Anisotropic Thermal Parameters

The occurrence of fluxional behaviour in CuL_6 chromophores of copper(II) complexes should be reflected in a marked anisotropy in the thermal motions of the six ligand atoms involved. As the differences are only small, care must be taken with the data collection[25] and the data, preferably, corrected for absorption in order to minimise the effect of any systematic errors in the data. The first observation of significant thermal motion, as evidence for dynamic Jahn-Teller effects, was in the data for $K_2Pb[Cu(NO_2)_6]$, [1][57] and $[Cu(en)_3][SO_4]$, [2],[25]. But a wider discussion of the effect[26] for the $M^{I}M^{II}[Cu(NO_2)_6]$ system has been given involving the comparison of the root-mean-square displacements

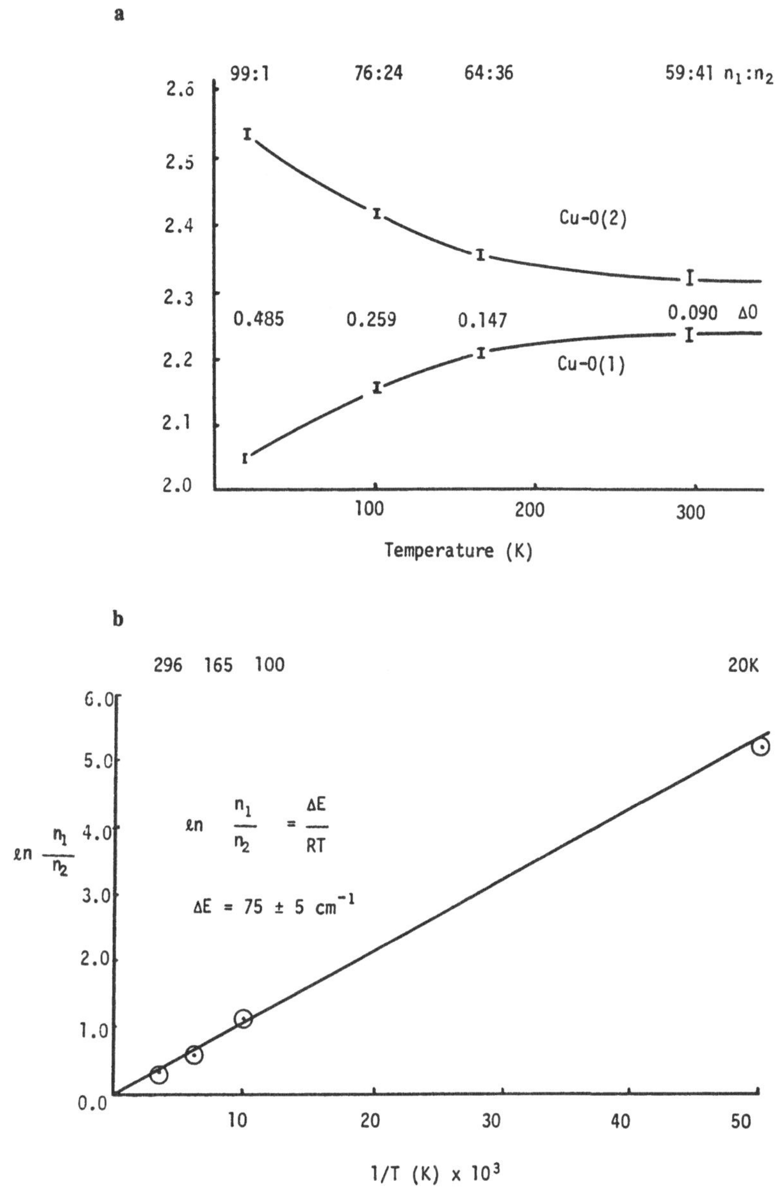

Fig. 7a, b. $[Cu(bipy)_2(ONO)][NO_3]$: **a** Cu-O distances v Temperature (K); **b** $\ell n\,(n_1/n_2)$ v 1/Temperature (K)

of the ligand atoms of the non-Jahn-Teller complexes $M^IM^{II}[Ni(NO_2)_6]$, Table 2. For non-dynamic systems, such as the nickel(II) complexes and the static copper(II) complexes, as the bending modes of an M-ligand bond are of lower energy than the stretching modes, higher root-mean-square displacements will occur perpendicular to the M-ligand bond directions, than along them, as shown in Table 2(a) and (b). For dynamic Jahn-

Table 2. A comparison of the nitrogen atom thermal motion[26)] for $M^{I}_2M^{II}[M(NO_2)_6]$ systems, root-mean-square displacements(Å), displacements along the M-N bonds are indicated by an asterisk (*)

		U_{11}	U_{22}	U_{33}
(a) Non-Jahn-Teller Complexes				
$K_2Pb[Ni(NO_2)_6]$	N(1)	0.130	0.133	0.115*
$K_2Sr[Ni(NO_2)_6]$	N(2)	0.120	0.122	0.110*
(b) Static Jahn-Teller Complexes				
$K_2Sr[Cu(NO_2)_6]$	N(1)	0.141	0.144	0.119*
	N(2)	0.113*	0.121	0.138
	N(3)	0.130	0.111*	0.129
(c) Dynamic Jahn-Teller Complexes (cubic)				
$K_2Pb[Cu(NO_2)_6]$	N(1)	0.170	0.164	0.182*
(d) Pseudo Compressed Jahn-Teller Complexes				
$K_2Pb[Cu(NO_2)_6]$ (276 °K)	N(1)	0.17	0.17	0.11*
	N(2)	0.15	0.16*	0.15
	N(3)	0.18*	0.11	0.25
$Rb_2Pb[Cu(NO_2)_6]$	N(1)	0.148	0.152	0.128*
	N(2)	0.182	0.190*	0.141
	N(3)	0.199*	0.151	0.144

Teller systems the opposite will apply, Table 2(c), and for the pseudo compressed Jahn-Teller systems, Table 2(d), which involve a static distortion along Cu-N(1), but a dynamic distortion along Cu-N(2) and Cu-N(3), smaller root-mean-square displacements are predicted to lie along the Cu-N bonds for N(1), and larger root-mean-square displacements for N(2) and N(3). These predictions are clearly born out for $A_2Pb[Cu(NO_2)_6]$, [A: Cs, Rb], but only partially for the less accurate data for $K_2Pb[Cu(NO_2)_6]$ (276 K).

The analysis of the root-mean-square displacements for CuL_6 chromophores in low symmetry systems[26)] can also be indicative of fluxional behaviour, but has ben expressed in the form $\Delta U^{1/2}$ (Cu-N), Table 3, where the nitrogen displacements are corrected for the displacements of the copper atom and the $\Delta Us'$ represent a bond displacement, rather than an isolated ligand atom displacement, directed along the Cu-ligand directions. Using this analysis for the $M^{I}M^{II}[Cu(NO_2)_6]$ systems of Table 2, identical conclusions in respect to a static or dynamic behaviour are deduced; the $\Delta U^{1/2}$ (Cu-N) values for static systems are < 0.10 Å, while for dynamic systems the values are > 0.10 Å; Table 3. Using the $\Delta U^{1/2}$ (Cu-ligand) values, Table 3, and the above criteria, this again suggests that $[Cu(en)_3][SO_4]$ and $K_2Pb[Cu(NO_2)_6]$ involve genuine dynamic Jahn-Teller effects and that $Cs_2Pb[Cu(NO_2)_6]$[23)], $Rb_2Pb[Cu(NO_2)_6]$[29)], $[Cu(dien)_2][NO_3]_2$[35)] and $[Cu(methoxyacetate)_2(OH_2)_2]$[36)] involve two dimensional fluxional behaviour consistent with their compressed rhombic octahedral stereochemistries, while $[NH_4]_2[Cu(OH_2)_6][SO_4]_2$[30)], is two dimensional fluxsional, consistent with its elongated rhombic octahedral stereochemistry with a high tetragonality, $T > 0.90$. The effect of

Table 3. Some $\Delta U^{1/2}$ (Cu-ligand) (Å)[26] for CuL_6 chromophores *along* the Cu-ligand bond directions

	Type	L(1)	L(2)	L(3)
(a) Fluxional				
$[Cu(en)_3][SO_4]_2$	(A)	0.176	0.176	0.176
$K_2Pb[Cu(NO_2)_6]$	(A)	0.144	0.144	0.144
$Rb_2Pb[Cu(NO_2)_6]$	(B)	0.062 (static)	0.145	0.147
$[Cu(dien)_2][NO_3]_2$[35]	(B)	0.248	0.062 (static)	0.186
$[Cu(methoxyacetate)_2(OH_2)_2]$[36]	(B)			
	298 K	0.098	0.093	0.048
	125 K	0.042	0.041	0.025
$[NH_4]_2[Cu(OH_2)_6][SO_4]_2$[30]	(C)			
	298 K	0.126	0.112	0.062
	203 K	0.141	0.132	0.107
	123 K	0.123	0.122	0.106
$[Cu(HBpz_3)_2]$[58] I	(C)	0.15	0.12	0.07
II	–	0.04	0.07	< 0.0
(b) Static				
$K_2Pb[Ni(NO_2)_6]$		0.047	0.047	0.047
$K_2Sr[Cu(NO_2)_6]$		0.030	0.040	0.053

reducing the temperature on the $\Delta U^{1/2}$ (Cu-L) values is to reduce their numerical values, Table 3, but still retains the differences appropriate to the fluxional behaviour at low temperature.

The visual asymmetry in the thermal ellipsoids of the oxygen atoms of the *cis*-distorted octahedral CuN_4O_2 chromophore when compared with the spherical symmetry of the corresponding ZnN_4O_2 chromophore, Fig. 8(a) and (b), first lead to the suggestion that these systems were fluxional and to the low temperature crystal structure determinations, that have established that the *cis*-distorted systems are indeed fluxional [12] and [13]. The effect of temperature on the thermal ellipsoids of [12], Fig. 8(c), shows that the fluxional behaviour is still present even at 100 K, and that 20 K is required for its removal. The usefulness of the $\Delta N^{1/2}$ (Cu-N) values is demonstrated for the two independent CuN_6 chromophores[58] of $[Cu(HBpz_3)_2]$, [15]. Both involve an elongated rhombic

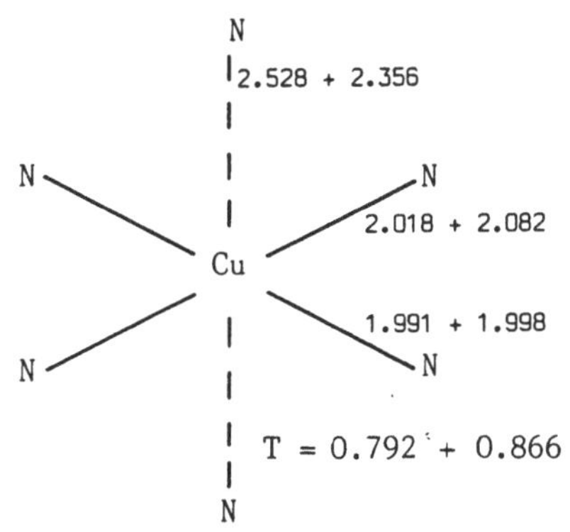

[15] $[Cu(HBpz_3)_2]$ forms I- and II-

octahedral CuN_6 chromophore for the same $[Cu(HBpz_3)_2]$ complex, but with significantly different tetragonalities, 0.792 and 0.866, respectively. The former suggests a static elongated Cu(1)N_6 chromophore with $\Delta N^{1/2}$ (Cu-N) values 0.05 Å, Table 3, while the latter has a significantly higher tetragonality consistent with the presence of some fluxional behaviour which is supported by the $\Delta U^{1/2}$ (Cu-N) values, Table 3, indicative of a two dimensional fluxional behaviour and consistent with the greater in-plane rhombic distortion of the Cu(2)N_6 chromophore in this lattice.

The room temperature data of Tables 2 and 3, not only establish the value of anisotropic temperature factors in indicating the presence of dynamic Jahn-Teller or pseudo Jahn-Teller effects in high symmetry crystals, but also for indicating the presence of fluxional behaviour[16] in the stereochemistry of elongated rhombic octahedral CuL_6 chromophores whose differing tetragonalities might otherwise be interpreted as arising from the plasticity effect[6, 7] of the copper(II) ion alone.

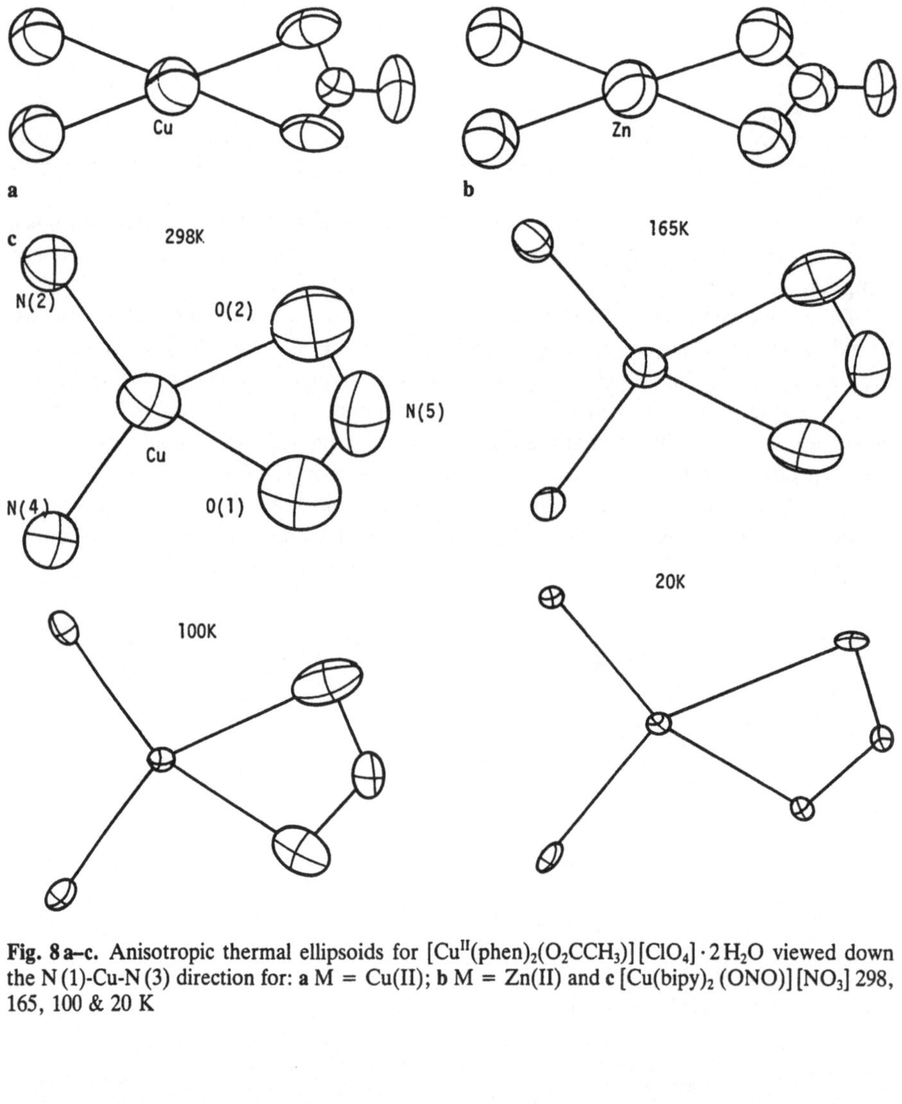

Fig. 8 a–c. Anisotropic thermal ellipsoids for $[Cu^{II}(phen)_2(O_2CCH_3)][ClO_4] \cdot 2H_2O$ viewed down the N(1)-Cu-N(3) direction for: **a** M = Cu(II); **b** M = Zn(II) and **c** $[Cu(bipy)_2(ONO)][NO_3]$ 298, 165, 100 & 20 K

7 Electron Spin Resonance Spectra

By virtue of an interaction time[43] of ca. 10^{-9} s e.s.r. measurements relate directly to static copper(II) environments, type Fig. 2 (C) behaviour ($\Delta E \gg kT$), but are time averaged in the event of fluxional behaviour[16], type Fig. 2 and 3, behaviour which occurs at a single copper centre. In static copper(II) environments the most important use[4, 16] of e.s.r. spectra is in the identification of those stereochemistries consistent with a local molecular d_{z^2} ground state, Table 4, namely, $g_3 \approx g_2 \gg g_1 \approx 2.0$, and those consistent with a local molecular $d_{x^2-y^2}$ ground state, namely, $g_3 \gg g_2 \approx g_1 > 2.0$ where $g_3 > g_2 > g_1$, Fig. 9 (1). For a fluxional pseudo octahedral copper environment, Fig. 2 (A), a time averaged isotropic g-value is obtained, as an average of the three local g-values; $g_3 \gg g_2 \approx g_1 > 2.0$, of the three 90° misaligned wells I, II and III (33.3 : 33.3 : 33.3%) of the single copper centre. For a fluxional pseudo compressed octahedral copper environment, Fig. 2 (B), a time averaged axial spectrum is obtained, with $g_3 \approx g_2 > g_1 > 2.0$, as an average of the two local g-values, $g_3 \gg g_2 \approx g_1 > 2.0$, of the two 90° misaligned wells I and II (50 : 50%) of the single copper centre, Fig. 10 (a). This spectrum, Fig. 9 (2) (b), is difficult to distinguish from the spectrum for a genuine static d_{z^2} ground state, Fig. 9 (1) (a), especially if the thermal population of wells I and II are not equal, Fig. 10 (b).

This diagnostic value of the e.s.r. spectra of non-dilute copper(II) complexes is further complicated by the presence of exchange coupling[4] between the crystallographically related local molecular copper environments, Fig. 11, and is dependent upon the relative orientation of the local molecular axes, which may be misaligned by the space group elements of symmetry. For aligned axes, Fig. 11 (a), (ferrodistortive order)

Table 4. The one electron orbital ground states for the known stereochemistries[4] of the copper(II) ion, Fig. 1 (for notation see Sect. 3)

1.	$d^1_{x^2-y^2}$	elongated tetragonal-octahedral elongated rhombic-octahedral square-coplanar square pyramidal
2.	$d^1_{z^2}$	compressed tetragonal-octahedral compressed rhombic-octahedral linear trigonal bipyramidal cis-distorted octahedral
3.	d^1_{xy}	compressed tetrahedral square-coplanar – $Cu(acac)_2$-type
4.[a]	$(d_{z^2}, d_{x^2-y^2})^3$	octahedral
5.[a]	$(d_{xz}, d_{yz})^3$	elongated tetrahedral
6.[a]	$(d_{xy}, d_{xz}, d_{yz})^5$	tetrahedral

[a] These configurations are orbitally degenerate; in 5 and 6, the degeneracy is removed by spin-orbit coupling (see Ref. 87 (b)), but it is not removed by this mechanism in 4 (see Sect. 3.1)

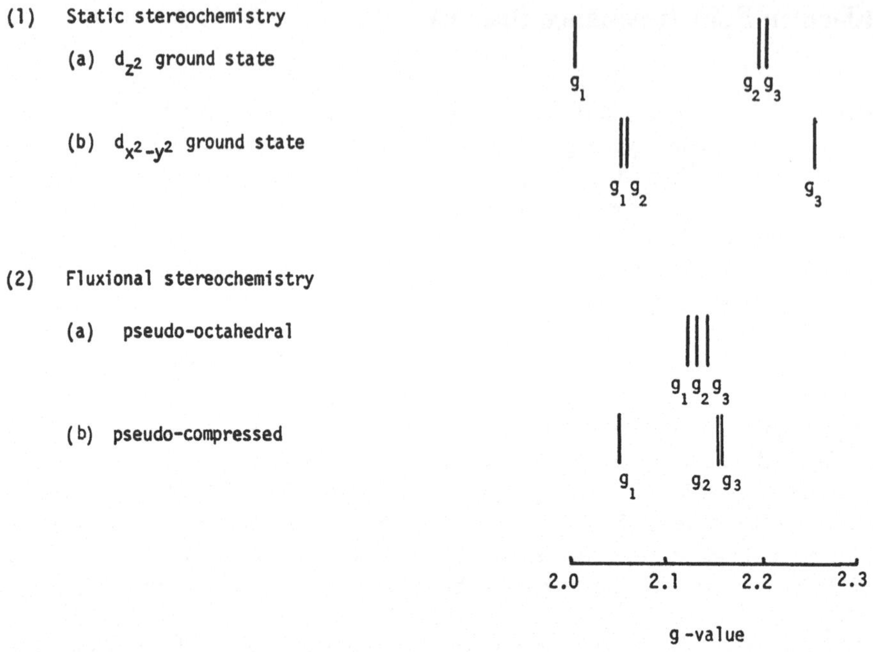

Fig. 9. Predicted types of e.s.r. spectra for (i) static and (ii) fluxional copper(II) environments

exchange narrowing occurs, the crystal g-values equate with the local molecular axes. For misaligned axes (2γ) the crystal g-values no longer equate with the local g-values, Fig. 11 (b), and only in the special case of $2\gamma = 90°$ (antiferrodistortive order), Fig. 11 (c), involving only two dimensional misalignment, can the resolution of the crystal g-values into the local molecular g-values be effected. Consequently, an "isotropic" e.s.r. spectrum may indicate, either, a pseudo octahedral stereochemistry or exchange coupled misaligned chromophores. Crystal g-values, $g_3 \approx g_2 \gg g_1 \approx 2.0$, are consistent with those static stereochemistries having a d_{z^2} ground state, Table 4 (2), or with a pseudo compressed stereochemistry, Fig. 10, or with antiferrodistortive ordering, Fig. 11 (c). The crystal g-values, $g_3 \gg g_2 \approx g_1 > 2.0$ only equate directly, with a $d_{x^2-y^2}$ ground state of the local molecular copper(II) environment in the special case of ferrodistortive ordering, Fig. 11 (a).

The problem of exchange coupling may be avoided if the copper complex is diluted in a diamagnetic host lattice, < 10% dilution, and while the observed g-values equate directly with the local molecular g-values, their appearance is complicated by the presence of copper hyperfine on the g-values, as illustrated in Fig. 12 (a)–(c). An additional value of the e.s.r. spectra of dilute systems is that they are more accurately measured, a clear advantage if their temperature variation is required. A disadvantage, is the lack of knowledge of the precise copper(II) ion environment present. Historically, electron spin resonance spectroscopy was first used[16, 17, 27] to demonstrate the temperature variability of fluxional CuL_6 chromophores and was then confirmed by the low temperature crystallographic evidence. The first high symmetry systems to be observed with isotropic

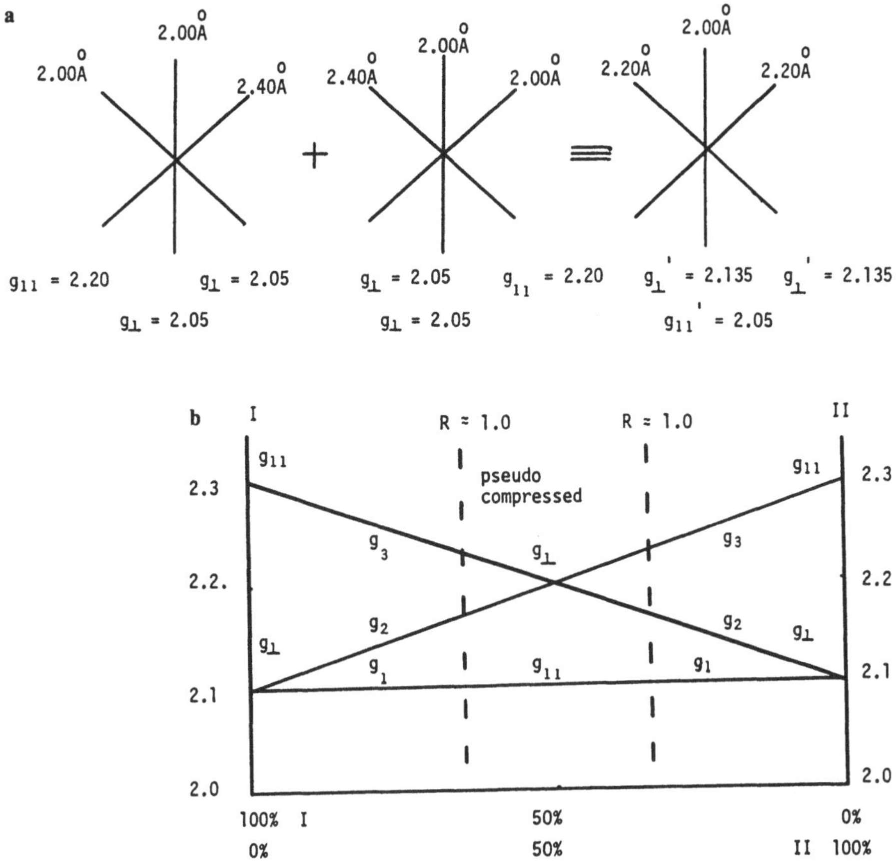

Fig. 10 a, b. Fluxional model at a single copper centre **a**; The consequence of two elongated tetragonal octahedral CuL_6 chromophores, ($g_3 \gg g_2 \approx g_1 > 2.0$) misaligned by 90° on the average Cu-L bond distances and the g-factors. **b** The effect of mixing two CuL_6 chromophores I and II misaligned by 90° as a function of the % composition of wells I and II, on the observed g-values ($R = g_2 - g_1/g_3 - g_2$ with $g_3 > g_2 > g_1$)

e.s.r. spectra were $K_2Pb[Cu(NO_2)_6]$ (cubic)[27] and $[Cu(en)_3][SO_4]$ (trigonal)[59], both change with temperature, Fig. 13 (a) and (c), to give an anisotropic polycrystalline e.s.r. spectrum which is reversed in form $g_\perp \gg g_{11} > 0$ relative to that observed for an elongated tetragonal octahedral CuL_6 chromophore, Fig. 13 (b), and $g_{11} \gg g_\perp > 2.0$, but with the lowest g-value (g_{11}) significantly above 2.0, the value predicted for a compressed rhombic octahedral stereochemistry. This "reversed" type e.s.r. spectrum[4] is understandable in terms of the two-dimensional averaging of two elongated rhombic octahedral CuL_6 chromophores misaligned by 90°, Fig. 2 (B), and as set out in Fig. 10 (a). The temperature variation of the polycrystalline e.s.r. spectrum[17,60] of $Cs_2Pb[Cu(NO_2)_6]$, Table 5 ranges from isotropic (cubic, Fig. 2 (A) behaviour, 420 K), reversed axial (orthorhombic – compressed rhombic octahedral CuN_6, Fig. 2 (B), behaviour, 298 K, and axial (monoclinic – elongated rhombic octahedral CuN_6, Fig. 2 (C), behaviour,

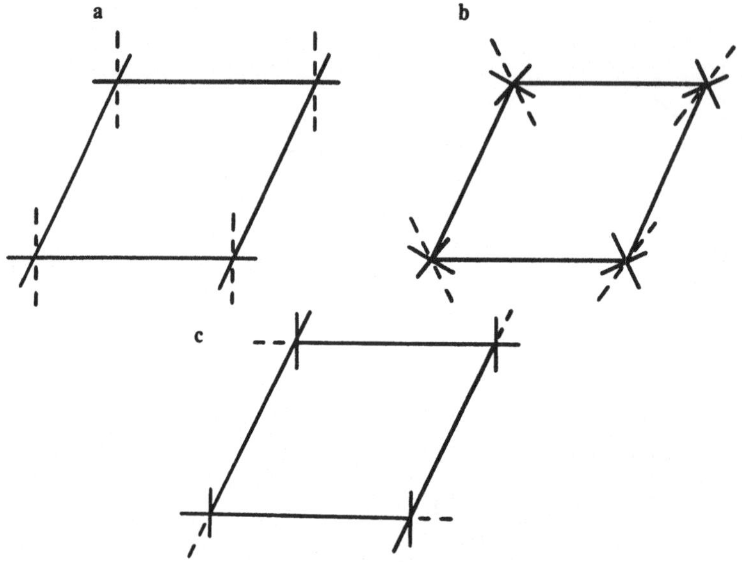

Fig. 11 a–c. Exchange coupling between crystallographic Cu centres; the types of crystallographic CuL_6 chromophore orientations: **a** aligned (ferrodistortive order); **b** misaligned ($2\gamma \approx 45°$); **c** 90° misaligned (antiferrodistortive order; $2\gamma = 90°$) (where 2γ = angle between elongation axes)

160 K), and in one complex contains all three types of e.s.r. behaviour associated with the Fig. 2 potential energy well system.

Temperature variable e.s.r. evidence for concentrated fluxional low symmetry copper(II) systems is not easily obtainable as exchange coupling of misaligned chromophores generate isotropic e.s.r. spectra, which are very insensitive to the temperature variation, and show no dramatic change in line shape (i.e. consistent with a change of phase) with decreasing temperature. This particularly applies to the e.s.r. spectra of $[Cu(dien)_2][NO_3]_2$ [10], g_i = 2.102–2.100, Table 5, but as no variation in the crystal structure down to 150 K was observed, this lack of change in the e.s.r. spectrum is not surprising. The single-crystal e.s.r. spectrum[4, 36] of $[Cu(methoxyacetate)_2(OH_2)_2]$ gives a very rhombic set of g-values 2.087, 2.185 and 2.360 at room temperature, which is rather surprising in view of the almost axially compressed CuO_6 chromophore structure Fig. 4. At low temperature the g's, Table 5, are much more axial, consistent with the change to

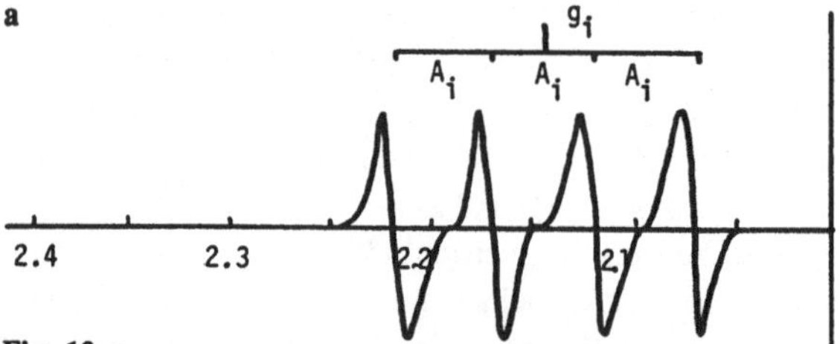

Fig. 12. a

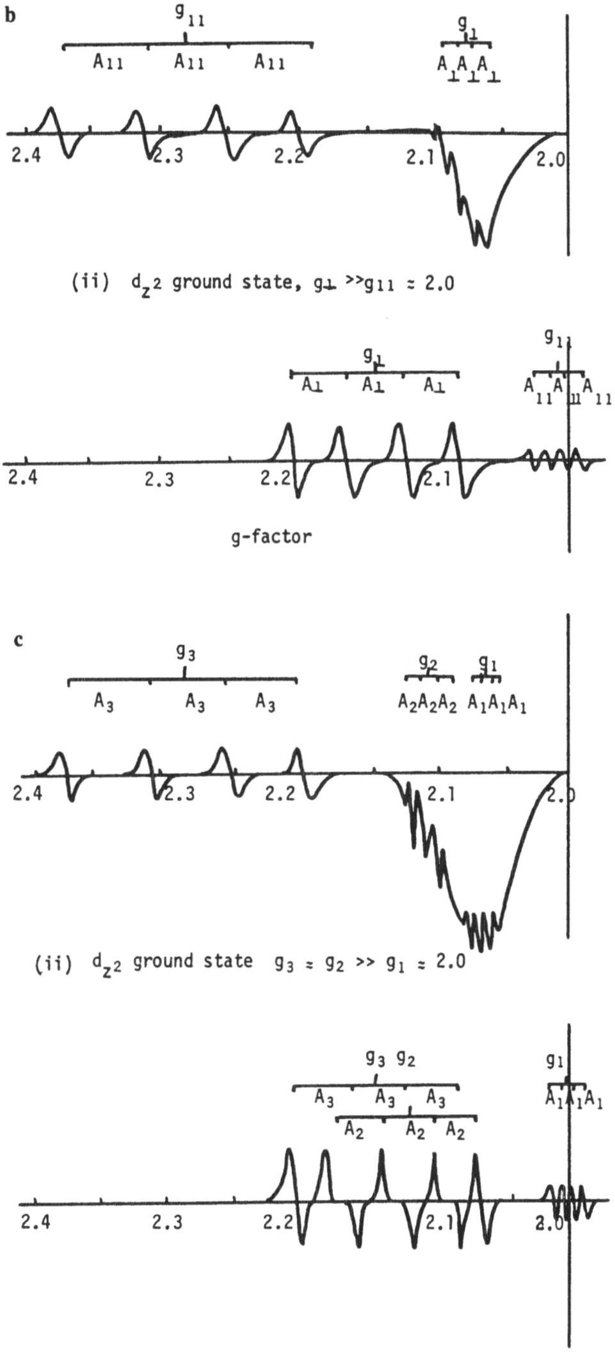

Fig. 12a–c. Polycrystalline e.s.r. spectra of a dilute copper(II) doped zinc(II) system: **a** isotopic g-factors, $d_{x^2-y^2}$ or d_{z^2} ground state; **b** axial g-factors (i) $d_{x^2-y^2}$ ground state, $g_{11} \gg g_{\perp} > 2.0$; (ii) d_{z^2} ground state, $g_{\perp} \gg g_{11} \approx 2.0$; **c** rhombic g-factors. (i) $d_{x^2-y^2}$ ground state $g_3 \gg g_2 \approx g_1 > 2.0$; (ii) d_{z^2} ground state, $g_3 \approx g_2 \gg g_1 \approx 2.0$

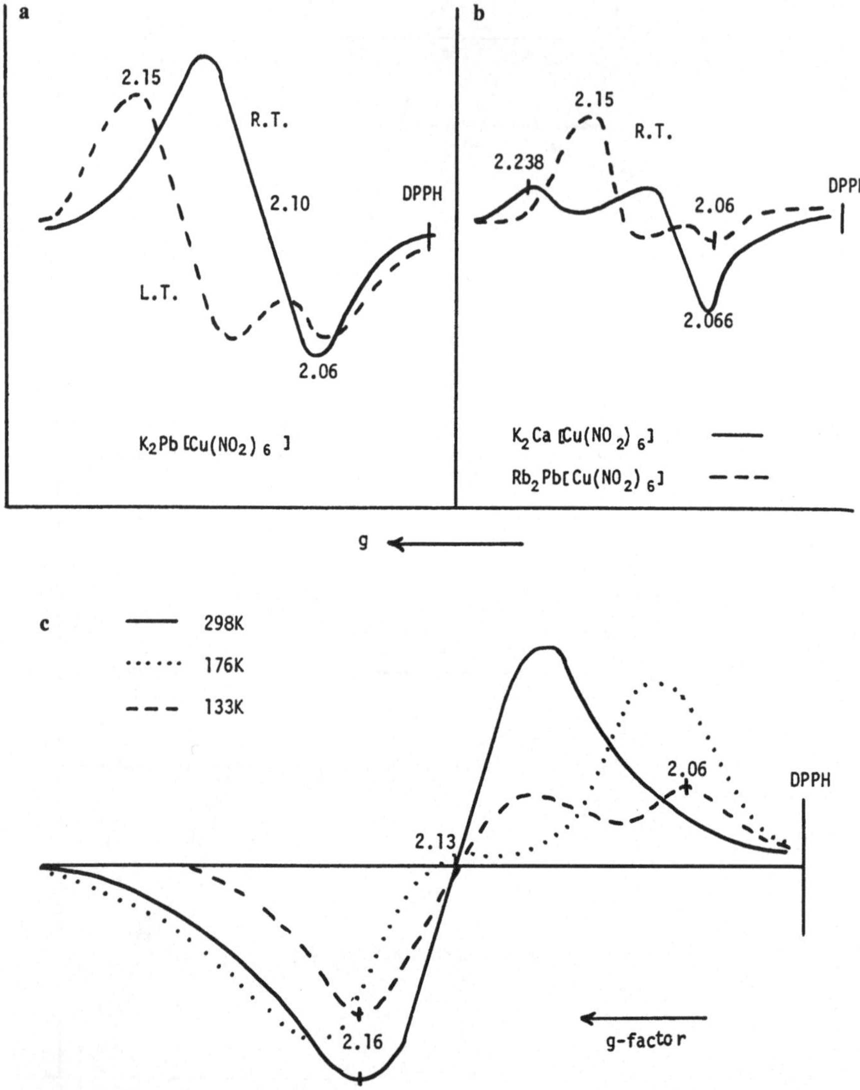

Fig. 13a–c. The powder e.s.r. spectra of **a** and **b** some $M^{I}M^{II}$ $[Cu(NO_2)_6]$ complexes and **c** of $[Cu(en)_3][SO_4]$

rhombic octahedral CuO_6 chromophore at 125 K. Although the CuO_6 chromophores of $[NH_4]_2[Cu(OH_2)_6][SO_4]_2$ are misaligned, the separate g's are resolved in the single-crystal e.s.r. spectrum Fig. 14, to show a variation that is characteristic of fluxional behaviour in the doped systems (see p. 83). The highest g-factor increases with decreasing temperature, the intermediate g-factor decreases and the lowest g-factor increases very slightly, changes that correspond directly with the increase in the Cu-O (7) distance, the decrease of the Cu-O (8) distance, with decreasing temperature, and the almost non-temperature variability of the Cu-O (9) distance[30], respectively. This corresponds to the

Table 5. Temperature variable e.s.r. spectra for some copper(II) complexes[16)]

	Temp. (K)	g_1	g_2	g_3
$K_2Pb[Cu(NO_2)_6]$	298	–	2.10	–
	77	2.06	–	2.15
$[Cu(en)_3][SO_4]$	298	–	2.13	–
	176	2.06	–	2.16
$Cs_2Pb[Cu(NO_2)_6]$	420	–	2.12	–
	298	2.066	–	2.156
	160	2.062	–	2.153
$[Cu(dien)_2][NO_3]_2$	298	–	2.102	–
	150	–	2.100	–
$[Cu(methoxyacetate)_2(OH_2)_2]$	298	2.087	2.185	2.360
	77	2.069	2.118	2.441
$[NH_4]_2[Cu(OH_2)_6][SO_4]_2$	298	2.083	2.215	2.365
	120	2.100	2.136	2.465

Fig. 2 (C) behaviour with a single low-energy potential well I, plus some thermal population of well II, and yields a ΔE value of $168 \pm 40\ cm^{-1}$.

Due to misalignment[38)] the single-crystal e.s.r. spectrum[61)] of the *cis*-distorted octahedral CuN_4O_2 chromophore of $[Cu(bipy)_2(ONO)][NO_3]$, [12], the polycrystalline e.s.r. spectrum shows little change with temperature, despite the temperature variable

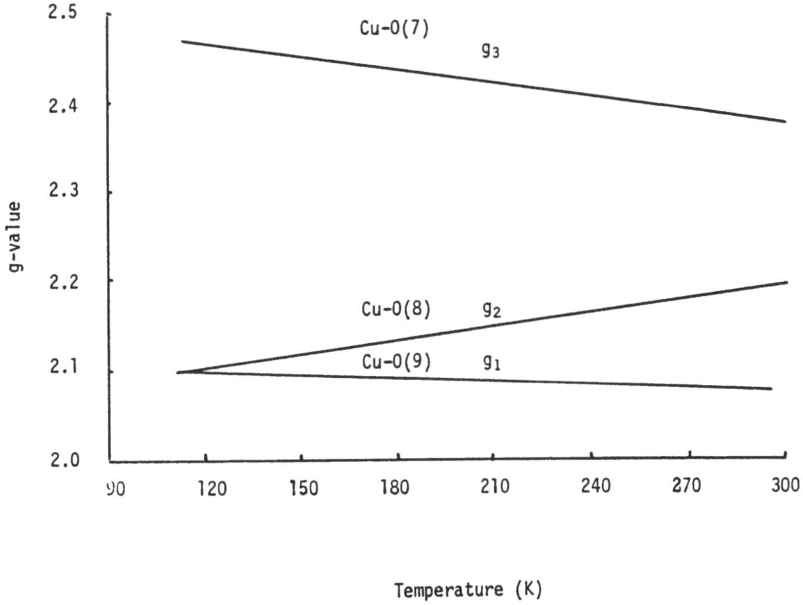

Fig. 14. Temperature variation of the single-crystal e.s.r. spectra of $[NH_4]_2[Cu(OH_2)_6][SO_4]_2$[30)]

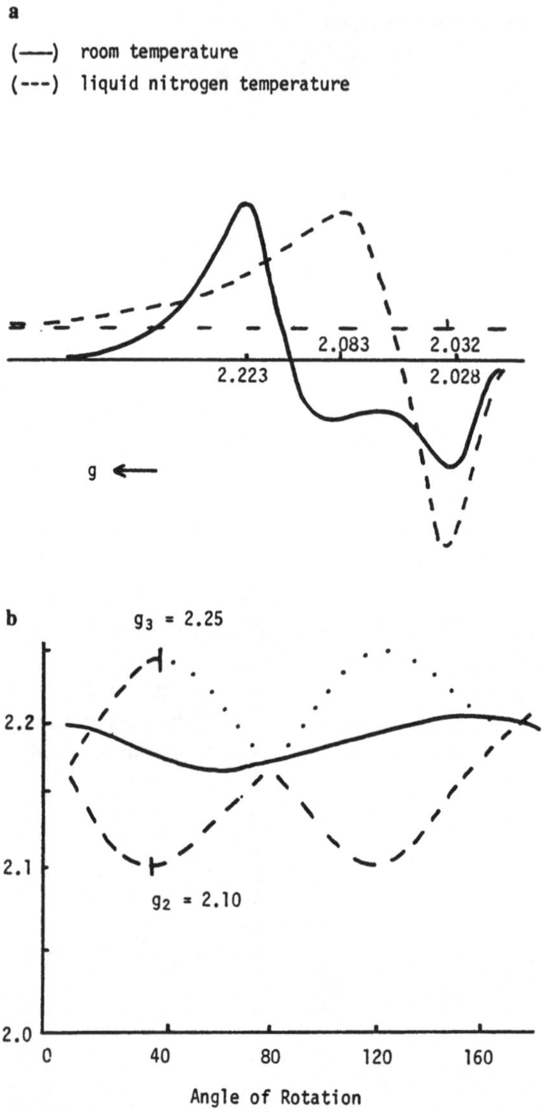

Fig. 15. **a** $[Cu(phen)_2(O_2CCH_3)][ClO_4] \cdot 2H_2O$, the polycrystalline e.s.r. spectra; **b** $[Cu(phen)_2(O_2CCH_3)][ClO_4]$, the single-crystal e.s.r. spectra

crystal structure [12][55]. In contrast, the polycrystalline e.s.r. spectrum[37] of $[Cu(phen)_2(O_2CCH_3)][ClO_4] \cdot 2H_2O$ [11] shows a marked change, Fig. 15(a), with decreasing temperature, suggesting a phase change, but otherwise is not very informative of the change in stereochemistry present. The single-crystal e.s.r. spectrum[37] of $[Cu(phen)_2(O_2CCH_3)][ClO_4]$ shows little temperature variation except in the CuN(2), N(4), O(1), O(2) plane, Fig. 15(b), where an isotropic spectrum is observed at room temperature which becomes anisotropic at the temperature of liquid nitrogen, consistent

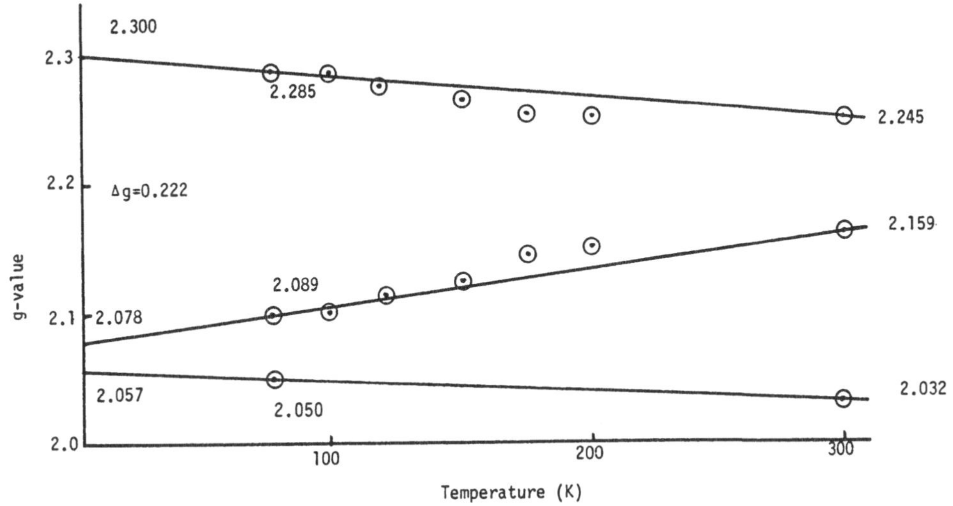

Fig. 16. Single-crystal e.s.r. spectra of $[Cu(phen)_2(O_2CCH_3)][NO_3]\cdot 2H_2O$ v temperature (K): calculated $\Delta E = 132 \pm 20\ cm^{-1}$

with a fluxional CuN_4O_2 chromophore behaviour. The polycrystalline e.s.r. spectrum of $[Cu(phen)_2(O_2CCH_3)][ClO_4]$, [13], shows a small change with temperature, consistent with the small change in structure on reducing the temperature from 298–173 K, [13], but gives no information on the change in stereochemistry involved. The best set of single-crystal g-values[37)] is for $[Cu(phen)_2(O_2CCH_3)][NO_3]\cdot 2H_2O$, Fig. 16, where the crystal g-values equate with the local molecular g-values in the triclinic unit cell, and yields a ΔE value of $132 \pm 20\ cm^{-1}$.

Historically the first evidence for fluxional copper(II) systems was obtained from the observation of the temperature variability[62)] of the e.s.r. spectra of the copper(II) doped $[K_2][Zn(OH_2)_6][SO_4]_2$ system, Fig. 17. The presence of a fluxional CuN_6 chromophore

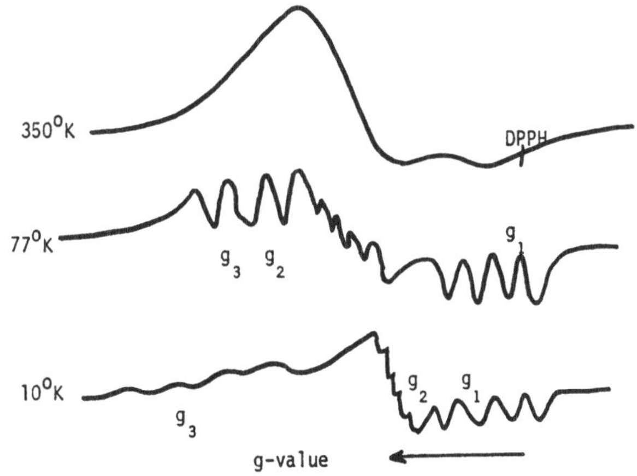

Fig. 17. The powder e.s.r. spectra of copper(II) doped $K_2[Zn(OH_2)_6][SO_4]_2$

Table 6. Copper(II) doped $[Zn(dien)_2]XY$ complexes, ΔA_3-values (%) between room temperature and liquid nitrogen temperature[63)]

X	Y	ΔA_3 (%)
Cl	ClO_4	8.9
Cl	BF_4	10.1
Cl	$Cl \cdot H_2O$	17.8
Br	$Br \cdot H_2O$	22.0
NO	NO_3 (monoclinic)	8.8
I	I	29.5
Cl	NO_3	31.4
Br	NO_3	36.0

in the $[Cu(dien)_2]XY$ complexes[63)] was first recognised from the variation of the A_3 values with temperature, Table 6, while the change in dilute e.s.r. spectrum[52)] of the copper(II) doped $[Zn(dien)_2][NO_3]_2$ system, Fig. 18, gave one of the most clear examples of a change from a d_{z^2} to a $d_{x^2-y^2}$-type e.s.r. spectra with decrease of temperature and consistent with the change from a compressed to an elongated octahedral CuN_6 chromophore stereochemistry. This is the only evidence that the stereochemistry of [10] is *pseudo*-compressed octahedral, as the corresponding concentrated $[Cu(dien)_2][NO_3]_2$, [10], complex[53)] showed no change in the isotropic e.s.r. spectrum with temperature (Table 5). Likewise, the effect of temperature on the polycrystalline e.s.r. spectrum[37)] of copper(II) doped $[Cu(phen)_2(O_2CCH_3)]X \cdot 2H_2O$, [11] complexes, Fig. 19, clearly demonstrate the change from a compressed CuN_4O_2-type chromophore stereochemistry to a more elongated stereochemistry. The temperature dependence of the polycrystalline g- and A-values for the copper(II) doped $K_2[Zn(OH_2)_6[SO_4]_2$ system, Fig. 20, established the general pattern of temperature variation for fluxional systems and produced the first

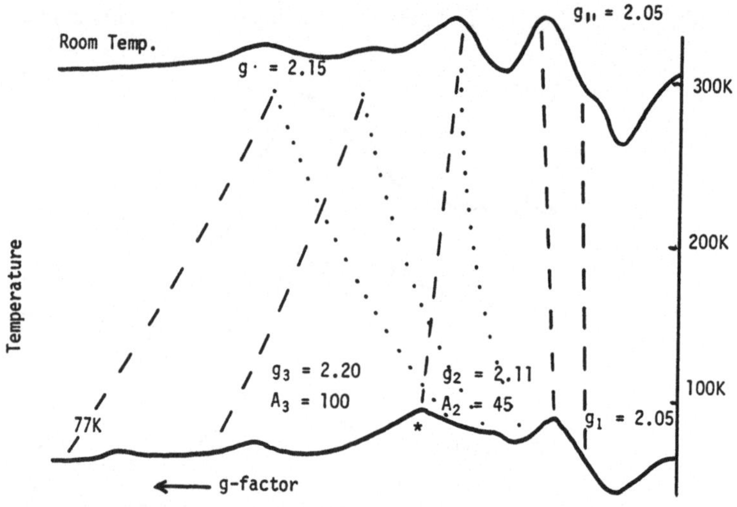

Fig. 18. Temperature variation of g-factors for copper(II) doped $[Zn(dien)_2][NO_3]_2$; ⋆ isotropic g-value of 2.13

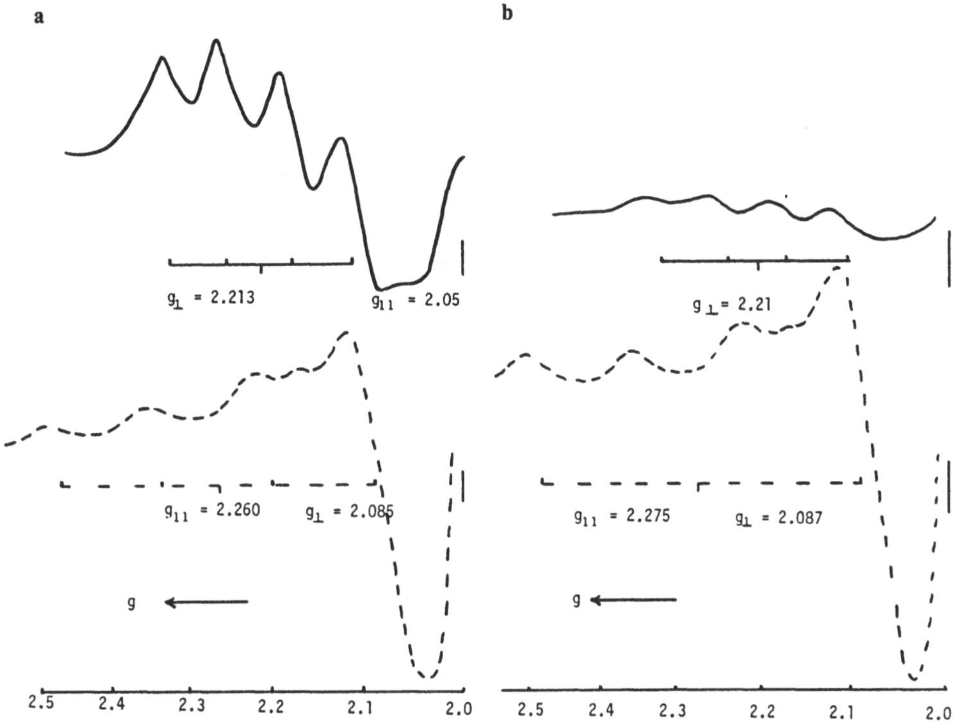

Fig. 19 a, b. The polycrystalline e.s.r. spectra of 10% copper doped $[Zn(phen)_2(O_2CCH_3)]X$; $[X = BF_4]^- \cdot 2H_2O$ – **a**; $X = [ClO_4]^- \cdot 2H_2O$ – **b** at room temperature (—) and at liquid nitrogen temperature (- - -)

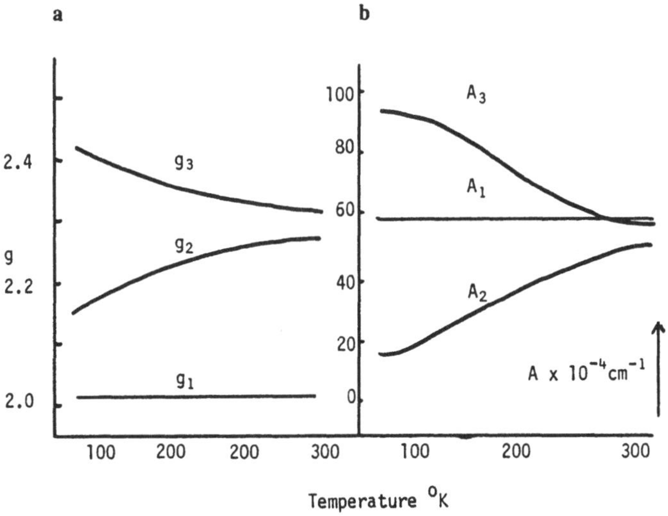

Fig. 20 a, b. Copper doped $K_2[Zn(OH_2)_6][SO_4]_2$ – Temperature variation of the **a** g-factors, and **b** A-values, powder data

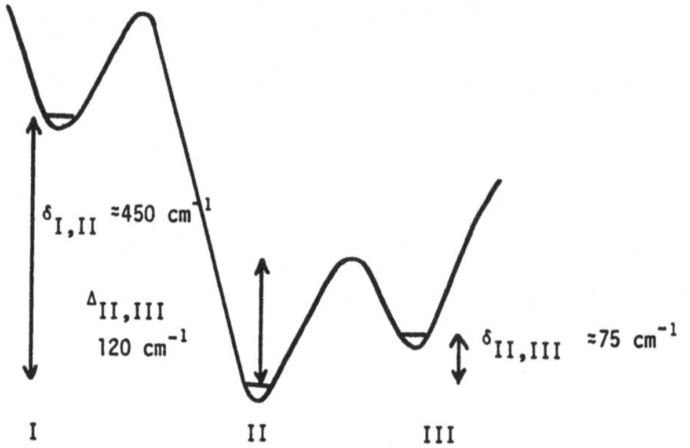

Fig. 21. Circular cross section of the potential energy surface associated with the three Jahn-Teller Wells I-III in copper doped $K_2[Zn(OH_2)_6][SO_4]_2$ (after Ref. 62)

set of definitive energy values for a Fig. 2 (C)-type potential energy well system, Fig. 21. While the single-crystal e.s.r. spectra of copper doped systems are more complicated, they yield clear evidence for fluxional behaviour[61] in copper(II) doped $[Zn(bipy)_2(ONO)][NO_3]$ and especially[37] $[Cu(phen)_2(O_2CCH_3)][BF_4] \cdot 2\,H_2O$, Fig. 22.

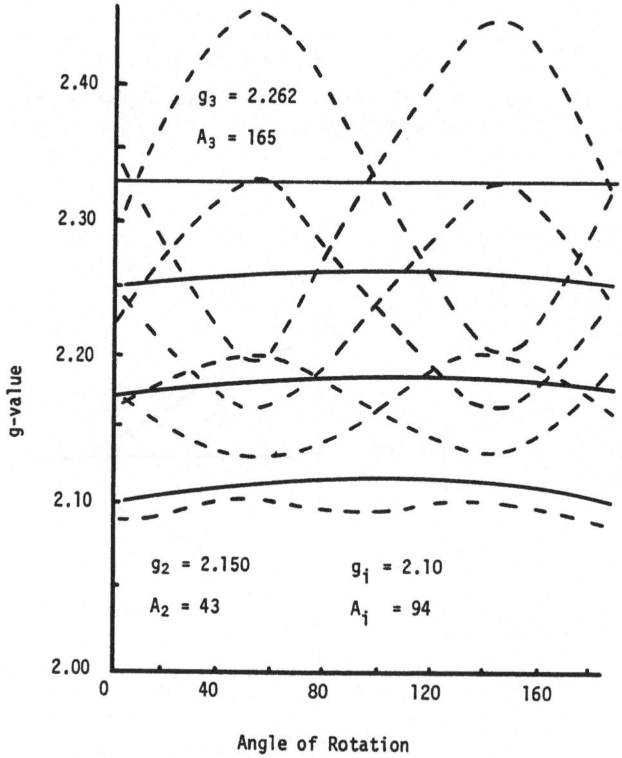

Fig. 22. Single-crystal e.s.r. spectra of 10% doped $[Zn(phen)_2(O_2CCH_3)][BF_4] \cdot 2\,H_2O$ measured in the Cu, N (2), N (4), O (1), O (2) plane

8 Electronic Spectra

As a consequence of the fluxional model the electronic energies of a fluxional CuL_6 chromophore can be interpreted using a cross-section of the Warped Mexican Hat potential energy surface,[16,17,26], Fig. 23(a), involving a low energy transition $\psi^- \rightarrow \psi^+$ of 4 $E_{J.T.}$, and a high energy transition, $\psi^- \rightarrow T_{2g}(d_{xz}, d_{yz}, d_{xy})$ of $\Delta + 2E_{J.T.}$ These transitions are not different from the one-electron energy level description,[4,11] except that a vibronic coupling description is used. Table 7 summarises[26] some structural and spectroscopic data for copper(II) systems involving six-equivalent ligands. The appearance of the electronic reflectance spectra of fluxional and static copper(II) systems do not differ significantly, Fig. 24. The three dimensional dynamic systems such as cubic $K_2Pb[Cu(NO_2)_6]$, [1], show no polarisation effects, the trigonal $[Cu(en)_3][SO_4]$, [2], only shows a change of intensity with polarisation, Fig. 25(a), but static systems, e.g. $K_2Ba[Cu(NO_2)_6]$, show marked polarisation[64], Fig. 25(b). The reason for this insensitivity of the electronic spectra to fluxional effects lies in the very short time scale[43] involved in electronic transitions of 10^{-15} s such that they relate to the underlying static distorted stereochemistry of a fluxional CuL_6 chromophore and not to the time average structure as determined by X-ray crystallographic and e.s.r. techniques. This insensitivity of the electronic spectra of copper(II) complexes has been a major factor in recognising the occurrence of fluxional copper(II) systems. Thus the "sameness" of the electronic spectra of the three complexes of Fig. 24, suggests essentially the same CuN_6 chromophore

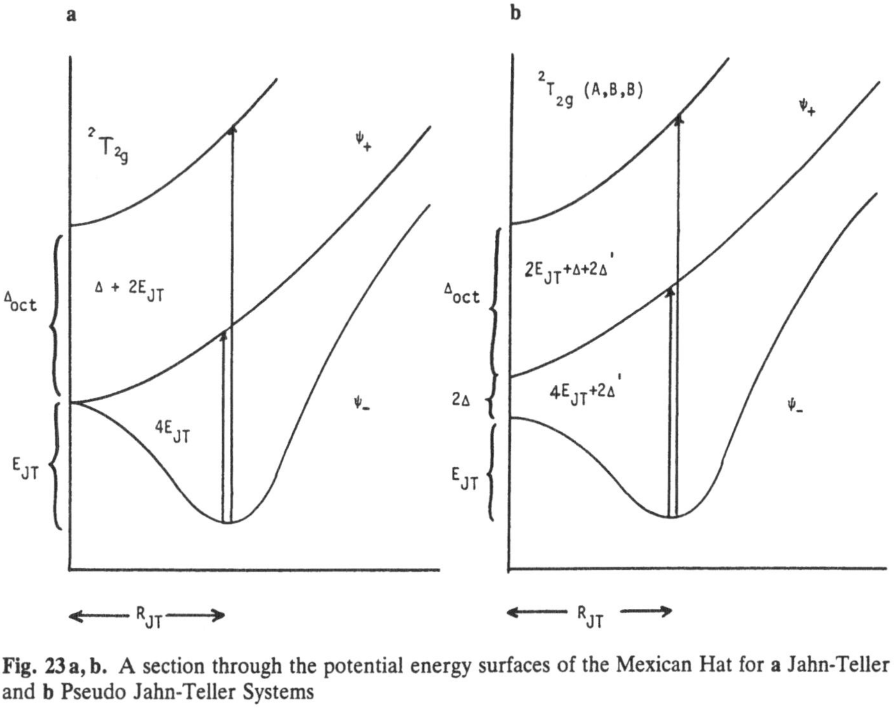

Fig. 23a, b. A section through the potential energy surfaces of the Mexican Hat for **a** Jahn-Teller and **b** Pseudo Jahn-Teller Systems

Table 7. Structural and spectroscopic data[26] for some copper(II) complexes involving six-equivalent ligands, room temperature

a) Dynamic (3D)	d_o, Å	R_{JT}Å	E_{JT}(kk)	E_2(kk)
$[Cu(en)_3][SO_4]$	2.150	0.358	2.18	15.7
$K_2Pb[Cu(NO_2)_6]$	2.111	0.333	1.75	16.5
b) Dynamic (2D)				
$Rb_2Pb[Cu(NO_2)_6]$	2.136	0.252	1.92	15.7
$Cs_2Pb[Cu(NO_2)_6]$	2.171	0.343	1.92	16.3
c) Static				
$K_2Ca[Cu(NO_2)_6]$	2.138	0.303	1.98	16.5
$K_2Sr[Cu(NO_2)_6]$	2.127	0.318	1.90	16.5
$K_2Ba[Cu(NO_2)_6]$	2.132	0.308	1.92	16.55

stereochemistry, modified by slight differences due to the different crystal lattices. Crystallographically, $K_2Pb[Cu(NO_3)_2]$ [1][22] has an octahedral CuN_6 structure, $Rb_2Pb[Cu(NO_2)_6]$, [4][29] has a compressed octahedral structure and $K_2Ba[Cu(NO_2)_6]$[64] has an elongated rhombic octahedral structure, which belie the fluxional properties of the first two complexes.

The electronic spectra of copper(II) complexes involving nonequivalent ligands and the potential energy surface of Fig. 3, can be similarly interpreted, Fig. 23 (b), but owing to the additional splitting of E_o by $2\Delta'$ it is not possible to evaluate $E_{J.T.}$. Nevertheless, the electronic spectra of these fluxional pseudo Jahn-Teller systems show the same insensitivity to the anions present. In a series $[Cu(bipy)_2(ONO)]X$ complexes, X = $[NO_3]^-$, [12], $[BF_4]^-$, [16][65] and $[PF_6]^-$ [17][66], the electronic spectra, Fig. 26 (a), consist of two broad peaks of comparable energy and intensity, despite the clear difference in inplane geometry of [12, 16 and 17]; $\Delta 0$ = {[Cu-O (2)] – [Cu-O (1)]} is 0.090,

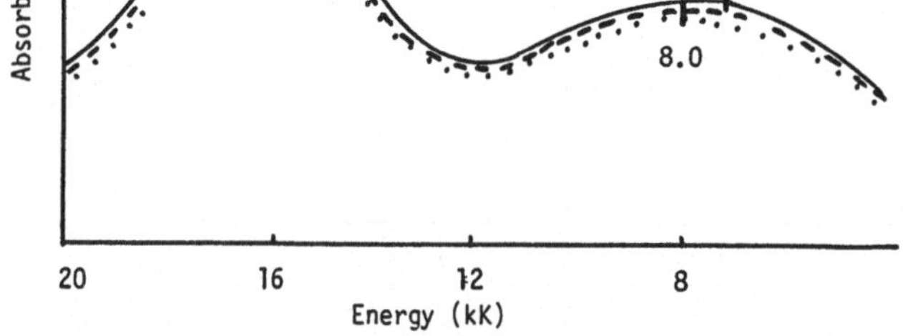

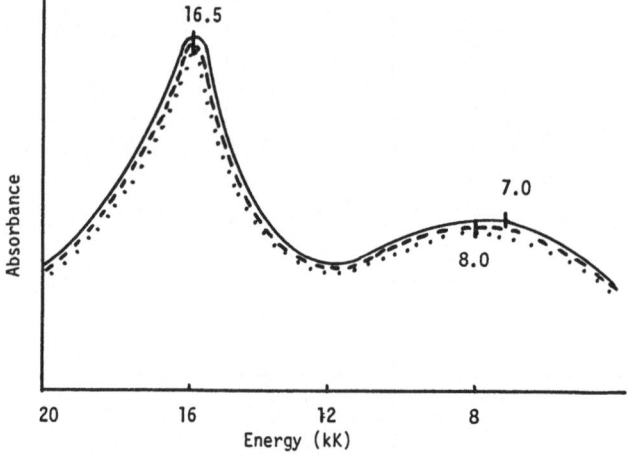

Fig. 24. The electronic reflectance of; a $K_2Pb[Cu(NO_2)_6]$ – three dimensional dynamic (—); b $Rb_2Pb[Cu(NO_2)_6]$ – two dimensional dynamic (- - -); c $K_2Ba[Cu(NO_2)_6]$ – static (- - -)

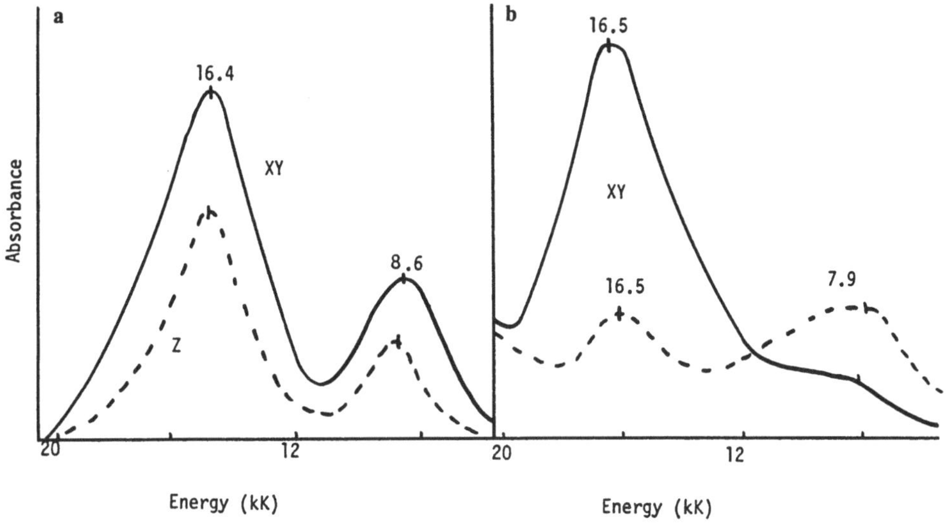

Fig. 25 a, b. The polarised single-crystal electronic spectra **a** $[Cu(en)_3][SO_4]$; **b** $K_2Ba[Cu(NO_2)_6]$

0.346 and 0.251, respectively. The comparable electronic spectra of [12] and [16] with bands at ca. 9 500 and 15 000 cm^{-1} suggest a closely comparable underlying *static* CuN_4O_2 chromophore stereochemistry, which is substantiated in the Cu-O (2) v Cu-O (1) plot of Fig. 26 (b). The low temperature data for [12] clearly give a linear correlation, and the data for [16] lie on this correlation, suggesting that [12] and [16] exhibit different degrees of fluxional behaviour, but with the *same* underlying static CuN_4O_2 chromophore stereochemistry. Using this correlation, the ΔE values for [12] and [16], are 75 and 125 cm^{-1}, respectively, corresponding, Fig. 7 (a), to a 59 and 64% thermal population for well I, respectively, at room temperature. In Fig. 26 (a), the spectrum of [17] is slightly different, with energies of 10, 200 and 14 000 cm^{-1} respectively, implying a different underlying static CuN_4O_2 geometry, support for which is found in the Cu-O data for [17]

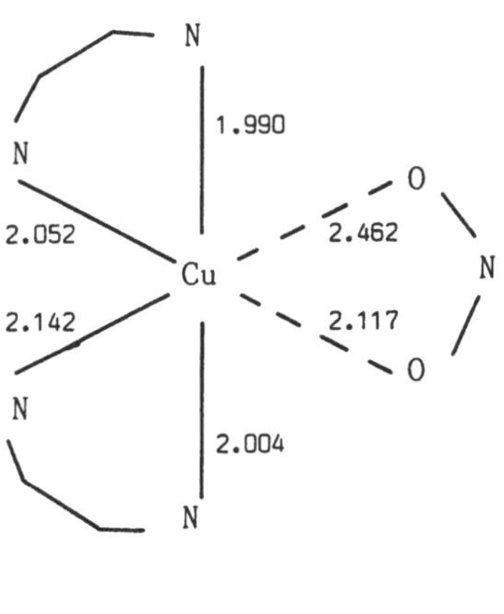

[16] $[Cu(bipy)_2(ONO)][BF_4]$

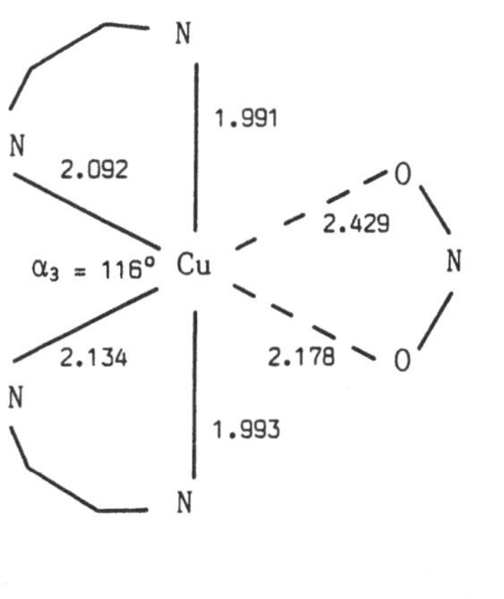

[17] $[Cu(bipy)_2(ONO)][PF_6]$

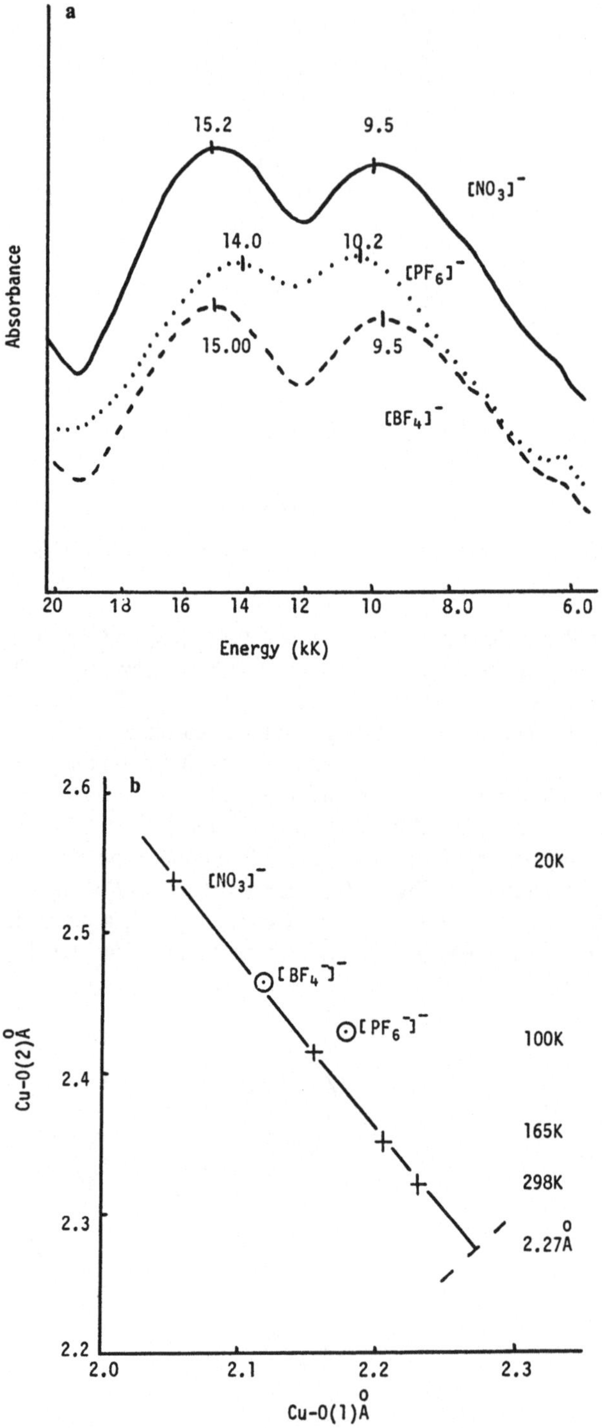

Fig. 26 a, b. [Cu(bipy)$_2$(ONO)] [Y]; **a** Electronic reflectance spectra; **b** Cu-O (1) v Cu-O (2)

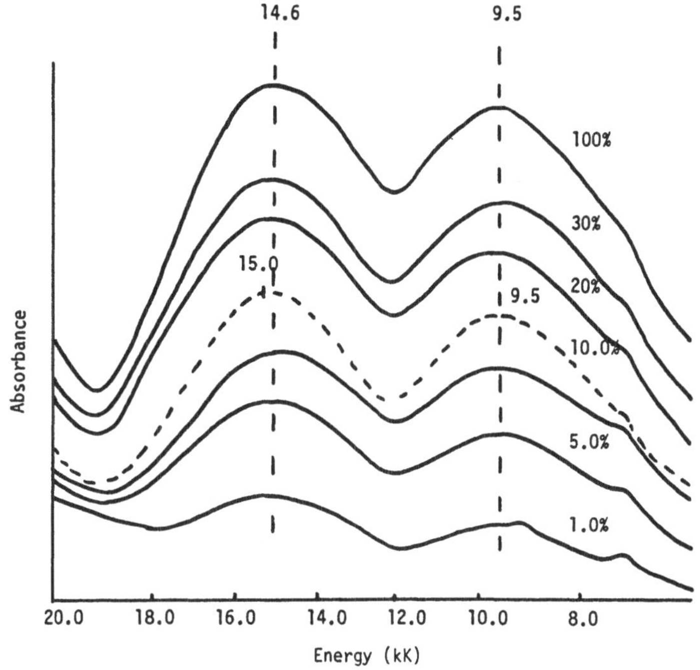

Fig. 27. The electronic reflectance spectra of 1–100% copper doped [Zn(bipy)$_2$(ONO)] [NO$_3$] (—) and [Cu(bipy)$_2$(ONO)] [BF$_4$] (---)

which lies well off the correlation of Fig. 26 (b). An examination of the room temperature structures of [12] shows that there is only a small shift (± 1°) in the N (2)-Cu-N (4) angle of 102° from 298 to 20 K, this angle is closely comparable in [16], 98.5°, but significantly different in [17], 116°, suggesting that the difference in the electronic spectrum of [17] lies in the difference in the N (2)-Cu-N (4) angle.

Another use of the apparent invariance of the electronic spectra is shown in Fig. 27, where the twin peaked electronic spectrum[61)] of the [Cu(bipy)$_2$(ONO)]$^+$ cation doped in the isomorphous lattice of [Zn(bipy)$_2$(ONO)] [NO$_3$] [18][65)] is shown to be independent of

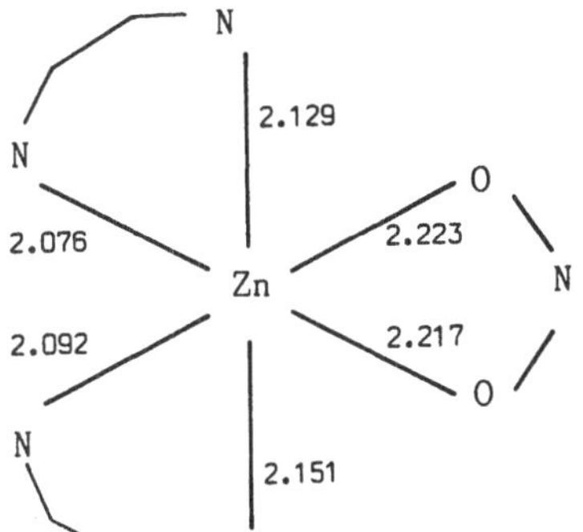

[18] [Zn(bipy)$_2$(ONO)][NO$_3$]

the copper concentration in this molecular type lattice[61], thus, notwithstanding the more nearly octahedral structure of the ZnN_4O_2 chromophore in [18] compared with that of the *cis*-distorted octahedral $[CuN_4O_2]$chromophore of [12]. This suggests that the structure of the CuN_4O_2 chromophore is the same in the concentrated copper(II) complex, and in the 1% copper(II) doped zinc(II) complex. Consequently, the more accurate g and A values determined in the > 10% copper(II) doped zinc(II) complex may be associated with the crystallographically determined CuN_4O_2 chromophore of the 100% copper(II) complex. In these molecular type lattices the structure of the CuN_4O_2 chromophore is independent of the structure of the host lattice, and the effect is referred to as the Non-Cooperative Jahn-Teller Effect[67].

9 Plasticity Effect

The previous sections have concentrated on a relatively small group of the six-coordinate complexes of the copper(II) ion, which exhibit a temperature variable or fluxional stereochemistry[16] due to the operation of the Jahn-Teller or pseudo Jahn-Teller effect[14]. It should be emphasised that the *bulk* of six-coordinate copper(II) complexes involve non-temperature variable structures[4], Fig. 2 (C), and that there also exist a wide range of four and five-coordinate stereochemistries, which are *not* susceptible to first-order Jahn-Teller effects. In addition, a significant number of complexes exist in which, due to the presence of relatively long bonds and off-the-z-axis bonding it is difficult to decide whether the geometry is best described as four, five, six, seven, eight or even nine coordinate[9]. Though all these geometries are connected by various aspects of the Jahn-Teller effect in one way or the other, the empirical concept referred to as the Plasticity Effect[6,7] has been introduced to classify the observed structures. For any non-regular stereochemistry the tetragonality, T, is not constant, but varies[6] as shown in Fig. 28. This implies that for a given stereochemistry the eccentricity of the copper(II) ion ellipsoid is not fixed, but *continuously* variable[6] (within certain limits), and the precise tetragonality observed will be determined by relatively minor factors, such as, lattice packing forces, including van der Waal forces, and hydrogen-bonding. It is this effect that is responsible for the apparent non-rigid or flexible stereochemistry[7] observed for the copper(II) ion in its complexes. These are of two types:

A-type. Distortion isomers in which complexes of the same empirical formula can exist in the radically different stereochemistries, such as in α- and β-$Cu(NH_3)_2Br_2$ [19][68] and [20][69], which involve the compressed and elongated octahedral $[CuN_2Br_2Br'_2]$ chromophore. In this system the β-form may be converted to the α-form by application of pressure.[70] The occurrence of this type of isomerism is much less common than the B-type.

B-type. Distortion isomers in which complexes of the same empirical formula can exist in only slightly different stereochemistries, namely, with slightly different degrees of distortion of the same basic stereochemistry, as in α- and β-[Cu(salicylaldehydrate)$_2$] [21][71], [22][72] which are rhombic coplanar, or the two forms of α- and β-[Cu(1,3-

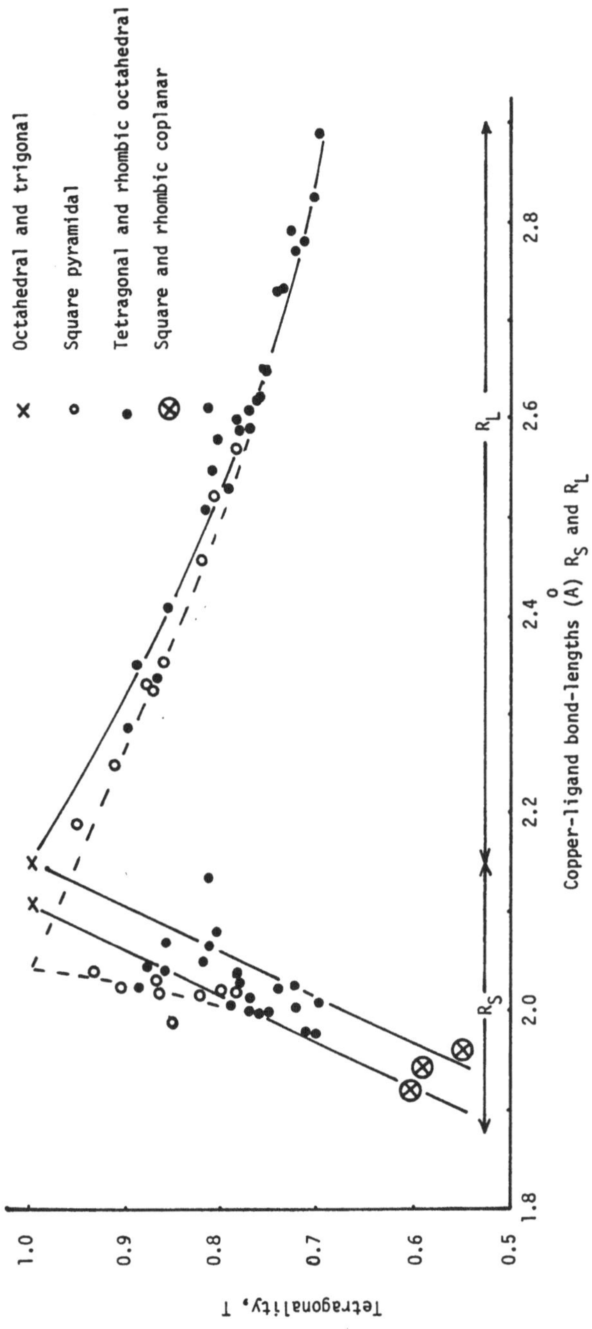

Fig. 28. R_S and R_L versus Tetragonality for nitrogen donor ligands

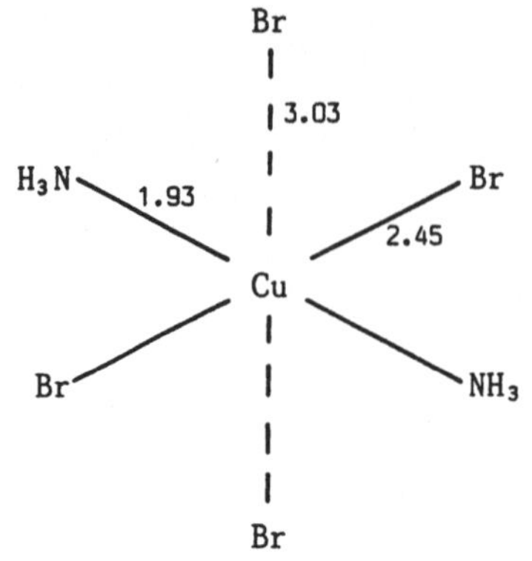

[19] α-[Cu(NH_3)$_2$Br$_2$]

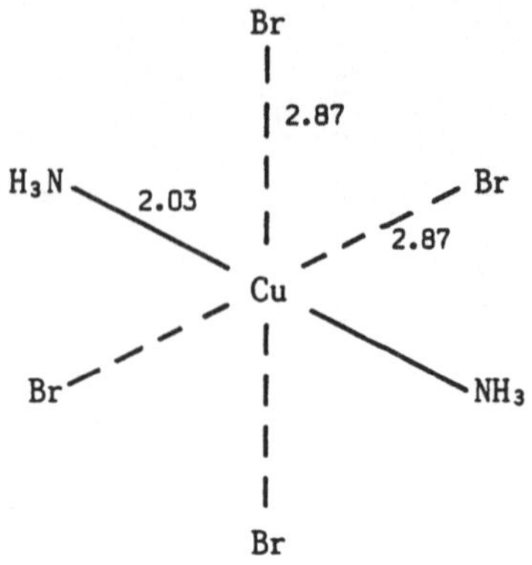

[20] β-[Cu(NH_3)$_2$Br$_2$]

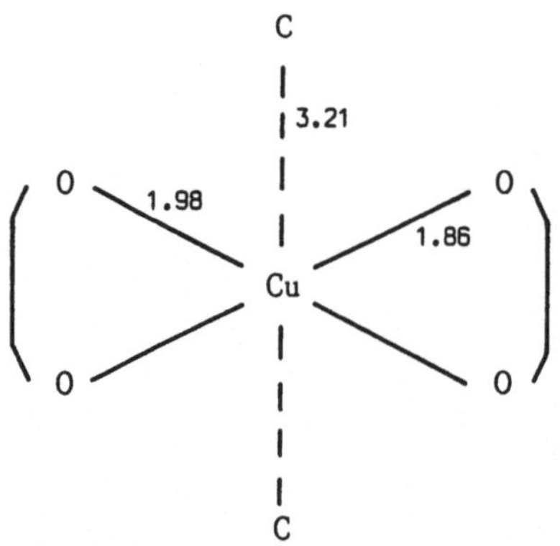

[21] α-[Cu(saliclaldehydate)$_2$]

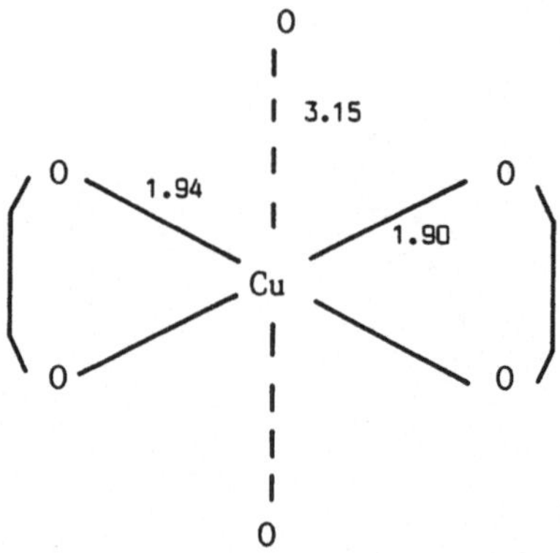

[22] β-[Cu(salicylaldehydate)$_2$]

diaminobutane)$_2$(ClO_4)$_2$] [23][73], [24][74], which are elongated tetragonal octahedral. In the latter the differences in bond-lengths and bond-angles are only small, i.e. < 0.1 Å or $< 5°$, but there are noticeable effects on the colour of the isomers, namely, blue-violet and red-violet, for the α and β forms. Also small differences in the polarised single-crystal electronic spectra[75] may occur, Fig. 29 for α- and β-[Cu(α-pic)$_2$(NO_3)$_2$] [25] and [26] and in measured g-factors[75], Table 8. If the term distortion isomer is extended slightly[76] to include:

C) anion distortion isomers; $M^{I}M^{II}$[Cu(NO_2)$_6$] and R_2[$CuCl_4$]

D) cation distortion isomers; [Cu(dien) (bipyam)] X_2 and [Cu(bipy)$_2$Cl] X

then a wide range of distortions are possible by varying the cations or anions in (C) and (D), respectively.

Examples of anion distortion isomers with different geometries have already been observed in the octahedral and compressed octahedral CuN_6 chromophores of K_2Pb[Cu(NO_2)$_6$] [1] and Rb_2Pb[Cu(NO_2)$_6$] [4] and in the compressed tetrahedral or

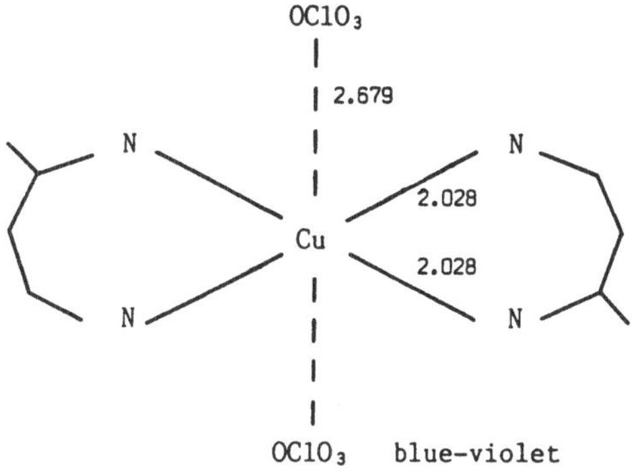

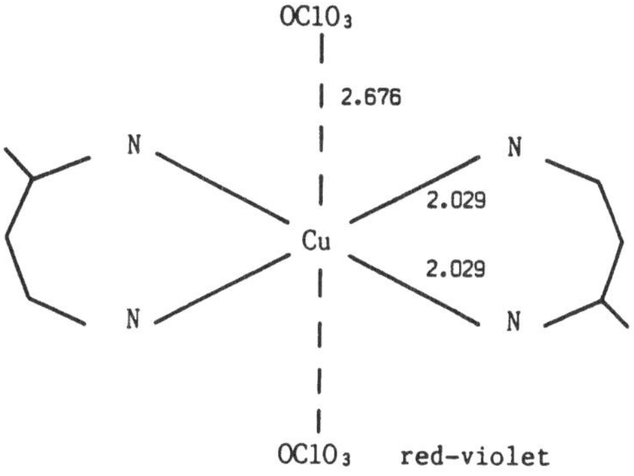

[23] α-[Cu(1,3-diaminobutane)$_2$(ClO$_4$)$_2$]

[24] β-[Cu(1,3-diaminobutane)$_2$(ClO$_4$)$_2$]

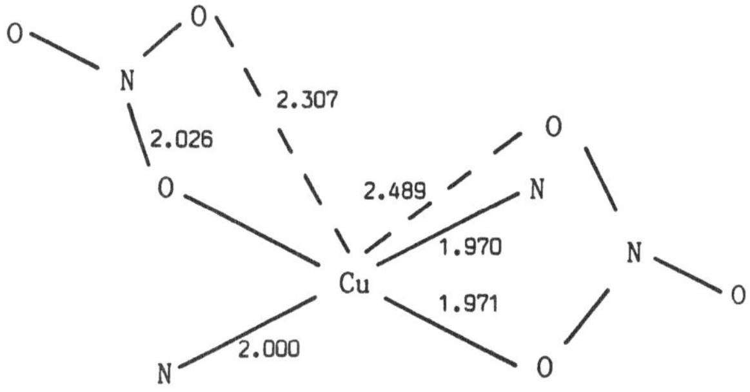

[25] α-[Cu(α-pic)$_2$(NO$_3$)$_2$]

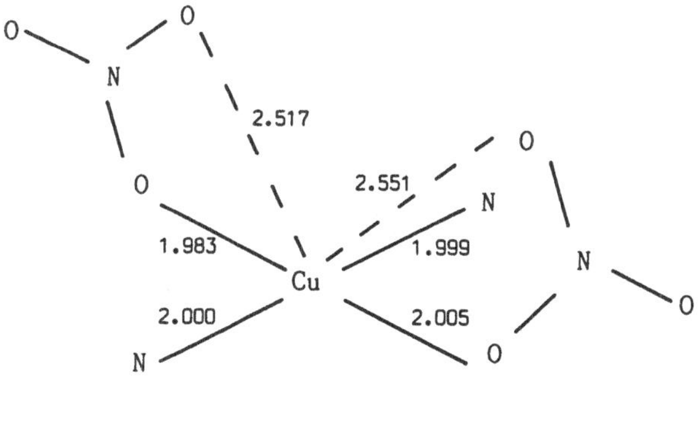

[26] β-[Cu(α-pic)$_2$(NO$_3$)$_2$]

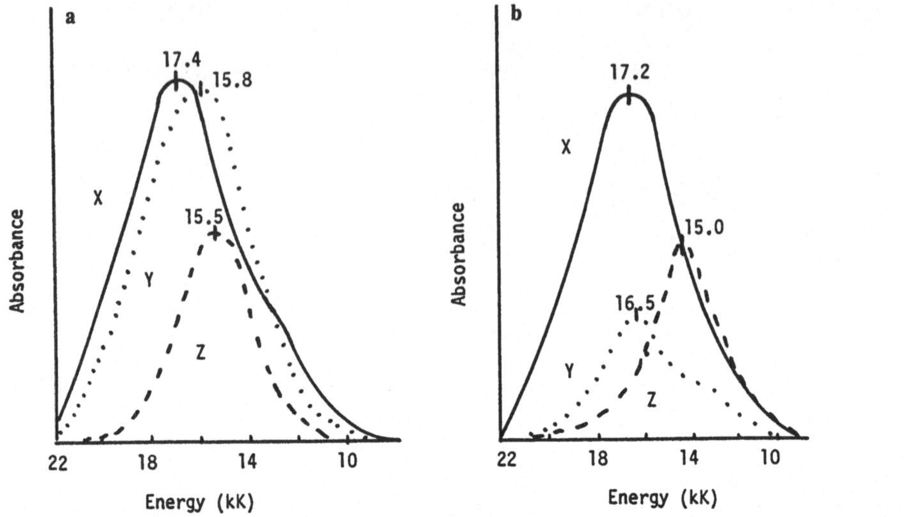

Fig. 29. **a** Pol.S.C. Electronic Spectra α-[Cu(α-pic)$_2$(NO$_3$)$_2$]; **b** Pol.S.C. Electronic Spectra β-[Cu(α-pic)$_2$(NO$_3$)$_2$]

square coplanar [$CuCl_4$] chromophores of Cs_2[$CuCl_4$] [27][77] and [$\phi CH_2CH_2NH_2CH_3$]$_2$[$CuCl_4$], [28][78], respectively. While examples of cation distortion isomers having different stereochemistries have been described in the compressed rhombic octahedral CuN_6 chromophore[35] of [Cu(dien)$_2$] [NO_3]$_2$, [10] and the elongated rhombic octahedral[40] chromophore in [Cu(dien)$_2$]$Br_2 \cdot H_2O$, [14].

The real significance of anion and cation distortion isomers is that for a given chromophone, small changes in geometry may be observed, which can then be correlated

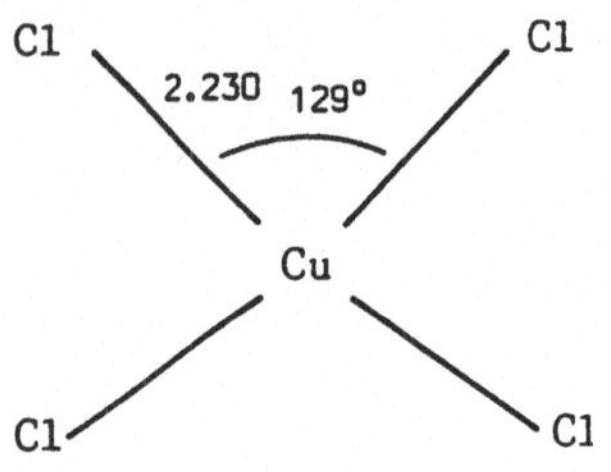

[27] Cs_2[$CuCl_4$]

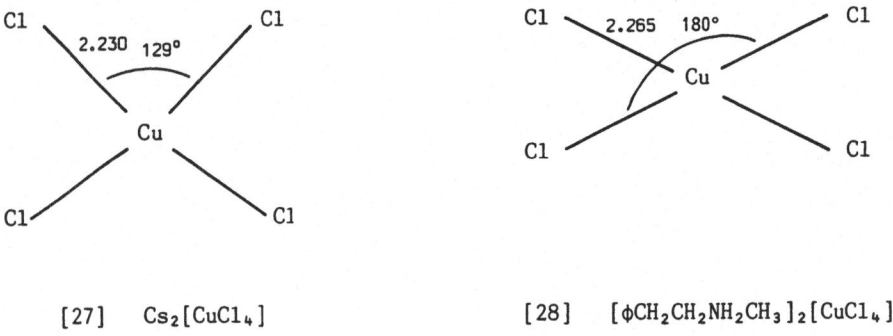

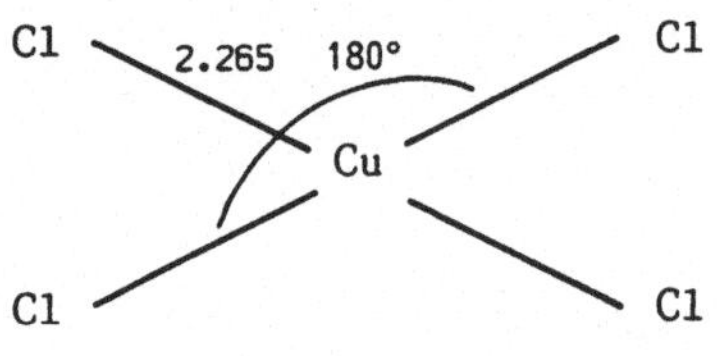

[28] [$\phi CH_2CH_2NH_2CH_3$]$_2$[$CuCl_4$]

Table 8. Single-crystal g-factors for α- and β-[Cu(α-pic)$_2$(NO$_3$)$_2$][75]

	g_x	g_y	g_z
α	2.0563	2.0742	2.2740
β	2.0619	2.0642	2.2830

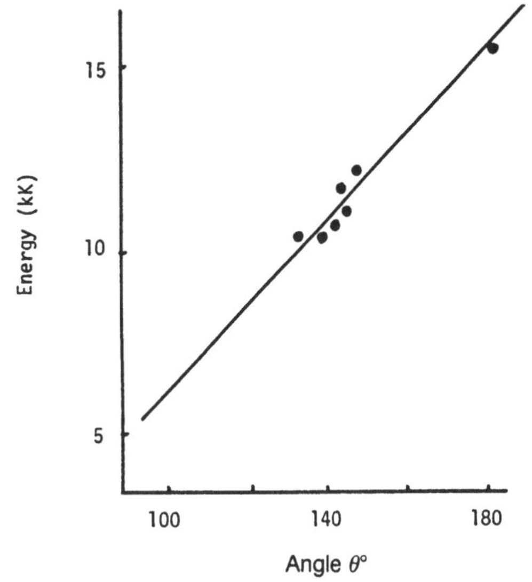

Fig. 30. Elect. Spectra of $[CuCl_4]^{=}$ v $\theta°$ angle (after Ref. 78)

with the observed changes in the corresponding electronic properties. Thus for the compressed tetrahedral $[CuCl_4]^{=}$ anion, there is a reasonable correlation[78, 79] between the maximum of the electronic reflectance spectra and the angular distortion θ of the anion, Fig. 30.

In the series of $[Cu(bipy)_2Cl]X$ complexes[80–82] the CuN_4Cl chromophore is basically trigonal bipyramidal [29], but with the angle α_3 varying from 97° to 124°, Table 9 (a), for the complexes where X = $[ClO_4]^-$; $1/2[S_5O_6] \cdot 3H_2O$, and $[PF_6]$, respectively. A range of angular distortions occur in these cation distortion isomers of 27°, a difference that is so large that the appearance of the electronic reflectance spectra noticeably changes in this series, Fig. 31 from two well resolved peaks (separated by ca. 4.0 kK) to a single peak at 11.9 kK[80]. Likewise in the series of cation distortion isomers[83] of $[Cu(dien)(bipyam)]X_2$ [30] the CuN_5 chromophore has a very distorted structure between trigonal bipyramidal

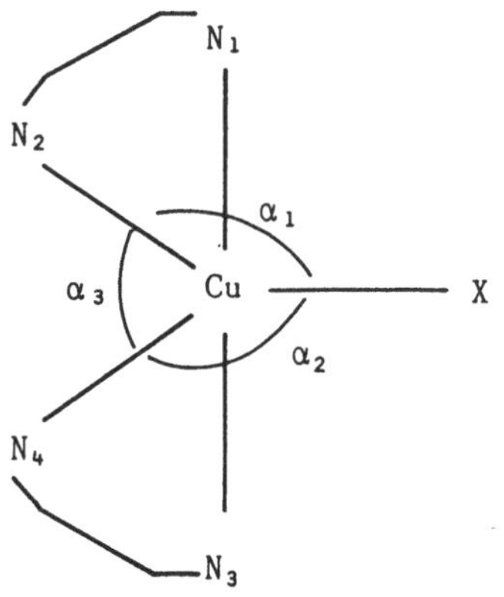

[29] $[Cu(bipy)_2Cl]X$

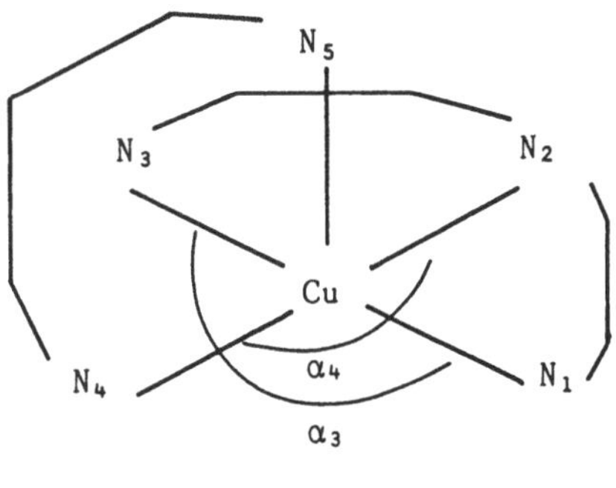

[30] $[Cu(dien)(bipyam)]X_2$

Table 9. X-ray crystallographic data for some cation distortion isomers

(a) $[Cu(bipy)_2Cl]X$ complexes[80]

X	$[ClO_4]^-$ [79]	$1/2\,[S_5O_6]^{=} \cdot 3\,H_2O$ [80]	$[PF_6]^-$ [81]
Cu-N (1)	1.993	1.998	1.996
Cu-N (2)	1.991	2.089	2.105
Cu-N (3)	2.076	1.988	2.005
Cu-N (4)	2.136	2.105	2.108
Cu-Cl	2.263	2.293	2.344
N (2)-Cu-Cl	137.1	130.1	115.7
N (4)-Cu-Cl	126.4	122.1	120.5
N (2)-Cu-N (4)	96.5	107.3	123.8

b) $[Cu(dien)(bipyam)]X_2$ complexes[83]

X	$[NO_3]^-$	$[ClO_4]^- \cdot 1/2\,H_2O$	$Cl^- \cdot H_2O$
Cu-N (1)	2.032 (19)	2.026 (5)	1.990 (12)
Cu-N (2)	2.039 (18)	2.023 (4)	2.013 (12)
Cu-N (3)	2.071 (18)	2.052 (5)	2.020 (12)
Cu-N (4)	1.991 (19)	1.993 (4)	1.998 (11)
Cu-N (5)	2.150 (19)	2.170 (3)	2.126 (11)
N (1)-Cu-N (3) (α_3)	135.5 (7)	151.9 (1)	159.1 (5)
N (2)-Cu-N (4) (α_4)	172.0 (8)	167.9 (1)	162.7 (5)

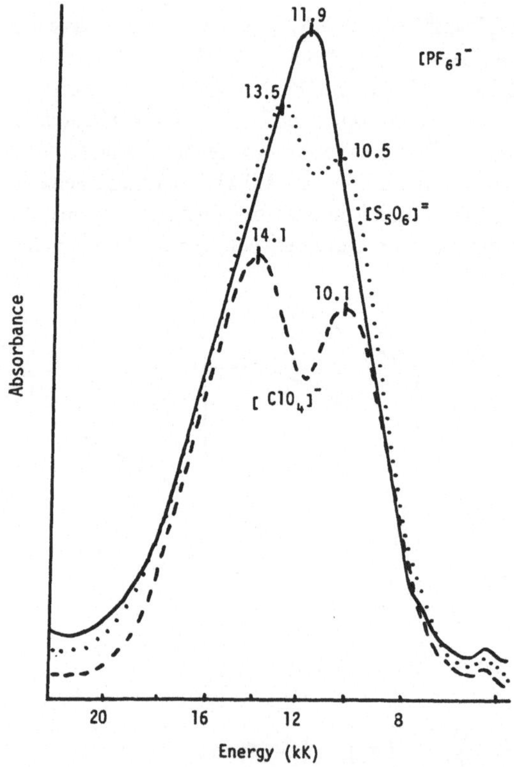

Fig. 31. Electronic reflectance spectra of $[Cu(bipy)_2Cl][Y]$ complexes

and square pyramidal, involving an angular change of 25° in α_3, Table 9 (b), from 135.5, 151.9 and 159.1° for X = $[NO_3]^-$, $[ClO_4]^- \cdot 0.5\ H_2O$ and $Cl^- \cdot H_2O$, respectively. These angular changes may be related to the distortion of the regular square pyramidal stereochemistry, α_3 = 165° in $K[Cu(NH_3)_5][PF_6]_3$ [31][84] to the trigonal bipyramidal stereochemistry, α_3 = 120° in $[Cu(tren)(NH_3)][ClO_4]_2$ [32][84], by the Berry Twist mechanism[85, 86], Fig. 32. The angular distortions of [30] are reflected in changes in the electronic spectra, Fig. 33 (a), which compare with changes in the electronic spectra[83] of [31] and [32], Fig. 33 (b).

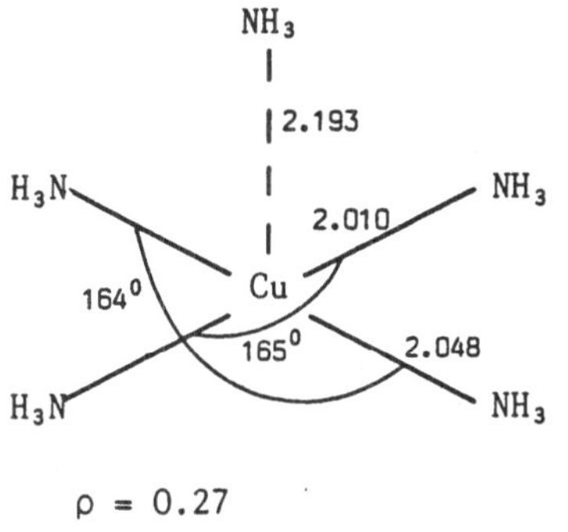

[31] $K[Cu(NH_3)_5][PF_6]_3$

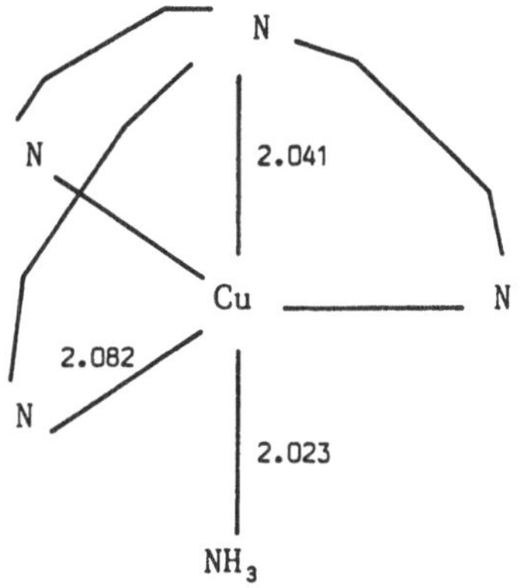

[32] $[Cu(tren)(NH_3)][ClO_4]_2$

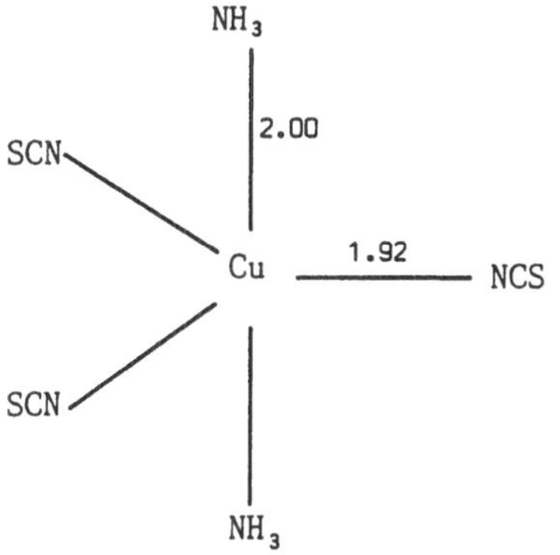

[33] $[Cu(NH_3)_2Ag(SCN)_3]$

The above examples demonstrate that it is not possible to predict a precise stereochemistry for a copper(II) complex in view of the existence of distortion isomers of Type A and B and of cation or anion distortion isomers. Each will have a definite stereochemistry, but a range of possible distortions (within certain limits) can arise due to the "Plasticity Effect" of the copper(II) ion and its variable eccentricity. This means that crystallographic data will be essential in determining the precise stereochemistry of the copper(II) ion in these distorted geometries. It also opens up the *possibility* that through

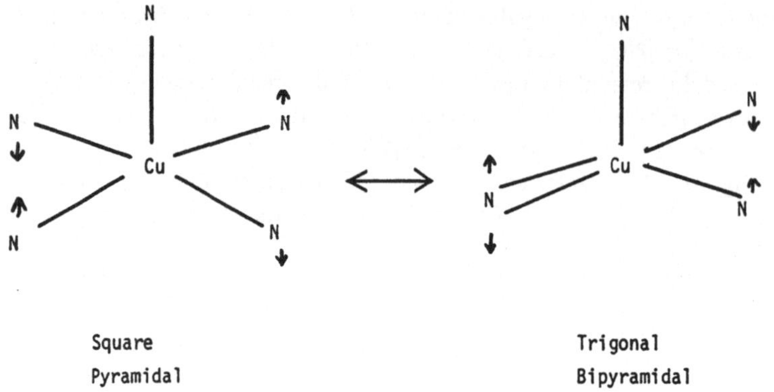

Fig. 32. Berry Twist Process[85]

these distortion isomers, especially cationic ones, it may be possible to map out how the electronic energies vary with stereochemistry and possibly establish an "electronic criteria of stereochemistry" for a given set of ligands in a particular chromophore, as in the [Cu(bipy)$_2$Cl] X, [29][80] and [Cu(dien) (bipyam)] X$_2$, [30][83] series of complexes above (see Sect. 12).

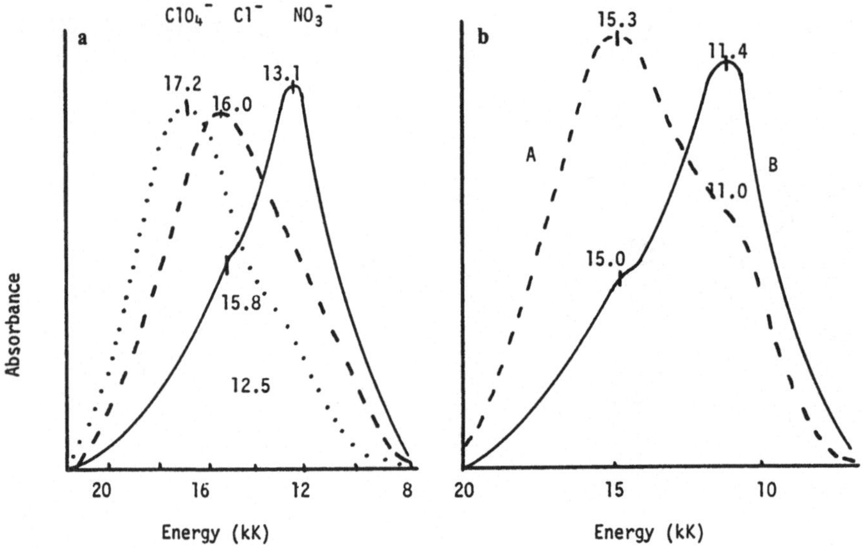

Fig. 33. a Electronic Spectra[83] [Cu(dien)(bipyam)] X$_2$ b Electronic Spectra[84] of A, K[Cu(NH$_3$)$_5$] [PF$_6$]$_3$ and B, [Cu(tren)(NH$_3$)] [ClO$_4$]$_2$

10 Stereochemistry of Copper(II) Complexes – Revisited

The stereochemistry of the copper(II) ion was reasonably described[4] in 1970 as in Fig. 34 with suitable examples to illustrate the structures. Alternative examples could readily be chosen but, with the possible exception[27] of $K_2Pb[Cu(NO_2)_6]$ and $[Cu(en)_3][SO_4]$, no doubt was ever expressed about considering these as static, non-temperature variable structures, all established by room temperature X-ray crystallography. As a consequence of the application of the Jahn-Teller Effect, [Sect. 2] particularly in the *three* potential well form of Fig. 2, the six-coordinate stereochemistries of the copper(II) ion can be very susceptible to the effect of variable temperature, along with their e.s.r. spectra. Thus, the structures of Fig. 34 may be listed, Table 10, according to their temperature variable and non-temperature variable types[20], in which 7 out of the 16 are potentially temperature variable, and *five* of these (1), (2), (5)–(7) only arise as artifacts of the fluxional model[16] and are best described as *pseudo*-stereochemistries of the copper(II) ion, Table 10. The elongated tetragonal and rhombic octahedral stereochemistries are genuine structures, which can occur with a static structure, but can also be modified by fluxional effects[30, 63]. Consequently, if Fig. 34 is up-dated in terms of more recent complexes, whose low temperature crystallography or e.s.r. properties have been established[6], an entirely different summary of the stereochemistry of the copper(II) ion is obtained. Only nine structures involve genuine static stereochemistries, two can be static or fluxional and five arise through fluxional effects and have no inherent existence as static structures, Fig. 34. This static/non-static stereochemical relationship is further complicated by the Plasticity Effect[6, 7] [Sect. 9], which due to the non-spherical symmetry of the copper(II) ion[4]

Table 10. Breakdown of Copper(II) Stereochemistries into their temperature variable (A) and non-temperature variable types (B)

A.	Temperature Variable	
(1)	pseudo-ocatahedral	$K_2Pb[Cu(NO_2)_6]$
(2)	pseudo-trigonal octahedral	$[Cu(en)_3][SO_4]$
(3)	elongated tetragonal octahedral	$[Cu(H_2O)_4(HCO_2)_2]$
(4)	elongated rhombic octahedral	$Ba_2[Cu(HCO_2)_6]\cdot 4H_2O$
(5)	pseudo compressed tetragonal octahedral	$Rb_2Pb[Cu(NO_2)_6]$
(6)	pseudo compressed rhombic octahedral	$[Cu(dien)_2][NO_3]_2$
(7)	pseudo-cis-distorted octahedral	$[Cu(bipy)_2(ONO)][NO_3]$
B.	Non-Temperature Variable	
(8)	linear	$CuCl_2$ (gaseous)
(9)	trigonal bipyramidal	$[Cu(NH_3)_2Ag(SCN)_3]$ [33]
(10)	square pyramidal	$K[Cu(NH_3)_5][PF_6]_3$
(11)	square coplanar	$Ca[CuSi_4O_{10}]$
(12)	rhombic coplanar	$[Cu(3\text{-}Me\text{-}acac)_2]$
(13)	eight coordinate	$Ca[Cu(CH_3CO)_4]\cdot 6H_2O$
(14)	compressed tetrahedral	$Cs_2[CuCl_4]$
(15)	distorted square pyramidal	$[Cu(dien)(C_2O_4)]\cdot 4H_2O$
(16)	assymmetric cis-distorted octahedral	$[Cu(bipy)_2(ONO)][BF_4]$

LINEAR – C_{2v}	TRIGONAL BIPYRAMIDAL – D_{3h}	DISTORTED SQUARE PYRAMIDAL	SQUARE PYRAMIDAL – C_{4v}
$CuCl_2$ (Gaseous)	$[Cu(NH_3)_2Ag(SCN)_3]$	$[Cu(dien)(O_2CH)][HCO_2]$	$K[Cu(NH_3)_5][PF_6]_3$
COMPRESSED TETRAGONAL OCTAHEDRAL – D_{4h}	OCTAHEDRAL – O_h	ELONGATED TETRAGONAL OCTAHEDRAL – D_{4h}	SQUARE COPLANAR – D_{4h}, D_{2h}
			$Na_4[Cu(NH_3)_4][Cu(S_2O_3)_2]_2 \cdot OH_2$
$Rb_2Pb[Cu(NO_2)_6]$	$K_2Pb[Cu(NO_2)_6]$	$[Cu(NH_3)_4(SCN)_2]$	$[Cu(3Me\text{-}acac)_2]$
COMPRESSED RHOMBIC OCTAHEDRAL – D_2	CIS-DISTORTED OCTAHEDRAL – C_2	TRIGONAL OCTAHEDRAL – D_3	ELONGATED RHOMBIC OCTAHEDRAL – D_{2h}
$[Cu(dien)_2][NO_3]_2$	$[Cu(phen)_2(O_2CCH_3)][ClO_4].2H_2O$	$[Cu(en)_3][SO_4]$	$Ba_2[Cu(HCO_2)_6] \cdot 4H_2O$
COMPRESSED TETRAHEDRAL – D_{2d}	ASYMMETRIC CIS-DISTORTED OCTAHEDRAL (4+1+1*)	SEVEN COORDINATE – C_{2v} (3+2+2*)	EIGHT COORDINATE – S_4 (4+4*)
$Cs_2[CuCl_4]$	$[Cu(bipy)_2(ONO)][BF_4]$	$[Cu(py)_3(O_2NO)_2]$	$Ca[Cu(CH_3CO_2)_4] \cdot 6H_2O$

═══ STATIC ONLY ═ ═ ═ STATIC + FLUXIONAL —_—_—_ PSEUDO ONLY

Fig. 34. A summary of the known stereochemistries of the Copper(II) ion, their idealised molecular symmetries, and the relationship between the regular and distorted geometries

enables it, for a given set of ligands, to give rise to distortion isomers, and to ranges of cation or anion distortion isomers, such as [Cu(dien)(bipyam)]X [30] and [Cu(bipy)$_2$Cl]X [29] or M^I_2Pb[Cu(NO$_2$)$_6$] [1], etc. This means that not only does the precise structure of a copper(II) cation or anion vary depending upon the counter ion present for a series of *static* copper(II) cation distortion isomers, but in fluxional systems, such as the Tutton salts[30)], the basic static structure, involved in the three-potential-well-systems, is not necessarily exactly the same from one complex to another. Consequently, the small differences in the $E_{J.T.}$ and E_2 energies of the $M^I_2M^{II}$[Cu(NO$_2$)$_6$] series of anion distortion isomers [Table 7] may be due to inherent differences in the underlying static CuN_6 chromophore stereochemistry, which are revealed in the electronic reflectance spectra due to their short lifetime, 10^{-15} s, relative to that of the X-ray structure measurements, ca. 1.0 s. This difference in the structure of the underlying static chromophore, has already been discussed[54, 65, 66)] for the cation distortion isomers of [Cu(bipy)$_2$-(ONO)]X [12, 16 and 17]. But an even more attractive illustration of the delicate effect of lattice packing factors, is that in the single crystal X-ray determination of the structure[30)] of the fluxional [NH$_4$]$_2$[Cu(OH$_2$)$_6$][SO$_4$]$_2$ [5], the elongation is along the Cu-O(7) direction. In the corresponding neutron diffraction structure[50)] of [ND$_4$]$_2$[Cu(OD$_2$)$_6$][SO$_4$]$_2$ [34] the elongation direction is along the Cu-O(8) direction. Thus, even the difference between hydrogen and deuterium bonding in a crystal, is sufficient to change the direction of the CuO_6 chromophore elongation, as it is determined by a ΔE value[30)] that is less than thermal energy, namely 160 ± 20 cm^{-1}.

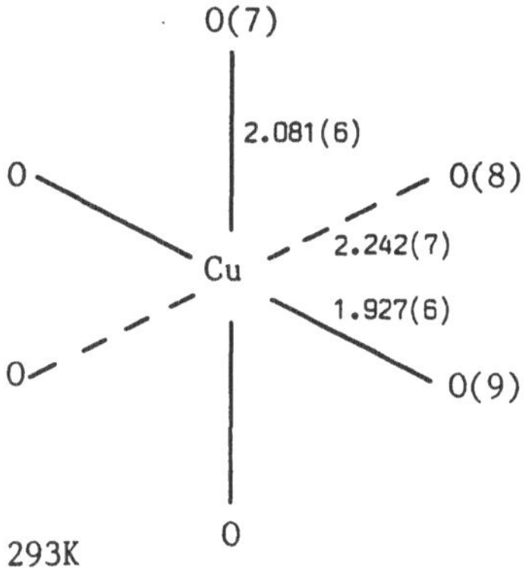

[34] [ND$_4$]$_2$[Cu(OD$_2$)$_6$][SO$_4$]$_2$

If the pseudo-structures of copper(II), Fig. 34 are now excluded from the discussion, the "genuine" stereochemistries of copper(II) ion are reduced to:
1. elongated tetragonal (or rhombic) octahedral,
2. square (or rhombic) coplanar,
3. compressed tetrahedral,
4. square pyramidal (including distorted),
5. trigonal bipyramidal (including distorted),

if the much less common stereochemistries of linear, seven-coordinate and eight-coordinate are also excluded[4, 9)]. Of these five, the occurrence of stereochemistries (2) to (5) are no different from any other first-row transition metal ion and only a regular octahedral stereochemistry is absent, but as the copper(II) ion in this structure is Jahn-Teller unstable, this static structure is replaced by the static elongated tetragonal octahedral stereochemistry. Thus, although the consequences of the Jahn-Teller effect on the stereochemistry of the six-coordinate copper(II) ion is paramount in respect to the fluxional pseudo-structures, it only accounts for *one* stereochemistry in the static structures, namely, elongated tetragonal octahedral. The absence of a regular tetrahedral structure with a $^2T_2$ degenerate ground state has been accounted for by the Jahn-Teller effect, but spin-orbit coupling is usually considered to remove the degeneracy of the electronic ground state of the copper(II) ion in this stereochemistry[87], although it has been suggested that the splitting of the $^2T_2$ ground state of Cu^{2+} in $ZnCr_2O_4$ is too large to be accounted for by spin-orbit coupling alone[87](b).

If the static stereochemistries (1) to (5) above are drawn out[20] as in Fig. 35, they can be arranged horizontally in terms of their basic coordination number, and vertically depending on whether the basic CuL_n chromophore is subject to a trigonal bipyramidal or tetrahedral sense of distortion. This generates four additional types of distorted stereochemistries namely, trigonal and tetrahedrally distorted, elongated tetragonal octahedral and square pyramidal. It is just these four types of distorted structures that predominate in copper(II) complexes involving chelate ligands, especially macrocyclic ligands, where significant bond-length and bond-angle distortions occur, but for which the trigonal distorted square pyramidal and tetrahedral distorted elongated rhombic octahedral predominate, Table 11. It is then unfortunate that the origin of these distorted structures (that are intermediate between the more regular structures of Fig. 35, expecially when all copper(II) stereochemistries are considered) are usually attributed to the *constraints* imposed by the chelate ligands in these complexes, rather than to the Plasticity[6,7] of the copper(II) ion.

Table 11. Copper(II) complexes involving (a) trigonal and (b) tetrahedral-type distortions

(a)	Trigonal distorted square pyramidal:	
	$[Cu(bipy)_2Cl]X$ [80]	$[Cu(dien)(bipyam)]X_2$ [83]
(b)	Tetrahedral Distorted Octahedral:	
	$[Cu(bipy)_2(OH_2)][S_2O_6]$ [81]	$[Cu(bipy)_2(O_2ClO_2)][ClO_4]$ [95]
	$[Cu(bipy)_2][PF_6]_2$ [92]	$[Cu(bipy)_2(O_2S_2O_6)]$ [96]
	$[Cu(bipy)_2(O_2NO)][NO_3]$ [90]	$[Cu(bipy)_2(O_2S_4O_4)]$ [97]

11 Structural Pathways in Copper(II) Stereochemistry

The question[7] must now be asked "how far do the arrows of Fig. 35 represent continuous transitions from one type of coordination to another"?; (i.e. square coplanar to tetrahedral) with *Structural Pathways*[86,88] not only between the regular static stereochemistries of Fig. 35, but between the intermediate distorted structures. Each structural pathway is

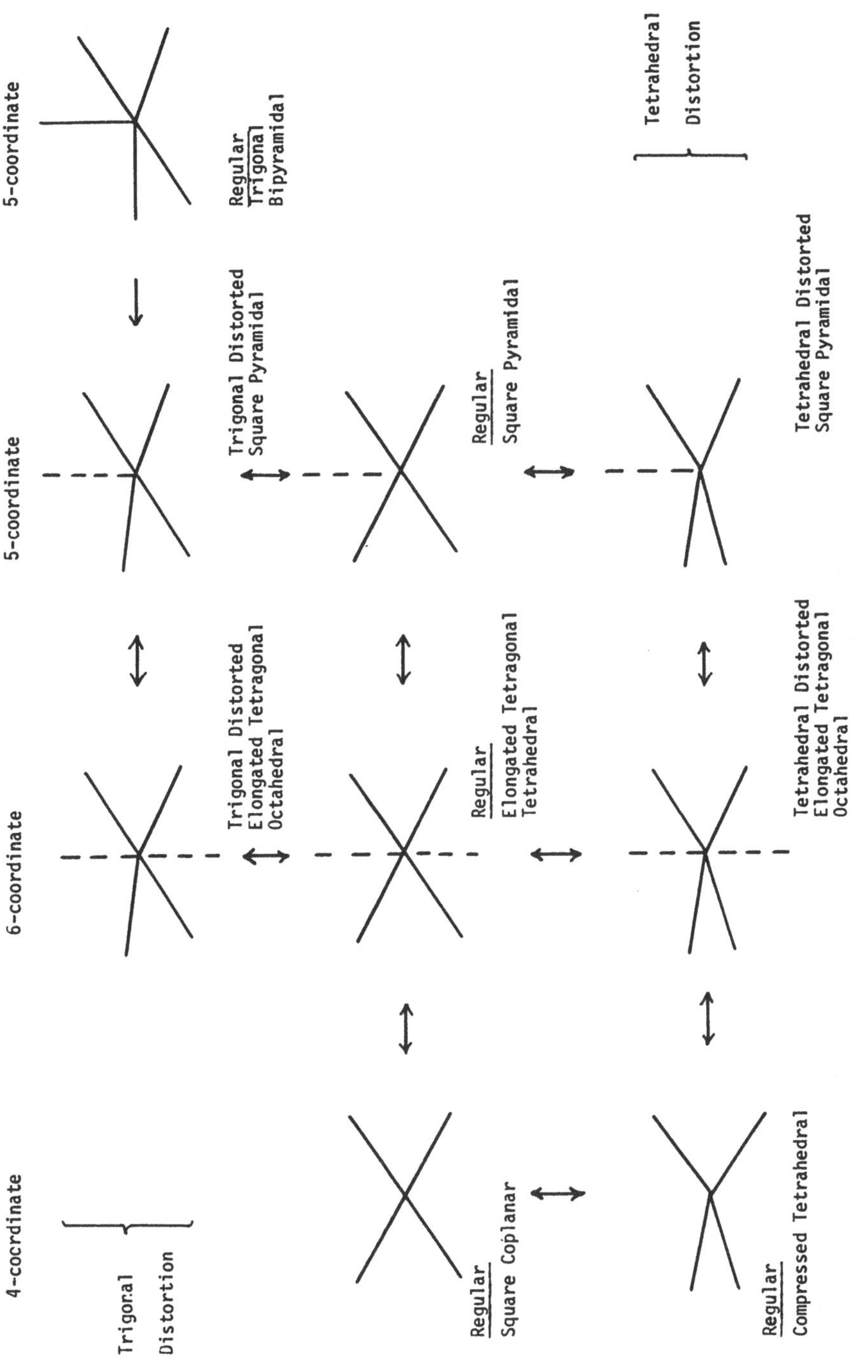

Fig. 35. The static stereochemistries of the copper(II) ion – regular and distorted structures

then determined by a "dominant" normal mode of vibration of the nuclear structure, Fig. 36 (or a linear combination of these) to yield a sequence of crystallographically established structures, which represent individual points in the structural pathway. The cation distortion isomers of the $[Cu(bipy)_2Cl]X$, nH_2O[80–82] and $[Cu(dien)(bipyam)]X_2$ series,[83] Table 9(a), and (b), then represent individual structures along the structural pathway of the regular trigonal bipyramidal to square pyramidal stereochemistries. As the constraints imposed by these chelate ligands are so different, compare [29] and [30], it is difficult to imagine that such different sets of ligands, determined the same individual nuclear CuN_5 or CuN_4Cl structures alone, but rather that these are determined by the e′ modes of vibration or a linear combination of these. Where non-equivalent ligands are involved, as in the $[Cu(bipy)_2Cl]X \cdot nH_2O$ series, then the alternative route A or route B (or C)-type distortions, Fig. 37, will be determined[89] by the appropriate linear combination of modes of vibration, Fig. 37, $\nu_4(e') + \nu_3(e')$ for route (A) and $\nu_7(e') + \nu_8(e')$ for

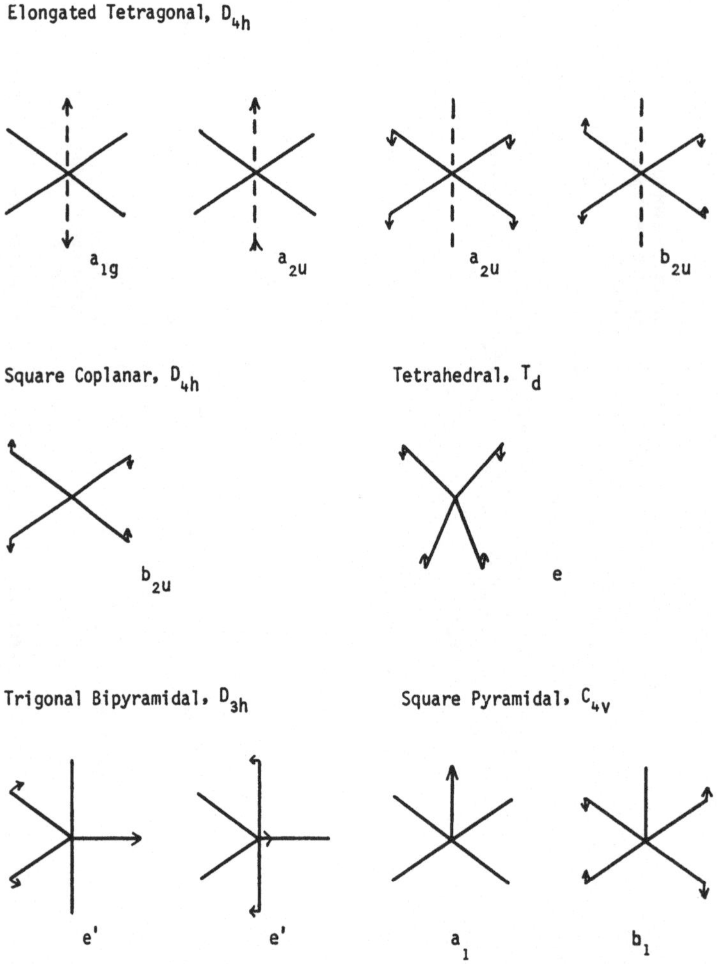

Fig. 36. Selected normal modes of vibration of regular stereochemistries

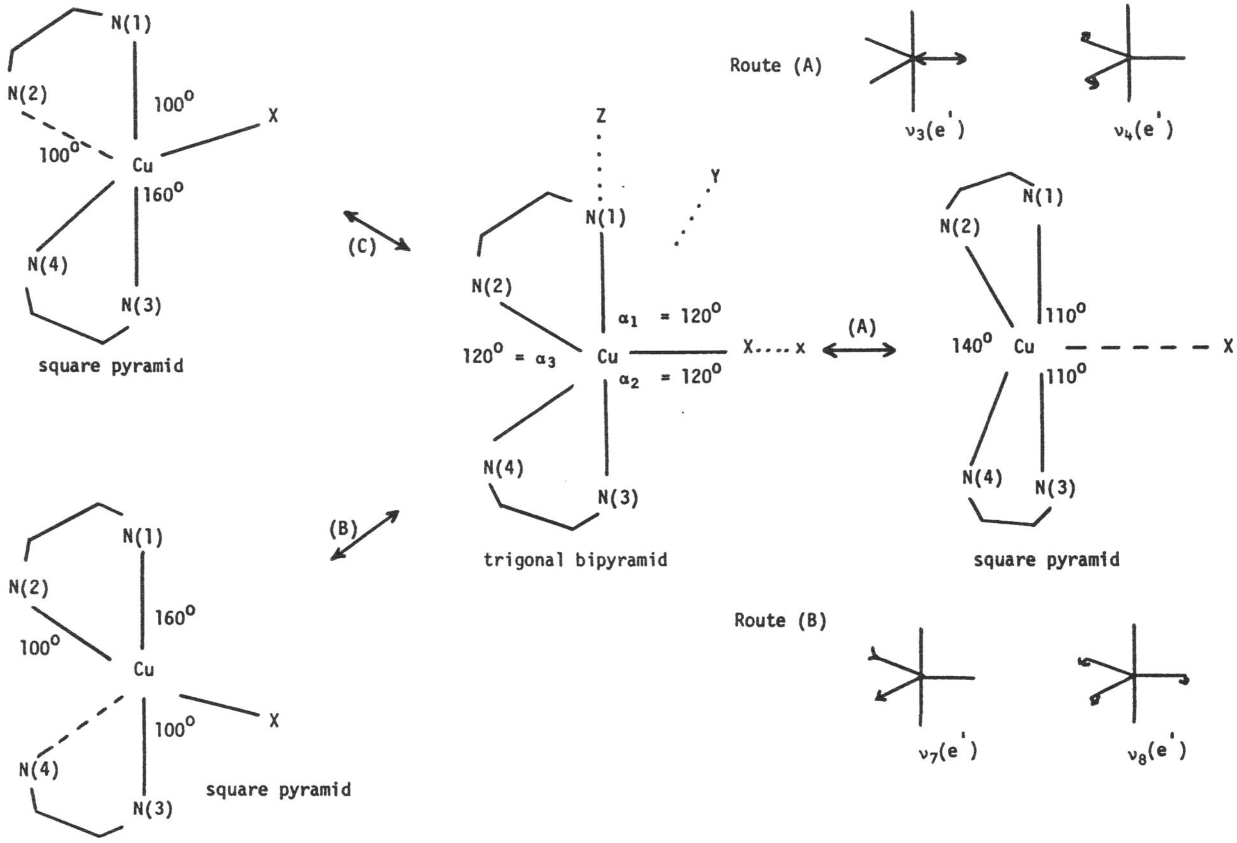

Fig. 37. The structural pathways for the distortion of the CuN_4Cl chromophore of the $[Cu(bipy)_2Cl]$ complexes, from trigonal bipyramid to square pyramid involving three alternative routes

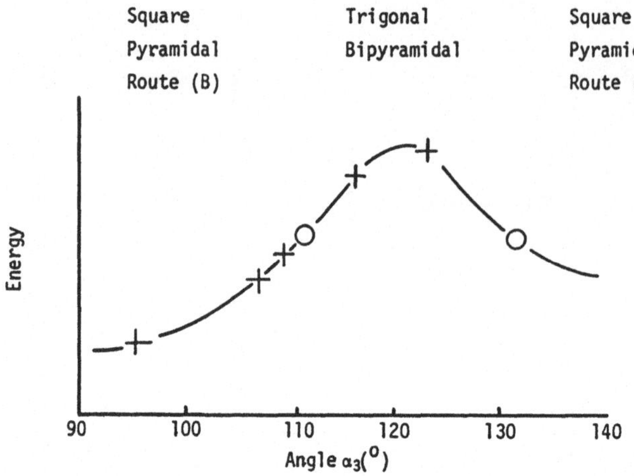

Fig. 38. Structural Profile $[Cu(bipy)_2X]^+$ cation: + $[Cu(bipy)_2Cl]^+$; O $[Cu(bipy)_2(OH_2)]^+$

route (B) or (C). The individual structures then map-out a structural pathway involving slightly different energies and may be used to draw a structural profile connecting the total energy E, and some structural parameter, such as the N(2)-Cu-N(4), α_3, angle in the $[Cu(bipy)_2Cl]X \cdot nH_2O$ series.[80–82] Extended Huckel calculations may then be used to estimate the relative total energies E, Fig. 38, for a $[Cu(NH_3)_4Cl]^+$ cation, using the molecular geometry of the individual complexes.

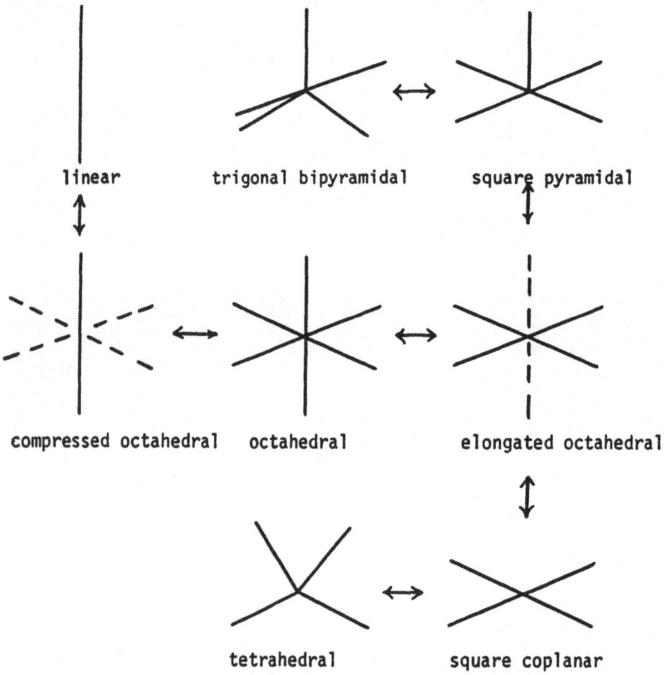

Fig. 39. Structural pathways between regular static and fluxional copper(II) stereochemistries

The structural pathway approach may then be suggested[61, 65] to link not only the regular and distorted *static* stereochemistries of the copper(II) ion, but extended, to include the *fluxional* structures, Fig. 39. Thus a structural pathway has been suggested[61], Fig. 40(a), to link the fluxional regular cis-distorted octahedral CuN_4O_2 chromophore to the square pyramidal (4 + 1 + 1*) stereochemistry via a linear combination of the ν_{1a} (a_1) and ν_{2a} (b_1) modes of vibration of a tris(chelate)copper(II) complex, Fig. 40(b) with the ν_{2a} (b_1) predominant, route E or F. But equally, to a bicapped square pyramidal (4 + 2) structure[90], route D with ν_{1a} (a_1) dominant, to a distorted bicapped square pyramid (4 + 1 + 1*) CuN_4O_2 chromophore, route G or H with ν_{2a} (b_1) again predominant. Although the regular bicapped square pyramidal structure has not been characterised in the [Cu(bipy)$_2$(O$_2$NO)][Y] systems, it has been established for [Cu(MeTAAB)(O$_2$NO)][NO$_3$], [35][91] where MeTAAB = tetrabenzo *b, f, j, n* [1, 5, 9, 13] tetraazacyclohexadecine, a macrocyclic ligand. The distorted bicapped square pyramid (4 + 1 + 1*) structure has been characterised in [Cu(bipy)$_2$(O$_2$NO)][NO$_3$] [36][90], which leads naturally, to the compressed tetrahedral CuN_4 chromophore of [Cu(bipy)$_2$][PF$_6$]$_2$ [37][92].

When put together as a series of [Cu(bipy)$_2$X]Y complexes, where X = Cl^-, OH_2, $[ONO]^-$ and $[ONO_2]^-$, an even more extensive Structural Pathway may be produced as in Fig. 41[92, 93].

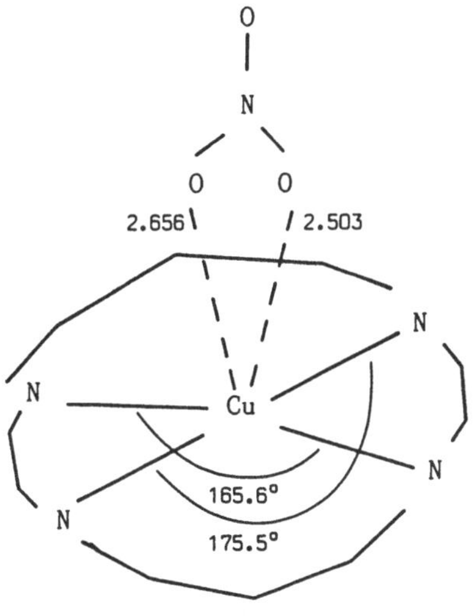

[35] [Cu(MeTAAB)(O$_2$NO)][NO$_3$]

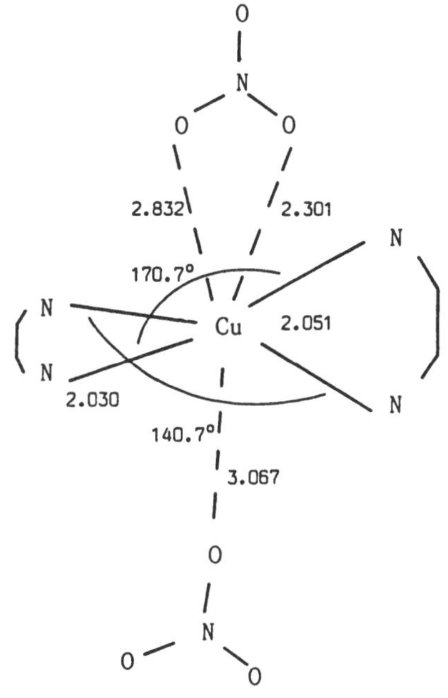

[36] [Cu(bipy)$_2$(O$_2$NO)][NO$_3$] · H_2O

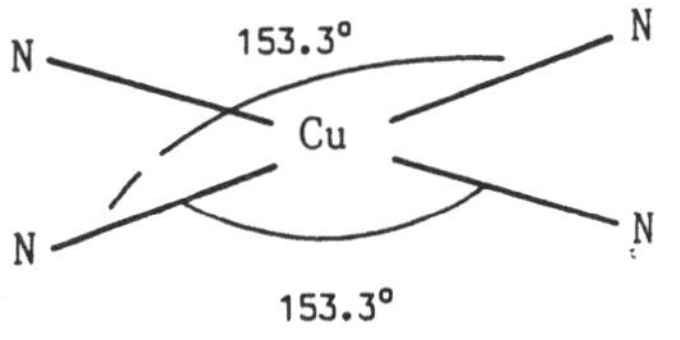

[37] [Cu(bipy)$_2$][PF$_6$]$_2$

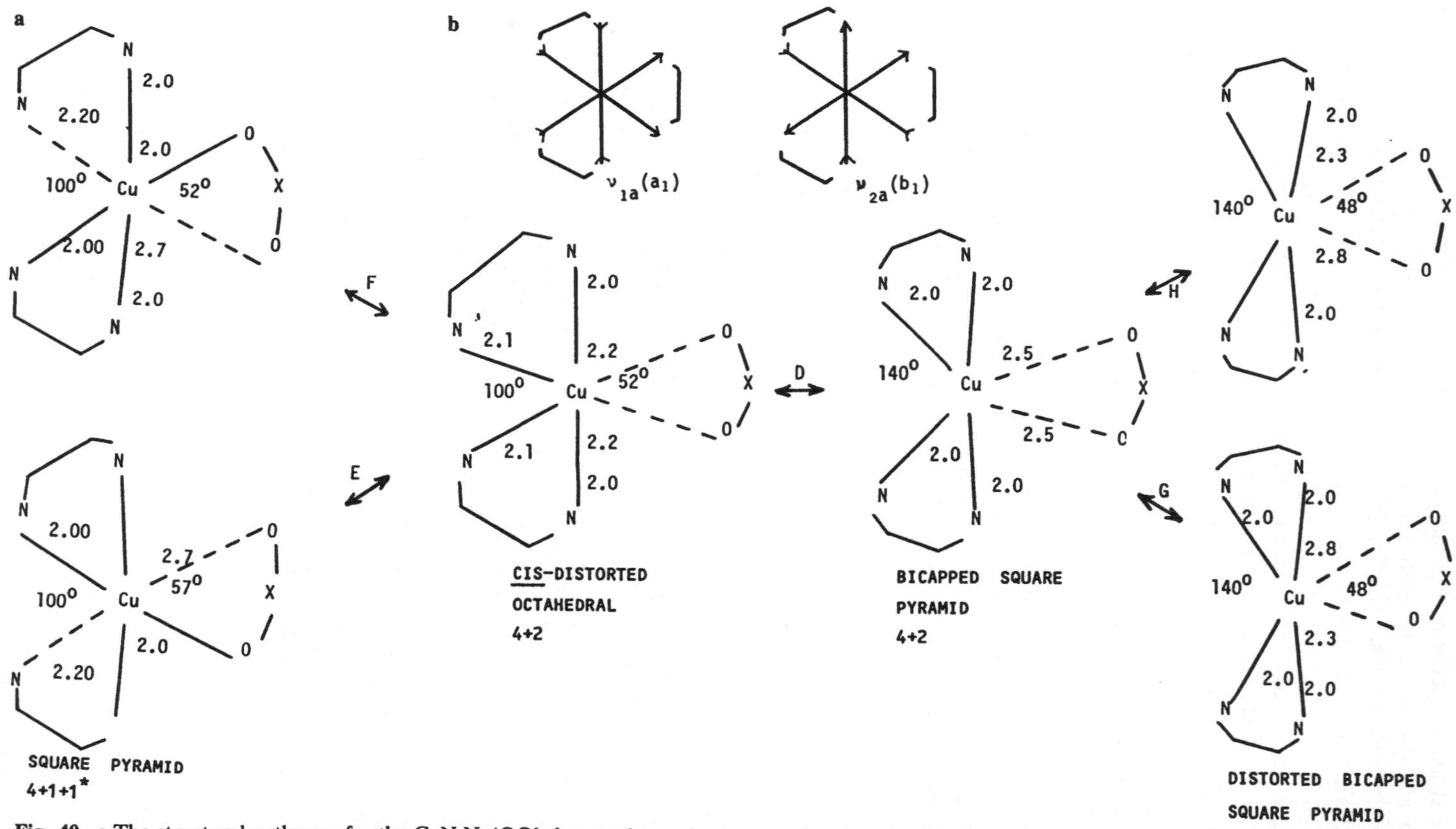

Fig. 40. **a** The structural pathways for the CuN_2N_2 'OO' chromophore of the $[Cu(bipy)_2(OXO)]^+$ cation for distortion from regular *cis*-distorted octahedral to distorted square pyramid (route E and F) and to bicapped square pyramid (route D and routes G or H); **b** modes of vibration of $[Cu(chelate)_3]X_2$ complexes

In this way a wide range of stereochemistries of the copper(II) ion may be understood if they are first separated in terms of fluxional behaviour and basic static stereochemistries and then the static structures linked by the normal modes of vibration to yield Structural Pathways connecting the regular structures via the alternative tetrahedral and trigonal distorted structures. The full potential and consequences of the Jahn-Teller effect are then realised in the Plasticity Effect and the Fluxional Behaviour in copper(II) stereochemistry as a whole.

12 Electronic Criteria of Copper(II) Stereochemistry

The previous sections have emphasised how much more complex the stereochemistry of the copper(II) ion now appears, than the simple static view[4)] of Fig. 34 in the light of both the Plasticity Effect[7)] and Fluxional Behaviour[16)], separately or together. Not only will the electronic properties vary with varying static stereochemistries, related as in the Structural Pathway[92)] of Fig. 42, to yield the correlations of the electronic properties[93)] of Fig. 43. A copper(II) complex containing the *same* cation or anion, can no longer be predicted to have a single possible structure, but that this can vary (within certain limits) depending upon the environment of the anion or cation in the crystal, see Table 9, for the $[Cu(bipy)_2Cl]X$ series[80–82)]. Nevertheless, as Fig. 31 showed, the electronic reflectance spectra do vary with the stereochemistry of the copper(II) chromophore CuL_x and the e.s.r. spectra, may be correlated with the ground state electron configuration, namely, a d_{z^2} or $d_{x^2-y^2}$ ground state ($g_3 \sim g_2 > g_1 \approx 2.0$) and $g_3 > g_2 \approx g_1 > 2.0$, respectively, Table 4, if only the Plasticity Effect is involved. If a fluxional copper(II) stereochemistry is involved, then the electronic reflectance spectra may be nearly invariant[65, 66)], as in the $[Cu(bipy)_2(ONO)][Y]$ series, Fig. 26(a), and the spectra will relate to the underlying static stereochemistry rather than the room temperature structures. In these fluxional systems the numerical values of the e.s.r. spectra will not be very informative, but their temperature variability should be the best pointer to the fluxional behaviour[16)].

Thus any attempt to establish an "electronic criteria of stereochemistry"[94)] for a copper(II) complex of unknown crystal structure should involve a number of considerations.

(a) the composition of the complex should be clearly established;

(b) from the number of ligands and their donor atoms an idea of the coordination number involved can be suggested, but the presence of non-coordinated water molecules and potentially coordinating polyatomic anions can introduce uncertainties. Only if 4, 5 or 6 equivalent ligands are present can regular 4, 5, and 6 coordinate stereochemistries occur and the operation of the Plasticity Effect introduces uncertainty into attempts to predict the stereochemistry by comparison with the structure of a complex involving a similar set of ligands, i.e. $[Cu(bipy)_2Cl]X$ complexes[80–82)], Table 9. With faculative polydentate ligands, as in many macrocyclic ligands, the prediction of stereochemistry is even more uncertain;

(c) from the polycrystalline e.s.r. spectra the approximate ground state can be determined, if exchange coupling is absent, Table 4 and Fig. 9, and the temperature variability of the e.s.r. spectra should indicate the possibility of a fluxional stereochemistry;

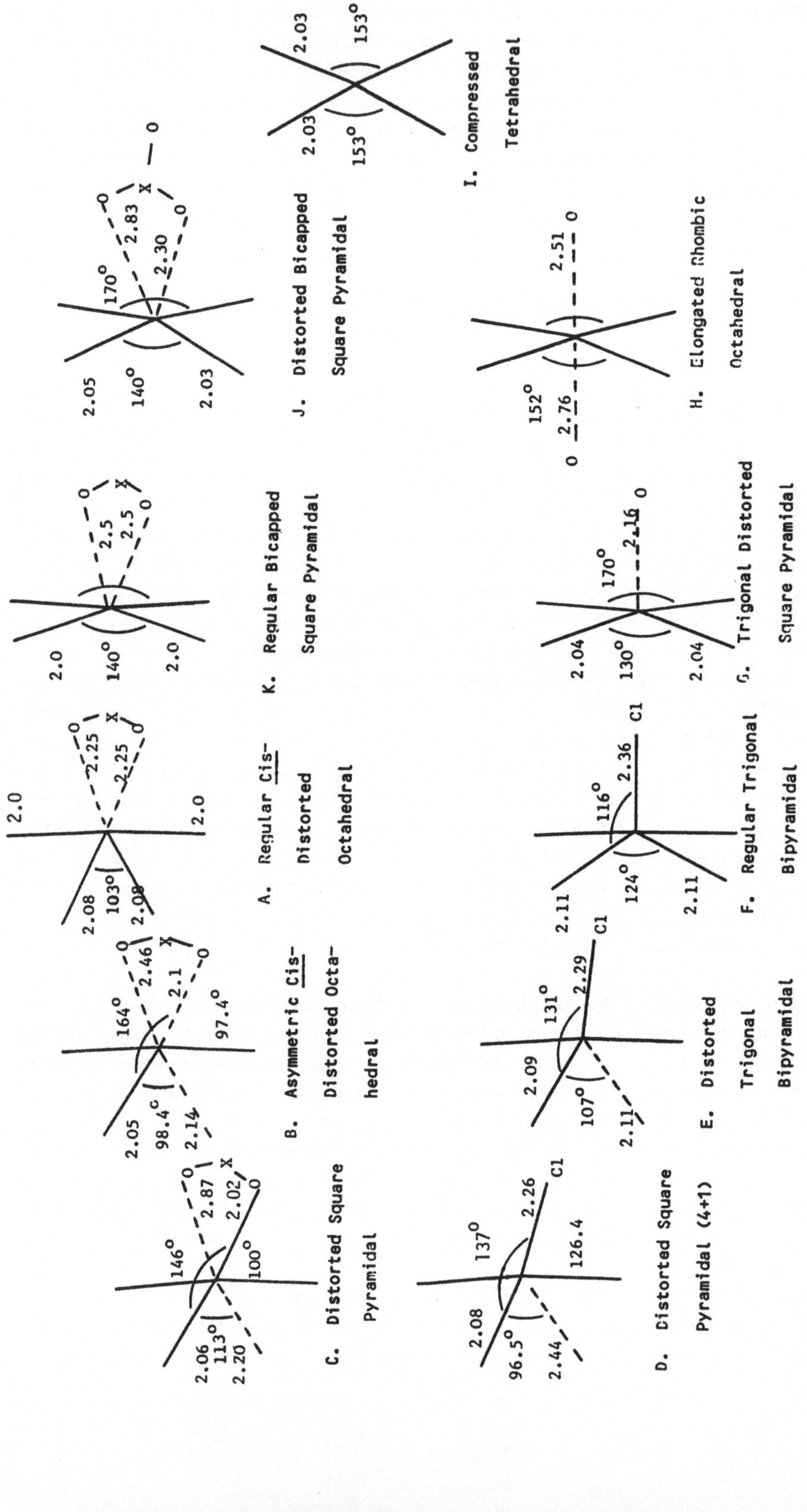

Fig. 41 A–K. $[Cu(bipy)_2X]Y$ Structural Pathway

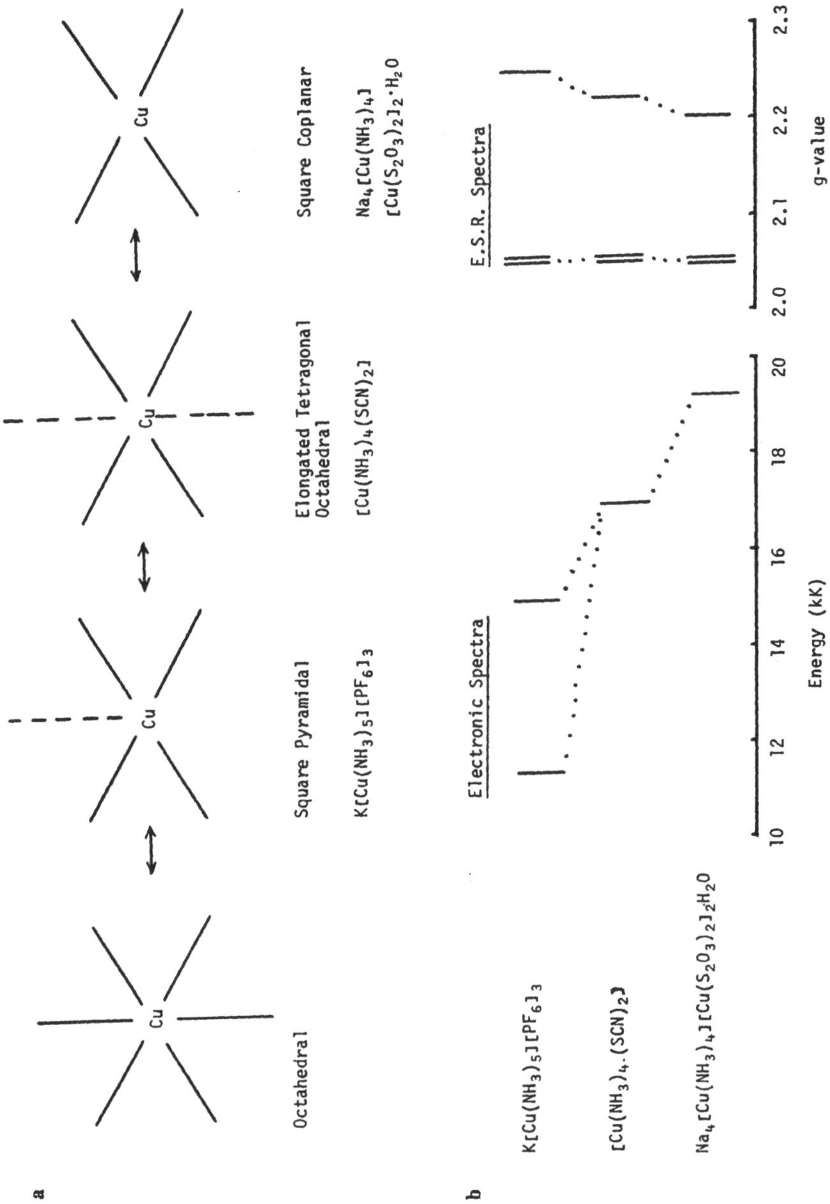

Fig. 42a, b. The correlation of a Structural Pathway and Electronic Properties

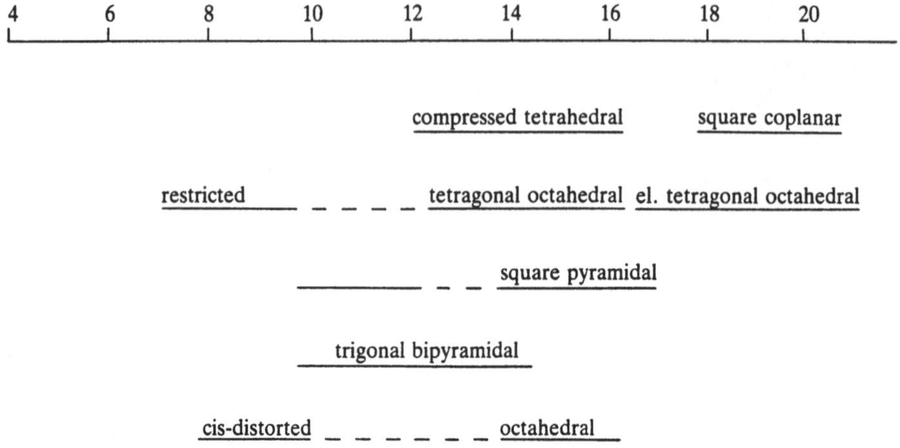

Fig. 43. A correlation of the range of energies for the d-d transitions for the CuN_x chromophore for different stereochemistries

(d) from the electronic reflectance some idea of the range of energy levels present can be determined and the number of possible stereochemistries reduced in number. Figure 43 summarises[94)] this approach for the CuN_x chromophores, with the limitation that it should be applied with caution and important results confirmed by X-ray crystallographic structure determination, where single crystals are available.

An alternative approach to the electronic criteria of stereochemistry is to examine the electronic properties of a series of complexes of known crystal structure[93)] involving a common set of ligands and see how these relate to the stereochemistry. Fig. 42, especially if such a set of ligands cover a wide range of copper(II) stereochemistries. An example of such a series of complexes are the $[Cu(bipy)_2X]Y$ complexes[93)], Fig. 41, for which there exist 34 complexes of known crystal structure. The series contain a wide range of stereochemistries, which may be represented in the Structural Pathway[92)] of Fig. 41, and their electronic properties have been reported in detail[65, 66, 80, 90, 92, 95)]. The electronic reflectance spectra are reported in Fig. 26 (a), Fig. 31 (a), Fig. 44 (a) and (b), Table 12, and the correlation of the g-factors (primarily single-crystal g-factors), and the elctronic energies of the band maximum are shown in Fig. 45, (i)–(iv). The electronic spectra vary from a single peak to twin peaks. In D to I the spectra relate to static stereochemistries, with D-F forming a series of cation distortion isomers, and twin peaked spectra relate to the more extreme distorted square pyramidal [4 + 1 + 1*] for D to compressed tetrahedral for I. The nearer trigonal bipyramidal stereochemistries of F and G are characterised by a single peak at ca. 12 000 cm^{-1}, and reversed type axial e.s.r. spectra consistent with a d_{z^2} ground state, Fig. 9, with $g_3 \approx g_2 > g_1 \approx 2.0$. The spectra of A-C show two clear peaks, Fig. 31, which are identical for A and C despite their clearly different room temperature structures [12 and 17] due to the fluxional behaviour of the CuN_4O_2 chromophore. In A-C the e.s.r. spectra are approximately axial, but reversed in type, Fig. 9. In the final quadrant of the Structural Pathway, Fig. 41, A, K, J, I, the large splitting of the electronic spectra of A, ca. 5000 cm^{-1} is reduced to ca. 2000 cm^{-1}, and shifted to significantly higher energy in I. In the same sequence the reversed type e.s.r. spectrum of A, $g_3 \approx g_2 > g_1 \approx 2.0$, changes to a normal type e.s.r. spectrum,

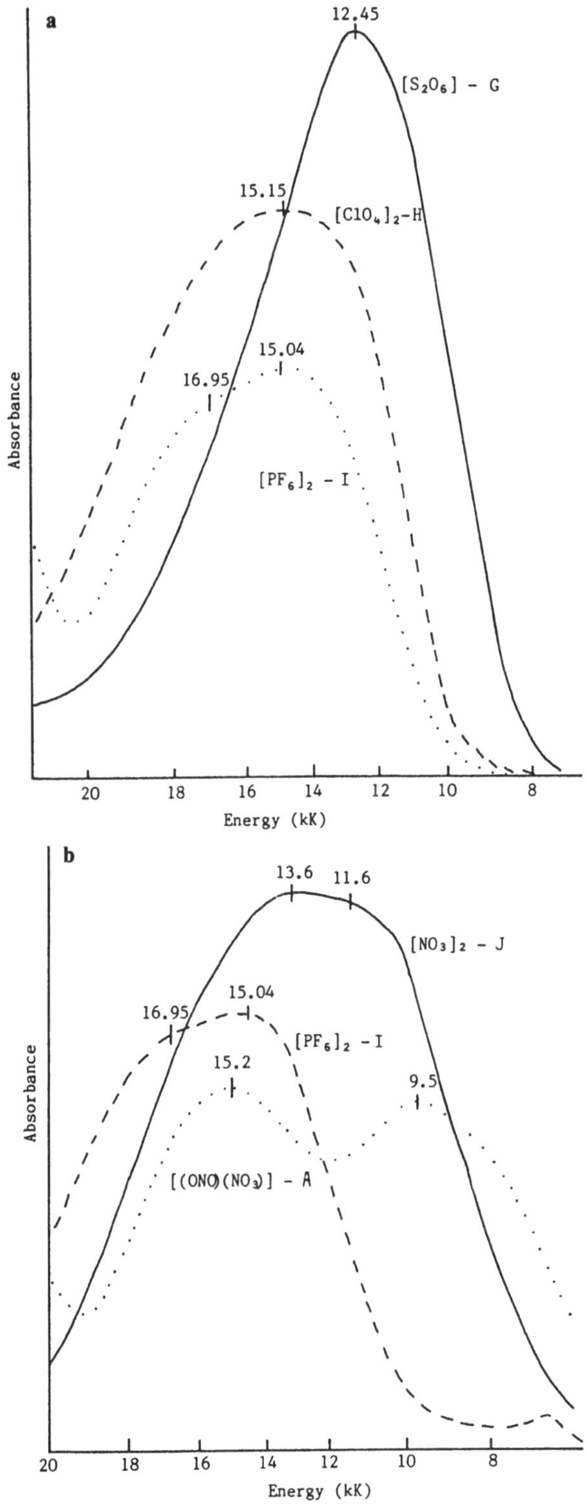

Fig. 44. **a** and **b** Electronic Reflectance Spectra of $[Cu(bipy)_2X]Y$ and $[Cu(bipy)_2]XY$ type complexes

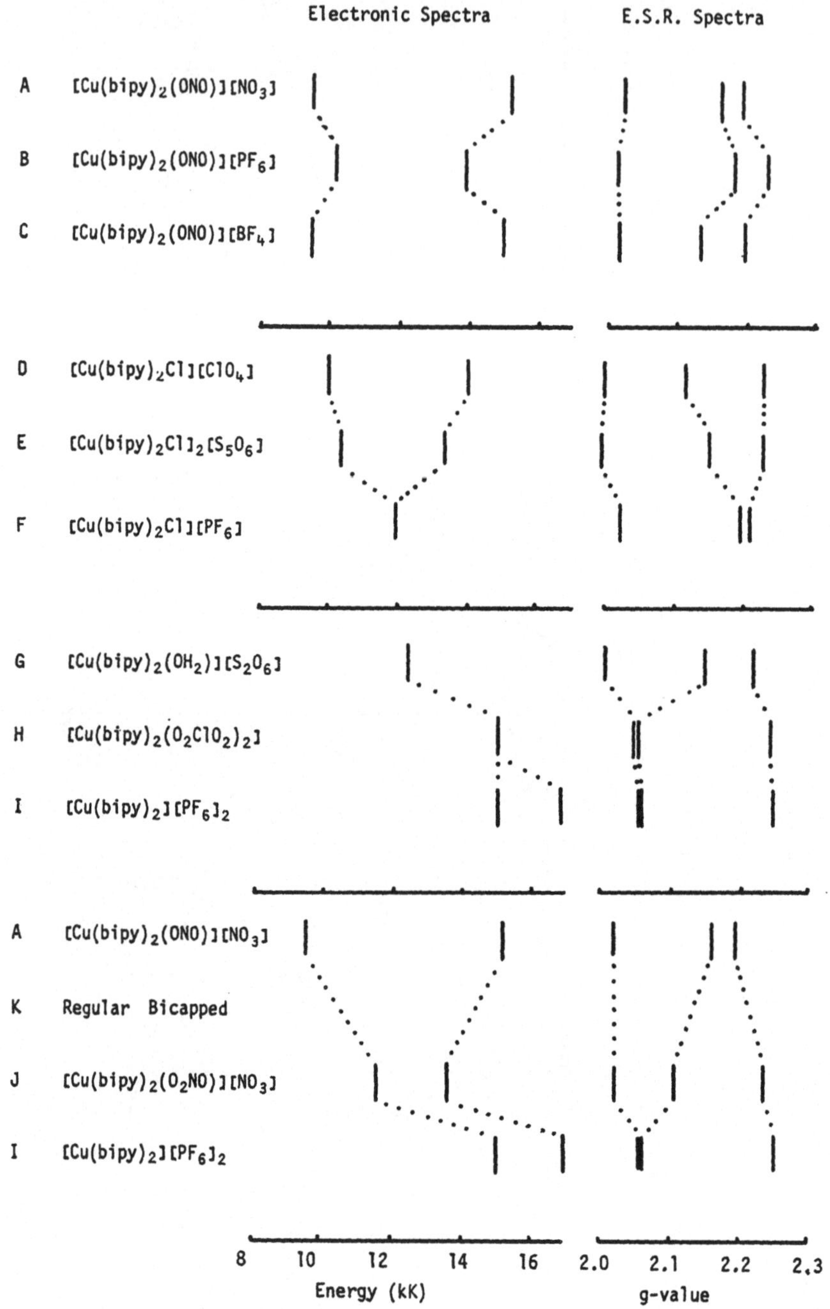

Fig. 45. Correlation of electronic spectral energies and g-values for $[Cu(bipy)_2X]$ Y complexes in a Structural Pathway (see Fig. 41)[93]

Table 12. The electronic properties of $[Cu(bipy)_2X]Y$ complexes

	Electronic Spectra (kk)			g-values		
A. $[Cu(bipy)_2(ONO)][NO_3]$	9.5		15.2	2.026	2.165	2.196
B. $[Cu(bipy)_2(ONO)][PF_6]$	10.2		14.0	2.016	2.186	2.233
C. $[Cu(bipy)_2(ONO)][BF_4]$	9.5		15.0	2.020	2.136	2.203
D. $[Cu(bipy)_2Cl][ClO_4]$	10.1		14.1	2.007	2.125	2.234
E. $[Cu(bipy)_2Cl]_2[S_5O_6] \cdot 6H_2O$	10.5		13.5	2.002	2.160	2.233
F. $[Cu(bipy)_2Cl][PF_6]$		11.9		2.029	2.203	2.216
G. $[Cu(bipy)_2(OH_2)][S_2O_6]$		12.45		2.011	2.158	2.225
H. $[Cu(bipy)_2(O_2ClO_2)][ClO_4]$		15.1		2.054	2.065	2.255
I. $[Cu(bipy)_2][PF_6]_2$	15.04		16.95	2.06	-	2.253
J. $[Cu(bipy)_2(O_2NO)][NO_3] \cdot H_2O$	11.6		13.6	2.023	2.110	2.240
K. Bicapped square pyramidal	-		-	-	-	-

$g_3 \gg g_2 \approx g_1 > 2.0$ in I, which together clearly distinguish the *cis*-distorted octahedral and compressed tetrahedral stereochemistries. Thus, not only does the Structural Pathway of the $[Cu(bipy)_2X]Y$ complexes[92)] of Fig. 41, nicely relate the wide range of stereochemistries in this series of complexes, the electronic reflectance spectra follow the changes in stereochemistry involved, the g-factors determine the electronic ground state, and identify fluxional behaviour, if present, and together they offer an "electronic criteria of stereochemistry" in this closely related series of complexes[98)].

13 Conclusions

From a detailed examination of the X-ray crystallographic data of copper(II) complexes at room and low temperature a more precise understanding of the Jahn-Teller and pseudo Jahn-Teller effects is being obtained. The characterisation of non-rigid geometries (fluxional behaviour) and ranges of distorted geometries (Plasticity Effect), suggests the linking of related stereochemistries through a series of Structural Pathways, determined by soft modes of vibration of the nuclear framework. Within the Structural Pathways the electronic and e.s.r. spectra may be used to establish an "Electronic Criteria of Stereochemistry" for a related series of complexes, such as the $[Cu(bipy)_2X][Y]$ complexes.

Acknowledgements. The author acknowledges helpful discussion on the pseudo Jahn-Teller Effect with Professor C. Simmons (Department of Chemistry, University of Puerto Rico), Dr. B. Boca (Department of Inorganic Chemistry, Slovak Technical University, Bratislava, Czechoslovakia), at the refereeing stage, Professor D. Reinen, Inorganic Chemistry Department, The Philips University, Marburg, Germany, and the essential help of research collaborators involved in the original publications.

14 Abbreviations

en	ethylenediamine	phen	1,10-phenanthroline
dien	diethylenetriamine	bipyam	2,2′-bipyridylamine
$HBPz_3$	hydrotris(pyrazolyl)borate	tren	tris(2-aminoethyl)amine
bipy	2,2-bipyridyl		methyl, 1,2-ethanediamine
H_4edta	ethylenediaminetetraacetic acid	Py	pyridine
		3-Me-acac	3-methyl-pentane-2,4-dione

15 References

1. Sidgwick, N. V.: The Chemical Elements and Their Compounds, Clarenden Press, Oxford 1950
2. Hatfield, W. E., Whyman, R.: Trans. Met. Chem. *5,* 47 (1969)
3. Massey, A.: Comprehensive Inorganic Chemistry, (Ex. Ed. Trotman-Dickenson, A. F.), Oxford, Pergamon Press, Vol. II (1973) p. 1
4. Hathaway, B. J., Billing, D. E.: Coord. Chem. Rev. *5,* 143 (1970)
5. Hathaway, B. J., Tomlinson, A. A. G.: ibid. *5,* 1 (1970)
6. Hathaway, B. J., Hodgson, P. G.: J. Inorg. Nucl. Chem. *35,* 4071 (1973)
7. Gazo, J., Bersuker, I. B., Garaj, J., Kabesova, M., Kohout, J., Langfelderova, H., Melnik, M., Serator, M., Valach, F.: Coord. Chem. Rev. *21,* 253 (1976)
8. Wood, J. S.: Prog. Inorg. Chem. *16,* 227 (1972)
9. Hathaway, B. J.: Struct. Bonding *14,* 49 (1973)
10. Smith, D. W.: ibid. *12,* 49 (1972); Smith, D. W.: Coord. Chem. Rev. *21,* 93 (1976)
11. Hathaway, B. J.: Essays in Chemistry *2,* 61 (1971)
12. Jahn, H. A., Teller E.: Proc. Roy. Soc., London *161,* 220 (1937)
13. Dunitz, J. D., Orgel, L. E.: J. Phys. Chem. Solids *3,* 20 (1957)
14. Bersuker, I. B.: Coord. Chem. Rev. *14,* 357 (1975)
15. Pradilla-Sorzano, J., Fackler, J. P.: Inorg. Chem. *12,* 1182 (1973)
16. Hathaway, B. J., Duggan, M., Murphy, A., Mullane, J., Power, P. C., Walsh, A., Walsh, B.: Coord. Chem. Rev. *36,* 267 (1981)
    Bencini, A., Gatteschi, D.: Trans. Met. Chem. *8,* 1 (1982)
    Bertini, I., Gatteschi, D., Scozzafava, A.: Coord. Chem. Rev. *29,* 67 (1979)
17. Reinen, D., Friebel, C.: Struct. Bonding *37,* 1 (1979)
18. Burdett, J. K.: 'Molecular Shapes', John Wiley and Sons, Inc., New York, 1980;
    Burdett, J. K.: Inorg. Chem. *20,* 1959 (1960)
19. Hathaway, B. J.: Coord. Chem. Rev. *35,* 211 (1981)
20. Hathaway, B. J.: ibid. *41,* 423 (1982)
21. Hathaway, B. J.: ibid. *52,* 87 (1983)
22. Isaacs, N. W., Kennard, C. H. L.: J. Chem. Soc. *A* 386 (1969);
    Joesten, M. D., Hussain, M. S., Lenhert, P. G.: Inorg. Chem. *16,* 2680 (1977)
23. Mullen, D., Heger, G., Reinen, D.: Solid State Commun. *17,* 1249 (1975)
24. Cola, M., Guiseppetti, G., Mazzi, F.: Atti Accad. Sci. Torino, Cl. Sci. Fis., Mat. Natur. *96,* 381 (1962)
25. Cullen, D. L., Lingafelter, E. C.: Inorg. Chem. *9,* 1858 (1970)
26. Ammeter, J. H., Burgi, H. B., Gamp, E., Meyer-Sandrin, V., Jensen, W. P.: ibid. *18,* 733 (1979)
27. Elliott, H., Hathaway, B. J., Slade, R. C.: ibid. *5,* 669 (1966)
28. Friebel, C.: Z. Anorg. Allg. Chem. *417,* 197 (1975)
29. Takagi, S., Joesten, M. D., Lenhert, P. G.: Acta Crystallogr. Sect. B, *32,* 1278 (1976)
30. Duggan, M., Murphy, A., Hathaway, B. J.: Inorg. Nucl. Chem. Lett. *15,* 103 (1979);
    Alcock, N. W., Duggan, M., Murphy, A., Tyagi, S., Hathaway, B. J., Hewat, A. W.: J. Chem. Soc., Dalton Trans. 7 (1984)

31. Alcock, N. W., Walsh, B., Hathaway, B. J.: unpublished results
32. Shields, K. G., Kennard, C. H. L.: Crys. Struct. Comm. *1*, 189 (1972)
33. Porai-Koshits, M. A.: Zhur. Strukt. Khim. *4*, 584 (1963)
34. Brown, D. S., Lee, J. D., Melson, B. G. A., Hathaway, B. J., Procter, I. M., Tomlinson, A. A. G.: Chem. Commun. 369 (1967);
Brown, D. S., Lee, J. D., Melson, B. G. A.: Acta Crystallogr. Sect. B, *24*, 730 (1968)
35. Stephens, F. S.: J. Chem. Soc. A, 883 (1969)
36. Mtetwa, V. S. B., Prout, K., Murphy, A., Hathaway, B. J.: Spring Crystallography Meeting, Royal Soc. of Chem., Univ. Durham, April 1982
37. Clifford, F., Counihan, E., Fitzgerald, W., Seff, K., Simmons, C., Tyagi, S., Hathaway, B. J.: J. Chem. Soc. Chem. Commun., 196 (1982);
Fitzgerald, W., Hathaway, B. J., Simmons, C.: J. Chem. Soc., Dalton Trans. 1984, accepted for publication
38. Procter, I. M., Stephens, F. S.: J. Chem. Soc. A, 1248 (1969)
39. Simmons, C., Lundeen, M., Alcock, N. W., Fitzgerald, W., Hathaway, B. J.: Acta Crystallogr. 1983, submitted for publication
40. Stephens, F. S.: J. Chem. Soc. A, 2233 (1969)
41. Boca, R.: privat communication
42. Ammeter, J. H.: Nouv. J. Chim. *4*, 631 (1980)
43. Muetterties, E. L.: Inorg. Chem. *4*, 795 (1965)
44. (a) *Ref.* 20 p. 448; (b) Pascal, J.-L., Potier, J., Jones, D. J., Roziere, J., Michalowicz, A.: ibid. 1984, accepted for publication
45. Klein, S., Reinen, D.: J. Solid State Chem. *25*, 295 (1978)
46. Klein, S., Reinen, D.: ibid *32*, 311 (1980)
47. Bertini, I., Dapporto, P., Gatteschi, D., Scozzafava, A.: J. Chem. Soc., Dalton Trans. 1409 (1979)
48. Robinson, D. J., Kennard, C. H. L.: Cryst. Struct. Commun. *1*, 185 (1972)
49. Van der Zee, J. J., Shields, K. G., Graham, A. J., Kennard, C. H. L.: ibid. *1*, 367 (1972)
50. Hathaway, B. J., Hewat, A. W.: J. Solid. State Chem. 1983, accepted for publication
51. Smith, G., Moore, F. H., Kennard, C. H. L.: Cryst. Struct. Commun. *4*, 407 (1975)
52. Murphy, A., Mullane, J., Hathaway, B. J.: Inorg. Nucl. Chem. Lett. *16*, 129 (1980)
53. Mtetwa, V., Prout, K., Fitzgerald, W., Hathaway, B. J.: 1983, unpublished results
54. Simmons, C., Clearfield, A., Fitzgerald, W., Tyagi, S., Hathaway, B. J.: J. Chem. Soc., Chem. Commun. 189 (1982)
55. Simmons, C. Clearfield, A., Fitzgerald, W., Tyagi, S., Hathaway, B. J.: Inorg. Chem. *22*, 2463 (1983)
56. Crama, W. J.: Acta Crystallogr. Sect. B, *37*, 2133 (1981)
57. Cullen, D. L., Lingafelter, E. C.: Inorg. Chem. *10*, 1264 (1971)
58. Muphy, A., Hathaway, B. J., King, T. J.: J. Chem. Soc., Dalton Trans., 1646 (1979)
59. Bertini, I., Gatteschi, D., Scozzafava, A.: Inorg. Chem. *16*, 1973 (1977)
60. Friebel, C.: Z. Anorg. Allg. Chem. *417*, 197 (1975)
61. Fitzgerald, W., Murphy, B., Tyagi, S., Walsh, B., Walsh, A., Hathaway, B. J.: J. Chem. Soc., Dalton Trans. 2271 (1981)
62. Silver, B. L., Getz, D.: J. Chem. Phys. *61*, 638 (1974)
63. Duggan, M., Hathaway, B. J., Mullane, J.: J. Chem. Soc., Dalton Trans., 690 (1980)
64. Hathaway, B. J., Dudley, R. J., Nicholls, P.: J. Chem. Soc. A, 1845 (1969);
Takagi, S., Joesten, M. D., Lenhert, P. G.: Acta Crystallogr., Sect. B, *31*, 596 (1975)
65. Walsh, A., Walsh, B., Murphy, B., Hathaway, B. J.: ibid. *37*, 1512 (1981)
66. Tyagi, S., Hathaway, B. J.: Proc. 9th Con. Coord. Chem., Bratislava, C.S.S.R., 417 (1983)
67. Duggan, M., Horgan, M., Mullane, J., Power, P. C., Ray, N., Walsh, A., Hathaway, B. J.: Inorg. Nucl. Chem. Lett. *16*, 407 (1980)
68. Cakajdova, I. A., Hanic, F.: Acta Crystallogr. *11*, 610 (1958)
69. Hanic, F.: ibid. *12*, 739 (1959)
70. Serator, M., Langfelderova, H., Gazo, J.: Inorg. Chim. Acta *30*, 267 (1978)
71. Hall, D., McKinnon, A. J., Waters, T. N.: J. Chem. Soc. 425 (1965)
72. McKinnon, A. J., Waters, T. N., Hall, T. N.: ibid. 3290 (1964)
73. Pajunen, A., Smolander, K., Belinskii, I.: Suom. Kemistilehti Sect. B, *45*, 317 (1972)

74. Cameron, A. F., Taylor, D. W., Nuttall, R. H.: J. Chem. Soc., Dalton Trans., 58 (1972)
75. Dudley, R. J., Fereday, R. J., Hathaway, B. J., Hodgson, P. G., Power, P. C.: ibid. 1044 (1973)
76. Ray, N., Hullett, L., Sheahan, R., Hathaway, B. J.: Inorg. Nucl. Chem. Lett. *14*, 305 (1978)
77. Morosin, B., Lingafelter, E. C.: J. Phys. Chem. *65*, 50 (1961);
McGinnety, J. A.: J. Amer. Chem. Soc. *94*, 8406 (1972)
78. Harlow, R. L., Wells, W. J., Watt, G. W., Simonsen, S. H.: Inorg. Chem. *14:* 1468 (1975)
79. Battaglia, L. P., Bonamartini Corradi, A., Marcotrigiano, G., Menabue, L., Pellacani, G. C.: ibid. *18*, 148 (1979)
80. Harrison, D., Kennedy, D. M., Power, M., Sheahan, R., Hathaway, B. J.: J. Chem. Soc., Dalton Trans. 1556 (1981)
81. Harrison, W. D., Hathaway, B. J., Kennedy, D.: Acta Crystallogr., Sect. B, *35*, 2301 (1979)
82. Tyagi, S., Hathaway, B. J., Kremer, S., Stratemeier, H., Reinen, D.: J. Chem. Soc., Dalton Trans., 1984, accepted for publication
83. Ray, N., Hulett, L., Sheahan, R., Hathaway, B. J.: ibid. 1463 (1981)
84. Duggan, M., Ray, N., Hathaway, B. J., Tomlinson, A. A. G., Brint, P., Pelin, K.: ibid. 1342 (1980)
85. Berry, S.: J. Chem. Phys. *32,* 933 (1960)
86. Burgi, H., Dunitz, J. D.: Acc. Chem. Res. *16*, 153 (1983)
87. Ballhausen, C. J.: 'Introduction to Ligand Field Theory' McGraw Hill, New York 1962, p. 272;
Reinen, D., Grefer, J.: Z. Naturforsch., Teil A, *28*, 1185 (1973);
Reinen, D.: Comments Inorg. Chem. *2,* 227 (1983)
88. Dunitz, J. D.: 'X-Ray Analysis and the Structure of Organic Molecules', Cornell University Press, London 1979, ch. 7
89. Nakamoto, K.: 'Infrared Spectra of Inorganic and Coordination Compounds', John Wiley and Sons, New York, 3nd Ed., 1978
90. Fereday, R. J., Hodgson, P. G., Tyagi, S., Hathaway, B. J.: J. Chem. Soc., Dalton Trans., 2070 (1981)
91. Jircitano, A. J., Sheldon, R. I., Mertes, K. B.: J. Amer. Chem. Soc. *105,* 3022 (1983)
92. Foley, J. Tyagi, S., Hathaway, B. J.: J. Chem. Soc., Dalton Trans., 1 (1984)
93. Hathaway, B. J.: Proc. 9th Con. Coord. Chem., Bratislava, C.S.S.R., 93 (1983)
94. Hathaway, B. J.: J. Chem. Soc., Dalton Trans., 1196 (1972)
95. Foley, J., Kennefick, D., Phelan, D., Tyagi, S., Hathaway, B. J.: ibid. 2333 (1983)
96. Harrison, W. D., Hathaway, B. J.: Acta Crystallogr. Sect. B, *36*, 1069 (1980)
97. Harrison, W. D., Hathaway, B. J.: ibid. *34*, 2843 (1978)
98. McKenzie, E. D.: J. Chem. Soc., Dalton Trans., 3095 (1970)

Calculations of the Jahn-Teller Coupling Constants for d^x Systems in Octahedral Symmetry via the Angular Overlap Model

Keith D. Warren

Department of Chemistry, University College, Cardiff, Wales, United Kingdom*

The Angular Overlap Model is applied to the calculation of the Jahn-Teller coupling constants for all possible ground states of ML_6d^x systems in O_h symmetry. In the weak field scheme the $|LSM_LM_S\rangle$ basis is treated by an operator equivalent technique and results are also derived for the $|LSJM_J\rangle$ basis to include the effects of spin-orbit coupling. In the strong field basis both high- and low-spin ground states are considered where these arise and in all cases the extent of quenching of Jahn-Teller activity due to spin-orbit coupling is evaluated. The derivation of the required angular overlap parameters and the general utility of the method are briefly considered.

* Work carried out as sabbatical visitor, Michaelmas Term 1982, School of Chemical Sciences, University of East Anglia, Norwich NR4 7TJ, England

1 Introduction

As recently as 1978 Bacci[1)] demonstrated that the Jahn-Teller coupling constants, hitherto only accessible theoretically with considerable difficulty, could be derived in a remarkably simple manner using the framework of the Angular Overlap Model. Bacci himself[1)] gave the results for d^1 (and equivalent) systems both for ML_6 O_h species and for a number of lower symmetries and more recently[2)] has listed his findings for various other d^x ground states in O_h symmetry, although these latter calculations were carried out only in the strong field basis and neglected the effects of spin-orbit coupling.

Nevertheless, although the strong field formalism lends itself most readily and naturally to the evaluation of the Jahn-Teller coupling constants, there may still arise situations in which a weak field basis might be more appropriate. Moreover, in either basis it is frequently desirable to determine the influence of spin-orbit interactions and the present work therefore presents results for all the d^x ground states of ML_6 O_h systems in both the weak and strong field bases and considers specifically the consequences of spin-orbit coupling in either formalism.

2 Theory

The necessary theory for the calculation of the linear Jahn-Teller coupling constants via the Angular Overlap Model has already been fully reported[1-4)] and is therefore only briefly rehearsed here. The coupling constant, c, is defined as

$$c = \langle \psi_a | \partial V / \partial Q_\gamma^\Gamma | \psi_b \rangle$$

where Q_γ^Γ is the Jahn-Teller active normal coordinate, the component γ of the Γ representation. The usual formulation for the angular overlap matrix element then gives

$$c = \sum_\lambda \sum_\omega \sum_{j=1}^{N} \frac{\partial[e_{\lambda\omega} F_{\lambda\omega}^l(\psi_a) F_{\lambda\omega}^l(\psi_b)]}{\partial(R_j, \theta_j, \varphi_j)} \frac{\partial(R_j, \theta_j, \varphi_j)}{\partial Q_\gamma^\Gamma}$$

from which the appropriate coupling constants are easily derived. Here the $F_{\lambda\omega}^l$ are the elements of the general angular transformation matrix, θ_j and φ_j are the polar angles defining the positions of the $j^{th.}$ of the N ligands, R_j being the metal-ligand distance, and the $e_{\lambda\omega}$ are the standard angular overlap parameters. For Jahn-Teller active stretching vibrations the c terms will be functions of $\partial e_\lambda / \partial R$, whilst for bending vibrations the matrix elements involve simply e_λ / R.

For O_h systems the active modes are ε_g (stretching) which may couple with either $E(\Gamma_3)$, $T_{1,2}(\Gamma_{4,5})$ or Γ_8 degeneracies, and τ_{2g} (bending), coupling only with $T_{1,2}(\Gamma_{4,5})$ or Γ_8 states. Within the manifolds of these degeneracies various symmetry determined relationships subsist between the matrix elements (see Table 1), from which the appropriate Jahn-Teller coupling constant is readily obtained as outlined by Bersuker[5)] and by Liehr[6)] (c.f. also Ref. 4), and these in turn are simply related to the appropriate Jahn-Teller

Table 1. Standard Forms of the Jahn-Teller Coupling Matrices[a]

$E \otimes \varepsilon_g$	$\vert\theta\rangle$	$\vert\varepsilon\rangle$
$\langle\theta\vert^*$	$+A(Q_{z^2})$	$-A(Q_{x^2-y^2})$
$\langle\varepsilon\vert^*$		$-A(Q_{z^2})$

$E_{JT} = A^2/2K_\varepsilon$

$T_{1,2} \otimes \varepsilon_g$	$\vert y\rangle$ $\vert xz\rangle$	$\vert z\rangle$ $\vert xy\rangle$	$\vert x\rangle$ $\vert yz\rangle$
$\langle x\vert^*$, $\langle xz\vert^*$	$-\frac{1}{2}C(Q_{z^2}) + \frac{1}{2}\sqrt{3}C(Q_{x^2-y^2})$	0	0
$\langle z\vert^*$, $\langle xy\vert^*$		$+C(Q_{z^2})$	0
$\langle y\vert^*$, $\langle yz\vert^*$			$-\frac{1}{2}C(Q_{z^2}) - \frac{1}{2}\sqrt{3}C(Q_{x^2-y^2})$

$E_{JT} = C^2/2K_\varepsilon$

$T_{1,2} \otimes \tau_{2g}$	$\vert y\rangle$ $\vert xz\rangle$	$\vert z\rangle$ $\vert xy\rangle$	$\vert x\rangle$ $\vert yz\rangle$
$\langle x\vert^*$, $\langle xz\vert^*$	0	$+B(Q_{yz})$	$+B(Q_{xy})$
$\langle z\vert^*$, $\langle xy\vert^*$		0	$+B(Q_{xz})$
$\langle y\vert^*$, $\langle yz\vert^*$			0

$E_{JT} = 2B^2/3K_\tau$

$T_{1,2} \otimes \varepsilon_g$	$\vert+1\rangle$	$\vert 0\rangle$	$\vert-1\rangle$
$\langle+1\vert^*$	$-\frac{1}{2}C(Q_{z^2})$	0	$-\frac{1}{2}\sqrt{3}C(Q_{x^2-y^2})$
$\langle 0\vert^*$		$+C(Q_{z^2})$	0
$\langle-1\vert^*$			$-\frac{1}{2}C(Q_{z^2})$

$E_{JT} = C^2/2K_\varepsilon$

$T_{1,2} \otimes \tau_{2g}$	$\vert+1\rangle$	$\vert 0\rangle$	$\vert-1\rangle$
$\langle+1\vert^*$	0	$-\frac{1}{2}\sqrt{2}B(Q_{xz} - iQ_{yz})$	$+iB(Q_{xy})$
$\langle 0\vert^*$		0	$+\frac{1}{2}\sqrt{2}B(Q_{xz} - iQ_{yz})$
$\langle-1\vert^*$			0

$E_{JT} = 2B^2/3K_\tau$

$\Gamma_8(T_{1,2}) \otimes \varepsilon_g$	$\vert\varkappa\rangle$	$\vert\lambda\rangle$	$\vert\mu\rangle$	$\vert\nu\rangle$
$\langle\varkappa\vert^*$	$+A_\varepsilon(Q_{z^2})$	0	$+A_\varepsilon(Q_{x^2-y^2})$	0
$\langle\lambda\vert^*$		$-A_\varepsilon(Q_{z^2})$	0	$+A_\varepsilon(Q_{x^2-y^2})$
$\langle\mu\vert^*$			$-A_\varepsilon(Q_{z^2})$	0
$\langle\nu\vert^*$				$+A_\varepsilon(Q_{z^2})$

$E_{JT} = A_\varepsilon^2/2K_\varepsilon$

$\Gamma_8(T_{1,2}) \otimes \tau_{2g}$	$\vert\varkappa\rangle$	$\vert\lambda\rangle$	$\vert\mu\rangle$	$\vert\nu\rangle$
$\langle\varkappa\vert^*$	0	$-A_\tau(Q_{xz} - iQ_{yz})$	$+iA_\tau(Q_{xy})$	0
$\langle\lambda\vert^*$		0	0	$+iA_\tau(Q_{xy})$
$\langle\mu\vert^*$			0	$+A_\tau(Q_{xz} + iQ_{yz})$
$\langle\nu\vert^*$				0

$E_{JT} = A_\tau^2/2K_\tau$

[a] The standard forms are listed for (i) E, T_1, and T_2 states in a real orbital basis, (ii) T_1 and T_2 states in a complex orbital basis, and (iii) Γ_8 levels arising from $T_{1,2}$ orbital degeneracies: Γ_8 levels arising from E states retain the same Jahn-Teller activity as for that E state.

stabilization energy, E_{JT}. Thus, for the coupling $E \otimes \varepsilon_g$, the coupling constant, A, is related to E_{JT} by $E_{JT} = A^2/2K_\varepsilon$, whilst for $T_{1,2} \otimes \tau_{2g}$ and $T_{1,2} \otimes \varepsilon_g$ respectively one finds $E_{JT} = 2B^2/3K_\tau$ and $E_{JT} = C^2/2K_\varepsilon$, where the K_Γ are the appropriate force constants for the ε_g and τ_{2g} vibrations, and B and C are the coupling constants for the τ_{2g} and ε_g modes respectively. For $\Gamma_8 \otimes \varepsilon_g$ or $\Gamma_8 \otimes \tau_{2g}$, with coupling constants A_ε and A_τ, the relationship in either case is $E_{JT} = A_\Gamma^2/2K_\Gamma$.

Table 2. Matrix Elements of the Jahn-Teller Operators in the Real d-Orbital Basis ML_6, O_h

$\partial V/\partial Q^\varepsilon_{z^2}$	$\lvert z^2\rangle$	$\lvert x^2-y^2\rangle$	$\lvert xz\rangle$	$\lvert xy\rangle$	$\lvert yz\rangle$
$\langle z^2\rvert^*$	$+\frac{1}{2}\sqrt{3}\dot{\sigma}$	0	0	0	0
$\langle x^2-y^2\rvert^*$		$-\frac{1}{2}\sqrt{3}\dot{\sigma}$	0	0	0
$\langle xz\rvert^*$			$+(1/3)\sqrt{3}\dot{\pi}$	0	0
$\langle xy\rvert^*$				$-(2/3)\sqrt{3}\dot{\pi}$	0
$\langle yz\rvert^*$					$+(1/3)\sqrt{3}\dot{\pi}$

$\partial V/\partial Q^\varepsilon_{x^2-y^2}$	$\lvert z^2\rangle$	$\lvert x^2-y^2\rangle$	$\lvert xz\rangle$	$\lvert xy\rangle$	$\lvert yz\rangle$
$\langle z^2\rvert^*$	0	$-\frac{1}{2}\sqrt{3}\dot{\sigma}$	0	0	0
$\langle x^2-y^2\rvert^*$		0	0	0	0
$\langle xz\rvert^*$			$+\dot{\pi}$	0	0
$\langle xy\rvert^*$				0	0
$\langle yz\rvert^*$					$-\dot{\pi}$

$\partial V/\partial Q^\tau_{xz}$	$\lvert z^2\rangle$	$\lvert x^2-y^2\rangle$	$\lvert xz\rangle$	$\lvert xy\rangle$	$\lvert yz\rangle$
$\langle z^2\rvert^*$	0	0	$+\frac{1}{2}\sqrt{3}\sigma$	0	0
$\langle x^2-y^2\rvert^*$		0	$+(3/2)\sigma$	0	0
$\langle xz\rvert^*$			0	0	0
$\langle xy\rvert^*$				0	$+2\pi$
$\langle yz\rvert^*$					0

$\partial V/\partial Q^\tau_{yz}$	$\lvert z^2\rangle$	$\lvert x^2-y^2\rangle$	$\lvert xz\rangle$	$\lvert xy\rangle$	$\lvert yz\rangle$
$\langle z^2\rvert^*$	0	0	0	0	$-\sqrt{3}\sigma + 2\sqrt{3}\pi$
$\langle x^2-y^2\rvert^*$		0	0	0	$-3\sigma + 2\pi$
$\langle xz\rvert^*$			0	$+2\pi$	0
$\langle xy\rvert^*$				0	0
$\langle yz\rvert^*$					0

$\partial V/\partial Q^\tau_{xy}$	$\lvert z^2\rangle$	$\lvert x^2-y^2\rangle$	$\lvert xz\rangle$	$\lvert xy\rangle$	$\lvert yz\rangle$
$\langle z^2\rvert^*$	0	0	0	$-\sqrt{3}\sigma$	0
$\langle x^2-y^2\rvert^*$		0	0	0	0
$\langle xz\rvert^*$			0	0	$+2\pi$
$\langle xy\rvert^*$				0	0
$\langle yz\rvert^*$					0

Table 3. Matrix Elements of the Jahn-Teller Operators in the Complex d-Orbital Basis, ML_6, O_h

$\partial V/\partial Q^{\varepsilon}_{z^2}$	$\|0\rangle$	$\|+1\rangle$	$\|-1\rangle$	$\|+2\rangle$	$\|-2\rangle$
$\langle 0\|^*$	$+\frac{1}{2}\sqrt{3}\dot{\sigma}$	0	0	0	0
$\langle +1\|^*$		$+(1/3)\sqrt{3}\dot{\pi}$	0	0	0
$\langle -1\|^*$			$+(1/3)\sqrt{3}\dot{\pi}$	0	0
$\langle +2\|^*$				$-(1/4)\sqrt{3}\dot{\sigma}$ $-(1/3)\sqrt{3}\dot{\pi}$	$-(1/4)\sqrt{3}\dot{\sigma}$ $+(1/3)\sqrt{3}\dot{\pi}$
$\langle -2\|^*$					$-(1/4)\sqrt{3}\dot{\sigma}$ $-(1/3)\sqrt{3}\dot{\pi}$

$\partial V/\partial Q^{\varepsilon}_{x^2-y^2}$	$\|0\rangle$	$\|+1\rangle$	$\|-1\rangle$	$\|+2\rangle$	$\|-2\rangle$
$\langle 0\|^*$	0	0	0	$-(1/4)\sqrt{6}\dot{\sigma}$	$-(1/4)\sqrt{6}\dot{\sigma}$
$\langle +1\|^*$		0	$-\dot{\pi}$	0	0
$\langle -1\|^*$			0	0	0
$\langle +2\|^*$				0	0
$\langle -2\|^*$					0

$\partial V/\partial Q^{\tau}_{xz}$	$\|0\rangle$	$\|+1\rangle$	$\|-1\rangle$	$\|+2\rangle$	$\|-2\rangle$
$\langle 0\|^*$	0	$-(1/4)\sqrt{6}\sigma$	$+(1/4)\sqrt{6}\sigma$	0	0
$\langle +1\|^*$		0	0	$-(3/4)\sigma$ $-\pi$	$-(3/4)\sigma$ $+\pi$
$\langle -1\|^*$			0	$+(3/4)\sigma$ $-\pi$	$+(3/4)\sigma$ $+\pi$
$\langle +2\|^*$				0	0
$\langle -2\|^*$					0

$\partial V/\partial Q^{\tau}_{yz}$	$\|0\rangle$	$\|+1\rangle$	$\|-1\rangle$	$\|+2\rangle$	$\|-2\rangle$
$\langle 0\|^*$	0	$+\frac{1}{2}\sqrt{6}i\sigma$ $-\sqrt{6}i\pi$	$+\frac{1}{2}\sqrt{6}i\sigma$ $-\sqrt{6}i\pi$	0	0
$\langle +1\|^*$		0	0	$-(3/2)i\sigma$	$-(3/2)i\sigma$ $+2i\pi$
$\langle -1\|^*$			0	$-(3/2)i\sigma$ $+2i\pi$	$-(3/2)i\sigma$
$\langle +2\|^*$				0	0
$\langle -2\|^*$					0

$\partial V/\partial Q^{\tau}_{xy}$	$\|0\rangle$	$\|+1\rangle$	$\|-1\rangle$	$\|+2\rangle$	$\|-2\rangle$
$\langle 0\|^*$	0	0	0	$-\frac{1}{2}\sqrt{6}i\sigma$	$+\frac{1}{2}\sqrt{6}i\sigma$
$\langle +1\|^*$		0	$+2i\pi$	0	0
$\langle -1\|^*$			0	0	0
$\langle +2\|^*$				0	0
$\langle -2\|^*$					0

For ML_6 O_h systems the real d-orbital set transforms as $e_g + t_{2g}$ and if only s and p orbitals of the ligands are considered (δ-bonding being neglected) the e_g level will be σ-anti-bonding and the t_{2g} level π-anti-bonding, assuming the absence of internal π-bonding systems from the ligands. The parameters which will arise are therefore limted to e_σ/R and e_π/R, written respectively simply as σ and π, and $\partial e_\sigma/\partial R$ and $\partial e_\pi/\partial R$, written respectively as $\dot{\sigma}$ and $\dot{\pi}$. The first step in the calculations is now therefore to derive all the non-vanishing matrix elements of $\partial V/\partial Q_\gamma^\Gamma$ for the γ components of $Q^{e_g}(Q_{z^2}, Q_{x^2-y^2})$ and $Q^{\tau_{2g}}(Q_{xz}, Q_{yz}, Q_{xy})$ in the real orbital basis, d_{z^2}, d_{xz}, d_{yz}, $d_{x^2-y^2}$, d_{xy}, and these are readily found to be as listed in Table 2. In turn these real orbital results easily yield the corresponding $\partial V/\partial Q_\gamma^\Gamma$ matrix elements in the complex orbital basis, $|m_l\rangle = 0, \pm 1, \pm 2$, and these are similarly given in Table 3.

In principle therefore the calculation of any of the Jahn-Teller coupling constants for the strong field and weak field d^x ground states could be carried out via the matrix elements of Tables 2 and 3, using the appropriate symmetry adapted d^x wave functions for the O_h point group (or for the O^* point group if spin-orbit coupling were to be included). As will be seen later this is a fairly simple matter using the strong field basis since the required O_h wave functions relate to relatively simple d configurations or their hole-equivalents, but in the weak field situation the necessary $O_h|LSM_LM_S\rangle$ (or $O^*|MLJM_J\rangle\rangle$) wave functions are frequently complicated enough to render the derivation of the matrix elements exceedingly tedious. Fortunately however there is an alternative approach to the weak field calculation via the operator equivalent formalism which eliminates most of the labour involved.

This technique was originally described by the author[7] as applied to f^x systems but the results are easily adapted to the d^x situation by truncating the expressions there listed since the O_6 terms now vanish. Thus one may write

$$\partial V/\partial Q_{z^2} = \alpha_{0\lambda} O_2^0 + \beta_{0\lambda}(O_4^0 - 7 O_4^4)$$
$$\partial V/\partial Q_{x^2-y^2} = \alpha_{2\lambda} O_2^2 + \beta_{2\lambda} O_4^2$$
$$\partial V/\partial Q_{xy} = \alpha_{\bar{2}\lambda} O_2^2 + \beta_{\bar{2}\lambda} O_4^2$$

where the O_m^n are appropriate operator equivalents as defined by Stevens[8]. Taking the $|m_l\rangle$ d-orbital set as a basis and using the $J = 2$ values for O_2^0, O_4^0, O_2^2, O_4^2, and O_4^4 listed by Hutchings[9], one easily finds for $\partial V/\partial Q_{z^2}$, in units of $\dot{\sigma}$ or $\dot{\pi}$

$$\alpha_{0\sigma} = -(1/21)\sqrt{3}\,; \quad \beta_{0\sigma} = +(1/336)\sqrt{3}\,; \quad \alpha_{0\pi} = -(1/21)\sqrt{3}\,;$$
$$\beta_{0\pi} = -(1/252)\sqrt{3}$$

from which, in terms of Stevens' operator equivalent coefficients, α and β, there follows

$$\alpha_{0\sigma} = +(1/2)\sqrt{3}\alpha\,; \quad \beta_{0\sigma} = +(3/32)\sqrt{3}\beta\,; \quad \alpha_{0\pi} = +(1/2)\sqrt{3}\alpha\,;$$
$$\beta_{0\pi} = -(1/8)\sqrt{3}\beta$$

Similarly, for $\partial V/\partial Q_{x^2-y^2}$ one obtains

$$\alpha_{2\sigma} = -(1/7)\,; \quad \beta_{2\sigma} = -(1/28)\,; \quad \alpha_{2\pi} = -(1/7)\,; \quad \beta_{2\pi} = +(1/21)$$

again in units of $\dot{\sigma}$ or $\dot{\pi}$, yielding in turn

$$\alpha_{2\sigma} = +(3/2)\,\alpha\ ; \qquad \beta_{2\sigma} = -(9/8)\,\beta\ ; \qquad \alpha_{2\pi} = +(3/2)\,\alpha\ ; \qquad \beta_{2\pi} = +(3/2)\,\beta$$

For $\partial V/\partial Q_{xy}$ the matrix elements are now in units of σ or π, rather than the derivative quantities, but apart from this the matrix elements of $\partial V/\partial Q_{xy}$ are easily obtained from those of $\partial V/\partial Q_{x^2-y^2}$ using the relationship

$$\langle M_J \,|\, \partial V/\partial Q_{xy} \,|\, M_J' \rangle = -2i\,\langle M_J \,|\, \partial V/\partial Q_{x^2-y^2} \,|\, M_J' \rangle\ , \qquad M_J > M_J'$$

Note that for all the operators, $\partial V/\partial Q_{z^2}$, $\partial V/\partial Q_{x^2-y^2}$, and $\partial V/\partial Q_{xy}$, the $\alpha_{n\pi}$ and $\beta_{n\pi}$ quantities are simply related to the $\alpha_{n\sigma}$ and $\beta_{n\sigma}$ terms such that in each case $\alpha_\pi/\alpha_\sigma = +1$ and $\beta_\pi/\beta_\sigma = -(4/3)$.

As demonstrated previously[7] the evaluation of the matrix elements of $\partial V/\partial Q_{z^2}$ (or $\partial V/\partial Q_{x^2-y^2}$) and $\partial V/\partial Q_{xy}$ is sufficient to determine all the Jahn-Teller coupling constants which may arise in any given case so that, provided the Stevens coefficients, α and β, are known for a particular LS state, the coupling constants can be found. Thus, since the α and β values have been listed by Hutchings[9] for ^2D (d^1 and d^9), ^5D (d^4 and d^6), ^3F (d^2 and d^8), and ^4F (d^3 and d^7), the results for all the weak field LS states are readily accessible.

The above procedure is therefore adequate to deal with all the ground states encountered in the $|LSM_LM_S\rangle$ basis but can readily be adapted when it is desired to take account of spin-orbit effects by the use of the $|LSJM_J\rangle$ basis. Thus any given $|LSJM_J\rangle$ component is related to the $|LSM_LM_S\rangle$ components of the parent LS state via the Clebsch-Gordan coefficients such that

$$|LSJM_J\rangle = \sum_{M_L+M_S \neq M_J} \langle LSM_LM_S \,|\, LSJM_J\rangle \cdot |LSM_LM_S\rangle$$

Consequently, since the matrix elements of the Jahn-Teller operator are functions only of the orbital terms, it easily follows that the required quantities in the $|LSJM_J\rangle$ basis are given by

$$\langle J, M_J \,|\, \partial V/\partial Q_\gamma^\Gamma \,|\, J, M_J' \rangle = \sum_{M_L+M_S=M_J}\ \sum_{M_L'+M_S'=M_J'} \langle LSM_LM_S \,|\, LSJM_J\rangle \cdot$$

$$\langle LSM_L'M_S' \,|\, LSJM_J'\rangle \cdot \langle LSM_LM_S \,|\, \partial V/\partial Q_\gamma^\Gamma \,|\, LSM_L'M_S'\rangle \cdot \delta_{M_S,M_S'}$$

In this way, once the results for any given LS state have been obtained via the operator equivalent technique, those for any of the J sub-levels thereof may easily be derived. This procedure could of course be extended to evaluate matrix elements between states of different J, arising from the *same* LS state, but for calculations, whether in the $|LSM_LM_S\rangle$ or the $|LSJM_J\rangle$ basis, in which it is required to incorporate mixing of *different* LS states the resulting cross-product matrix elements cannot be found by the operator equivalent method but must be determined directly from the wave functions.

The general method using the weak field basis may best be illustrated by considering as an example the $^3T_{1g}$ ground state of ^3F, d^2, working first with the $|LSM_LMS\rangle$ set. Following Hutchings[9] the required operator equivalents for ^3F, d^2, are those for $J = 3$ for the various O_m^n, together with $\alpha = -(2/105)$ and $\beta = -(2/315)$. The insertion of these values into the operator equivalent formulae for the matrix elements of $\partial V/\partial Q_{z^2}$ and $\partial V/\partial Q_{x^2-y^2}$ then yields the $\langle M_L \| M_L' \rangle$ quantities for $\partial V/\partial Q_{z^2}$

$$\langle\ 0\|\ 0\rangle = -(1/10)\sqrt{3}\dot{\sigma} + (2/5)\sqrt{3}\dot{\pi}$$
$$\langle\pm 1\|\pm 1\rangle = +(1/20)\sqrt{3}\dot{\sigma} + (2/15)\sqrt{3}\dot{\pi}$$
$$\langle\pm 2\|\pm 2\rangle = +(1/4)\sqrt{3}\dot{\sigma} - (1/3)\sqrt{3}\dot{\pi}$$
$$\langle\pm 3\|\pm 3\rangle = -(1/4)\sqrt{3}\dot{\sigma}$$
$$\langle\pm 2\|\mp 2\rangle = +(1/4)\sqrt{3}\dot{\sigma} - (1/3)\sqrt{3}\dot{\pi}$$
$$\langle\pm 3\|\mp 1\rangle = +(3/20)\sqrt{5}\dot{\sigma} - (1/5)\sqrt{5}\dot{\pi}$$

and for $\partial V/\partial Q_{x^2-y^2}$

$$\langle\pm 3\|\pm 1\rangle = +(1/10)\sqrt{15}\dot{\sigma} - (1/5)\sqrt{15}\dot{\pi}$$
$$\langle\ 0\|\pm 2\rangle = -(1/20)\sqrt{30}\dot{\sigma}$$
$$\langle\pm 1\|\mp 1\rangle = -(3/5)\dot{\sigma} + (2/5)\dot{\pi}$$

Following Griffith[10] (Table A 19) the orbital parts of the $|LSM_LM_S\rangle$ functions for $^3T_{1g}$ of ^3F, d^2, are, in terms of $|L, M_L\rangle$,

$$|T_1 + 1\rangle = -(1/2\sqrt{2})\{\sqrt{5}\,|3, -3\rangle + \sqrt{3}\,|3, +1\rangle\}$$
$$|T_1\quad 0\rangle = |3, 0\rangle$$
$$|T_1 - 1\rangle = -(1/2\sqrt{2})\{\sqrt{5}\,|3, +3\rangle + \sqrt{3}\,|3, -1\rangle\}$$

so that, working for simplicity with $M_S = +1$, the non-zero matrix elements for $^3T_{1g} \otimes \varepsilon_g$ prove to be

$$\langle T_1 + 1|\partial V/\partial Q_{z^2}|T_1 + 1\rangle = \langle T_1 - 1|\partial V/\partial Q_{z^2}|T_1 - 1\rangle = +(1/20)\sqrt{3}\dot{\sigma} - (1/5)\sqrt{3}\dot{\pi}\,;$$
$$\langle T_1 0|\partial V/\partial Q_{z^2}|T_1 0\rangle = -(1/10)\sqrt{3}\dot{\sigma} + (2/5)\sqrt{3}\dot{\pi}\,;$$
$$\langle T_1 + 1|\partial V/\partial Q_{x^2-y^2}|T_1 - 1\rangle = +(3/20)\dot{\sigma} - (3/5)\dot{\pi}\,.$$

Thus for $^3T_{1g} \otimes \varepsilon_g$ the coupling constant, C, is given by

$$C = -(1/10)\sqrt{3}\dot{\sigma} + (2/5)\sqrt{3}\dot{\pi}\,.$$

The constant, B, for the coupling $^3T_{1g} \otimes \tau_{2g}$ is now easily derived, noting that $\langle T_1 + 1|\partial V/\partial Q_{xy}|T_1 - 1\rangle = iB$, by use of the relationship between the matrix elements of $\partial V/\partial Q_{xy}$ and of $\partial V/\partial Q_{x^2-y^2}$ given above. Thus one finds

$$\langle T_1 + 1|\partial V/\partial Q_{xy}|T_1 - 1\rangle = (1/8)\{\sqrt{15}\,\langle -3\|-1\rangle + \sqrt{15}\,\langle +1\|+3\rangle + 3\langle +1\|-1\rangle\}$$

so that, since the elements of $\partial V/\partial Q_{xy}$ in the complex basis are Hermitian,

$$\langle T_1 + 1|\partial V/\partial Q_{xy}|T_1 - 1\rangle = i\{(6/5)\sigma - (9/5)\pi\} \text{ or } B = (3/5)(2\sigma - 3\pi)$$

The calculation of the necessary matrix elements in the spin-orbit, $|LSJM_J\rangle$, basis is also quite straightforward and $^3F_2$ of ^3F of d^2 is treated as an example. Thus, using the Clebsch-Gordan coefficients, the $|J, M_J\rangle$ components of $^3F_2$ may be expressed in terms of the $|M_L, M_S\rangle$ components of ^3F, yielding

$$|2, +2\rangle = (1/\sqrt{21})\{\sqrt{15}|+3, -1\rangle - \sqrt{5}|+2, 0\rangle + |+1, +1\rangle\}$$
$$|2, +1\rangle = (1/\sqrt{21})\{\sqrt{10}|+2, -1\rangle - 2\sqrt{2}|+1, 0\rangle + \sqrt{3}|0, +1\rangle\}$$
$$|2, \ \ 0\rangle = (1/\sqrt{7})\{\sqrt{2}|+1, -1\rangle - \sqrt{3}|0, 0\rangle + \sqrt{2}|-1, +1\rangle\}$$
$$|2, -1\rangle = (1/\sqrt{21})\{\sqrt{3}|\ 0, -1\rangle - 2\sqrt{2}|-1, 0\rangle + \sqrt{10}|-2, +1\rangle\}$$
$$|2, -2\rangle = (1/\sqrt{21})\{|-1, -1\rangle - \sqrt{5}|-2, 0\rangle + \sqrt{15}|-3, +1\rangle\}$$

Since all spin-orthogonal terms vanish, the evaluation of the $\langle M_J\|M_J'\rangle$ matrix elements in terms of the $\langle M_L\|M_L'\rangle$ quantities is very simple and yields, for $\partial V/\partial Q_{z^2}$,

$$\langle \pm 2\|\pm 2\rangle = (1/21)\{15\langle\pm 3\|\pm 3\rangle + 5\langle 0\|0\rangle + \langle\pm 1\|\pm 1\rangle\}$$
$$\langle \pm 1\|\pm 1\rangle = (1/21)\{3\langle 0\|0\rangle + 8\langle\pm 1\|\pm 1\rangle + 10\langle\pm 2\|\pm 2\rangle\}$$
$$\langle\ 0\|\ 0\rangle = (1/7)\{4\langle\pm 1\|\pm 1\rangle + 3\langle 0\|0\rangle\}$$
$$\langle \pm 2\|\mp 2\rangle = (1/21)\{\sqrt{15}\langle\pm 3\|\mp 1\rangle + \sqrt{15}\langle\pm 1\|\mp 3\rangle + 5\langle\mp 1\|\pm 1\rangle\}$$

and for $\partial V/\partial Q_{x^2-y^2}$

$$\langle\ 0\|\pm 2\rangle = (1/7\sqrt{3})\{\sqrt{30}\ \langle\pm 1\|\pm 3\rangle + \sqrt{15}\ \langle 0\|\pm 2\rangle + \sqrt{2}\ \langle\mp 1\|\pm 1\rangle\}$$
$$\langle \pm 1\|\mp 1\rangle = (1/21)\{\sqrt{30}\ \langle 0\|\pm 2\rangle + 8\langle\mp 1\|\pm 1\rangle + \sqrt{30}\ \langle 0\|\mp 2\rangle\}$$

When evaluated these afford, for $\partial V/\partial Q_{z^2}$,

$$\langle\pm 2\|\pm 2\rangle = -(7/60)\sqrt{3}\dot{\sigma} - (23/315)\sqrt{3}\dot{\pi}\ ;$$
$$\langle\pm 1\|\pm 1\rangle = +(13/105)\sqrt{3}\dot{\sigma} - (16/315)\sqrt{3}\dot{\pi}\ ;$$
$$\langle 0\|0\rangle = -(1/70)\sqrt{3}\dot{\sigma} + (26/105)\sqrt{3}\dot{\pi}\ ;$$
$$\langle\pm 2\|\mp 2\rangle = +(11/84)\sqrt{3}\dot{\sigma} - (11/63)\sqrt{3}\dot{\pi}\ ,$$

and for $\partial V/\partial Q_{x^2-y^2}$,

$$\langle 0\|\pm 2\rangle = (1/140)\sqrt{6}\dot{\sigma} - (13/105)\sqrt{6}\dot{\pi}\ ; \qquad \langle\pm 1\|\mp 1\rangle = -(13/35)\dot{\sigma} + (16/105)\dot{\pi}\ .$$

In O* symmetry the J = 2 $^3F_2$ state gives rise to a $\Gamma_3(E)$ and a $\Gamma_5(T_2)$ level, the components of which are given (in terms of $|J, M_J\rangle$) by Griffith[10] (Table A 19) as

$$|E\theta\rangle = |2, 0\rangle\ ; \qquad |E\varepsilon\rangle = (1/\sqrt{2})\{|2, +2\rangle + |2, -2\rangle\} \quad \text{and}$$
$$|T_2 + 1\rangle = |2, -1\rangle\ ; \qquad |T_2\, 0\rangle = (1/\sqrt{2})\{|2, +2\rangle - |2, -2\rangle\}\ ;$$
$$|T_2 - 1\rangle = -|2, +1\rangle$$

For the coupling $E \otimes \varepsilon_g$ the matrix elements are then $\langle E\,\theta|\partial V/\partial Q_{z^2}|E\,\theta\rangle =$

$$-\langle E\,\varepsilon|\partial V/\partial Q_{z^2}|E\,\varepsilon\rangle = -(1/70)\sqrt{3}\,\dot{\sigma} + (26/105)\sqrt{3}\,\dot{\pi} \quad \text{and}$$
$$\langle E\,\theta|\partial V/\partial Q_{x^2-y^2}|E\,\varepsilon\rangle = +(1/70)\sqrt{3}\,\dot{\sigma} - (26/105)\sqrt{3}\,\dot{\pi}\,,$$

thus yielding the coupling constant, A, as

$$A = -(1/70)\sqrt{3}\,\dot{\sigma} + (26/105)\sqrt{3}\,\dot{\pi}\,.$$

Similarly, for the coupling $T_2 \otimes \varepsilon_g$ the matrix elements are easily found to be

$$\langle T_2+1|\partial V/\partial Q_{z^2}|T_2+1\rangle = \langle T_2-1|\partial V/\partial Q_{z^2}|T_2-1\rangle = +(13/105)\sqrt{3}\,\dot{\sigma} - (16/315)\sqrt{3}\,\dot{\pi}\,;$$
$$\langle T_2\,0|\partial V/\partial Q_{z^2}|T_2\,0\rangle = -(26/105)\sqrt{3}\,\dot{\sigma} + (32/315)\sqrt{3}\,\dot{\pi}\,;$$
$$\langle T_2+1|\partial V/\partial Q_{x^2-y^2}|T_2-1\rangle = +(13/35)\,\dot{\sigma} - (16/105)\,\dot{\pi}\,.$$

The coupling constant, C, is therefore given by

$$C = -(26/105)\sqrt{3}\,\sigma + (32/315)\sqrt{3}\,\dot{\pi}\,.$$

For the coupling $T_2 \otimes \tau_{2g}$ it is as before necessary to determine the matrix element $\langle T_2+1|\partial V/\partial Q_{xy}|T_2-1\rangle$, and this is readily accomplished once the $\langle M_J||M_J'\rangle$ matrix elements for $\partial V/\partial Q_{xy}$ have been derived (as previously) from the $\partial V/\partial Q_{x^2-y^2}$ quantities. In this way one obtains the coupling constant, B, as

$$B = +(26/35)\,\sigma - (32/105)\,\pi\,.$$

Table 4. Jahn-Teller Coupling Constants for ML_6, O_h, Systems in the Weak Field $|LSM_LM_S\rangle$ Basis

$d^1\,(^2D)$	$^2T_{2g} \otimes \varepsilon_g$	$C = -(2/3)\sqrt{3}\,\dot{\pi}$
	$^2T_{2g} \otimes \tau_{2g}$	$B = +2\pi$
$d^2\,(^3F)$	$^3T_{1g} \otimes \varepsilon_g$	$C = -(1/10)\sqrt{3}\,(\dot{\sigma} - 4\dot{\pi})$
	$^3T_{1g} \otimes \tau_{2g}$	$B = +(3/5)\,(2\sigma - 3\pi)$
$d^3\,(^4F)$	$^4A_{2g}$	Jahn-Teller impotent
$d^4\,(^5D)$	$^5E_g \otimes \varepsilon_g$	$A = -(1/2)\sqrt{3}\,\dot{\sigma}$
$d^5\,(^6S)$	$^6A_{1g}$	Jahn-Teller impotent
$d^6\,(^5D)$	$^5T_{2g} \otimes \varepsilon_g$	$C = -(2/3)\sqrt{3}\,\dot{\pi}$
	$^5T_{2g} \otimes \tau_{2g}$	$B = +2\pi$
$d^7\,(^4F)$	$^4T_{1g} \otimes \varepsilon_g$	$C = -(1/10)\sqrt{3}\,(\dot{\sigma} - 4\dot{\pi})$
	$^4T_{1g} \otimes \tau_{2g}$	$B = +(3/5)\,(2\sigma - 3\pi)$
$d^8\,(^3F)$	$^3A_{2g}$	Jahn-Teller impotent
$d^9\,(^2D)$	$^2E_g \otimes \varepsilon_g$	$A = -(1/2)\sqrt{3}\,\dot{\sigma}$

Table 5. Jahn-Teller Coupling Constants for ML_6, O^*, Systems in the Weak Field $|LSJM_J\rangle$ Basis

$d^1\ (^2D_{3/2})$	$\Gamma_8 \otimes \varepsilon_g$ $\Gamma_8 \otimes \tau_{2g}$	$A_\varepsilon = -(1/5)\sqrt{3}(\dot{\sigma} + \dot{\pi})$ $A_\tau = +(2/5)\sqrt{3}(\sigma + \pi)$
$d^2\ (^3F_2)$	$\Gamma_3 \otimes \varepsilon_g$ $\Gamma_5 \otimes \varepsilon_g$ $\Gamma_5 \otimes \tau_{2g}$	$A = -(1/210)\sqrt{3}(3\dot{\sigma} - 52\dot{\pi})$ $C = -(2/315)\sqrt{3}(39\dot{\sigma} - 16\dot{\pi})$ $B = +(2/105)\ (39\sigma - 16\pi)$
$d^3\ (^4F_{3/2})$	$\Gamma_8 \otimes \varepsilon_g$ $\Gamma_8 \otimes \tau_{2g}$	$A_\varepsilon = +(12/175)\sqrt{3}(\dot{\sigma} + \dot{\pi})$ $A_\tau = -(24/175)\sqrt{3}(\sigma + \pi)$
$d^4\ (^5D_0)$	Γ_1	Jahn-Teller impotent
$d^5\ (^6S_{1/2})$	Γ_6	Jahn-Teller impotent
$d^6\ (^5D_4)$	$\Gamma_5 \otimes \varepsilon_g$ $\Gamma_5 \otimes \tau_{2g}$	$C = +(1/84)\sqrt{3}(9\dot{\sigma} + 4\dot{\pi})$ $B = +(1/14)\ (6\sigma + 5\pi)$
$d^7\ (^4F_{9/2})$	Γ_6	Jahn-Teller impotent
$d^8\ (^3F_4)$	$\Gamma_5 \otimes \varepsilon_g$ $\Gamma_5 \otimes \tau_{2g}$	$C = +(1/28)\sqrt{3}(\dot{\sigma} - 4\dot{\pi})$ $B = -(3/28)\ (\sigma + 3\pi)$
$d^9\ (^2D_{5/2})$	$\Gamma_8 \otimes \varepsilon_g$ $\Gamma_8 \otimes \tau_{2g}$	$A_\varepsilon = +(1/30)\sqrt{3}(9\dot{\sigma} + 4\dot{\pi})$ $A_\tau = +(2/15)\sqrt{3}(3\sigma - 2\pi)$

Before leaving the weak field calculation it is however important to consider under what circumstances it may be necessary or desirable to take specific account of spin-orbit effects. Thus, when the spin-orbit splitting of a given LS state is small, the separation between the various J components will often be minimal with respect to E_{JT} that is $E_{JT} \gg \zeta_{nd}$, where ζ_{nd} is the spin-orbit coupling constant for the appropriate d-shell, using the Condon-Shortley definition. Under these conditions the J components may to a good approximation be regarded as equienergetic and the E_{JT} value derived in the simple $|LSM_LM_S\rangle$ basis. On the other hand, should the spin-orbit coupling be large with respect to E_{JT} ($\zeta \gg E_{JT}$), then each J component may in contrast be treated separately, using the $|LSJM_J\rangle$ basis, neglecting those matrix elements connecting levels arising from different J components. When E_{JT} and ζ are of comparable magnitude the situation becomes much more complex and requires the whole LS manifold to be treated in the $|LSJM_J\rangle$ basis; the situations in which either the simple $|LSM_LM_S\rangle$ basis is applicable or in which a single J component may be treated in the $|LSJM_J\rangle$ basis merely represent the limiting cases of this more general case when, respectively, either $\zeta/E_{JT} \rightarrow 0$ or $\zeta/E_{JT} \rightarrow \infty$. This particular problem also of course arises in calculations using the strong field basis and is considered there too as well as in Sect. 3.

Turning now to the strong field basis calculations the situation is somewhat complicated by the fact that for the d^4, d^5, d^6, and d^7 cases it is now necessary to consider both high-spin and low-spin ground states. Nevertheless, working now simply in O_h symmetry and neglecting spin-orbit interactions, it proves to be unnecessary at any time to deal with anything more complex than a two-electron wave function in order to derive the desired Jahn-Teller coupling constants. To achieve this it is necessary only to remember that the coupling constants for any system and its hole-equivalent differ merely by an insignificant phase factor of -1, and that any symmetrically half-filled shell (e_g or t_{2g}) contributes nothing to the Jahn-Teller activity. Thus one requires in fact only the results for the $^2T_{2g}(t^1_{2g})$, $^3T_{1g}(t^2_{2g})$, and $^2E_g(e^1_g)$ cases: for the t_{2g} orbitally degenerate states one may work either in a real or in a complex basis (c.f. Table 1 and Table A 24 of Griffith[10]), although for the inclusion of spin-orbit effects the latter is more convenient (vide infra).

Thus the results in O_h symmetry for $^2T_{2g}(t^1_{2g})$ will also apply to the hole-equivalent $^2T_{2g}(t^5_{2g})$, the low-spin d^5 ground state, whilst those for $^3T_{1g}(t^2_{2g})$ will similarly hold for $^3T_{1g}(t^4_{2g})$, the low-spin d^4 ground state. For d^3 ($^4A_{2g}$) and d^8 ($^3A_{2g}$) the orbital singlet ground states are Jahn-Teller impotent and remain so the the first order even when spin-orbit coupling is included. For high-spin d^4 the ground state is $^5E_g(t^3_{2g}e^1_g)$ in which an e^1_g degeneracy is supplemented only by an inactive symmetrically half-filled t^3_{2g} shell so that the same coupling constant arises as for the $^2E_g(t^6_{2g}e^3_g)$ situation, a single hole in the e_g shell. For high-spin d^5 the orbital singlet ground state, $^6A_{1g}(t^3_{2g}e^2_g)$, is Jahn-Teller impotent and remains so, except to very high order, even when spin-orbit effects are considered. For the d^6 case the low-spin $^1A_{1g}(t^6_{2g})$ ground level is totally Jahn-Teller impotent whilst the high-spin $^5T_{2g}(t^4_{2g}e^2_g)$ ground state represents simply a $^3T_{1g}(t^2_{2g})$ hole-equivalent supplemented by an inactive symmetrically half-filled e_g shell. For d^7 systems the low-spin ground state, $^2E_g(t^6_{2g}e^1_g)$, represents merely an e^1_g, $^2E_g(e^1_g)$, system together with an inactive t^6_{2g} closed shell whilst the high-spin ground level, $^4T_{1g}(t^5_{2g}e^2_g)$, corresponds to a $^2T_{2g}(t^1_{2g})$ hole-equivalent plus an inactive symmetrically half-filled e shell. Consequently, in simple O_h symmetry, it is only required to evaluate the matrix elements of the one-electron $\partial V/\partial Q^\Gamma_\gamma$ operators for the three simple cases listed above and to derive therefrom the coupling constants in the usual way.

As in the weak field situation however the inclusion of spin-orbit coupling complicates this simple position. Again, just as in the weak field case, when ζ is small compared with E_{JT} the simple O_h results constitute a good approximation and the differences in spin-orbit contributions between the various O* components of the O_h ground state may be neglected. When however ζ is comparable to or greater than E_{JT} specific account must be taken of this splitting which arises from the coupling of the orbital and spin angular momenta contributions to the ground state. Once more, in the general case, it would be necessary to evaluate all the $\partial V/\partial Q^\Gamma_\gamma$ matrix elements within and between the various O* components arising from the original O_h ground level but again, when $\zeta \gg E_{JT}$, it is permissible to treat the individual O* components separately and to determine the Jahn-Teller coupling constants within them alone. Nevertheless, were this to necessitate in all cases the construction of the appropriate O* wave functions and the evaluation of the $\partial V/\partial Q^\Gamma_\gamma$ matrix elements therefrom, this would, even in simple cases, be a tedious process and would become more so as the value of the coupled spin, S, increased. Happily however this labour can be avoided.

As noted earlier, the matrix elements of the Jahn-Teller operators are functions of the orbital parts of the wave functions only. Consequently a given matrix element will be independent of the associated spin contribution and will vanish unless the spin components of the terms connected are identical. Moreover, for all the cases here of interest, the coupling constants required to couple together the spin and orbital parts to produce the O* wave functions have been tabulated by Griffith[10] (Table A 20). (Note that here Griffith gives complete Tables only in the complex orbital basis, hence the preference for this form.) Thus, for any given O* component, the wave functions may be written down in terms of linear combinations of products of the spin components and of the orbital components of the original O_h state. As a consequence therefore, the matrix elements of the $\partial V/\partial Q_\gamma^\Gamma$ within any O* component may readily be expressed in terms of the matrix elements of the same $\partial V/\partial Q_\gamma^\Gamma$ within the original O_h ground state. Thus the relationship between the coupling constants in the O* basis and the coupling constants in the O_h basis will be symmetry determined and can be expressed in terms of a simple numerical ratio, as can the associated values of E_{JT}. Moreover, the technique is in fact of even more general application and could be used in exactly the same fashion to evaluate matrix elements between *different* O* components arising from the original O_h state.

In general, when the O* components are well separated energetically so that E_{JT} within each may be determined, the values thus obtained exhibit a marked quenching of Jahn-Teller activity relative to that of the original O_h ground level manifold, and this quenching tends to become more pronounced as the magnitude of the spin contribution, S, increases. Some illustrative examples are therefore appropriate at this point.

Consider first therefore the simplest case arising, the ${}^2T_{2g}(t_{2g}^1)$ system. Here the spin, S, is given by $S = \frac{1}{2}$, which in O* symmetry transforms as Γ_6. The spin x orbital product in O* is thus $\Gamma_6 \times \Gamma_5 = \Gamma_7 + \Gamma_8$, of which the former, a Kramers' doublet, is inactive, so that any residual Jahn-Teller activity resides in the Γ_8 quartet only, assuming the Γ_7 and Γ_8 components to be well separated. Table A 20 of Griffith[10] then gives the Γ_8 wave functions as

$$|\varkappa\rangle = -(1/\sqrt{3})\{|+\tfrac{1}{2}, T_2\,0\rangle + \sqrt{2}|-\tfrac{1}{2}, T_2 + 1\rangle\}$$

$$|\lambda\rangle = |-\tfrac{1}{2}, T_2 - 1\rangle$$

$$|\mu\rangle = |+\tfrac{1}{2}, T_2 + 1\rangle$$

$$|\nu\rangle = -(1/\sqrt{3})\{\sqrt{2}|+\tfrac{1}{2}, T_2\,0\rangle + |-\tfrac{1}{2}, T_2 + 1\rangle\}$$

where the spin is denoted by $|\pm\frac{1}{2}\rangle$ and the orbital parts by $|T_2 \pm 1, 0\rangle$. For the coupling $\Gamma_8 \otimes \varepsilon_g$ the matrix element $\langle\varkappa|\partial V/\partial Q_{x^2-y^2}|\mu\rangle$ is equal to $+A_\varepsilon(Q_{x^2-y^2})$, but from the above wave functions

$$\langle\varkappa|\partial V/\partial Q_{x^2-y^2}|\mu\rangle = -(1/\sqrt{3})\{\langle T_2 - 1|\partial V/\partial Q_{x^2-y^2}|T_2 + 1\rangle\} =$$
$$-(1/\sqrt{3})\cdot(-\tfrac{1}{2}\sqrt{3}C) \quad \text{or} \quad A_\varepsilon = +\tfrac{1}{2}C\,.$$

Similarly, for the coupling $\Gamma_8 \otimes \tau_{2g}$, the matrix element $\langle\varkappa|\partial V/\partial Q_{xy}|\mu\rangle$ is equal to $+iA_\tau(Q_{xy})$, and from the wave functions

$$\langle\varkappa|\partial V/\partial Q_{xy}|\mu\rangle = -(1/\sqrt{3})\{\langle T_2 - 1|\partial V/\partial Q_{xy}|T_2 + 1\rangle\} =$$
$$-(1/\sqrt{3})\cdot(-iB) \quad \text{or} \quad A_\tau = +(1/\sqrt{3})\,B\,.$$

Thus, using the standard forms of the various coupling matrices listed in Table 1, simple expressions have been deduced relating the coupling constants within the Γ_8 O* level to those of the ${}^2T_{2g}(t_{2g}^1)$ state in O_h symmetry. In a similar way all the matrix elements within the Γ_8 level, or even within the complete manifold of $\Gamma_7 + \Gamma_8$, could be derived. For the Γ_8 state, for either ε_g or τ_{2g} coupling, $E_{JT} = A_\Gamma^2/2K_\Gamma$. Thus, since for $T_{2g} \otimes \varepsilon_g\, E_{JT} = C^2/2K_\varepsilon$, and for $T_{2g} \otimes \tau_{2g}\, E_{JT} = 2B^2/3K_\tau$, the Jahn-Teller stabilization energy is seen in both cases to be reduced to one-quarter of that for the parent ${}^2T_{2g}$ O_h state.

The d^2 ground state, $^3T_{1g}(t_{2g}^2)$, is also easily treated by the method described above. Here the spin, S, is S = 1, which transforms as Γ_4 in O*. The direct product is thus $\Gamma_4 \times \Gamma_4 = \Gamma_1(A_1) + \Gamma_3(E) + \Gamma_4(T_1) + \Gamma_5(T_2)$. For the E state Table A 20 of Griffith[10] gives the spin × orbital wave functions

$$|E\theta\rangle = (1/\sqrt{6})\{|+1, T_1-1\rangle + 2|0, T_1 0\rangle + |-1, T_1+1\rangle\}$$

$$|E\varepsilon\rangle = (1/\sqrt{2})\{|+1, T_1+1\rangle + |-1, T_1-1\rangle\}$$

For the coupling $E \otimes \varepsilon_g$ the matrix element $\langle E\theta|\partial V/\partial Q_{x^2-y^2}|E\varepsilon\rangle$ is equal to $-\frac{1}{2}\sqrt{3}A(Q_{x^2-y^2})$, but from the wave functions

$$\langle E\theta|\partial V/\partial Q_{x^2-y^2}|E\varepsilon\rangle = (1/2\sqrt{3}) \cdot 2\langle T_1+1|\partial V/\partial Q_{x^2-y^2}|T_1-1\rangle$$

whilst the $\partial V/\partial Q_{xy}$ terms, being Hermitian, vanish. From Table 1 it is seen that

$$\langle T_1+1|\partial V/\partial Q_{x^2-y^2}|T_1-1\rangle = -\tfrac{1}{2}\sqrt{3}C(Q_{x^2-y^2})$$

and also that

$$\langle E\theta|\partial V/\partial Q_{x^2-y^2}|E\varepsilon\rangle = -A(Q_{x^2-y^2}), \qquad \text{so that}$$

$$(1/2\sqrt{3}) \cdot 2(-\tfrac{1}{2}\sqrt{3}C) = -A, \quad \text{or simply} \quad A = +\tfrac{1}{2}C.$$

Consequently the Jahn-Teller stabilization energy, E_{JT}, is reduced in the E O* component to one-quarter of the $^3T_{1g}(t_{2g}^2)$ O_h value. In an exactly similar fashion it can also be shown that for the T_1 and T_2 O* components the coupling constants, B and C, are both reduced to one-half of their values in the parent $^3T_{1g}(t_{2g}^2)$ O_h state so that once again the value of E_{JT} will in each case be only one-quarter of the O_h value.

Finally, the well known case of the $^5T_{2g}(t_{2g}^4e_g^2)$ high-spin d^6 ground state, originally treated by Van Vleck[11] by the pseudo-angular momentum technique, affords a good example of the present approach. Here the spin, S = 2, transforms as E + T_2 so that the direct product, $(E + T_2) \times T_2$, yields the components $\Gamma_1(A_1) + \Gamma_3(E) + 2\Gamma_4(T_1) + 2\Gamma_5(T_2)$. Of these the T_2 level, characterised by the label J = 1 in Griffith's[10] Table A 20, lies lowest, the spin × orbital functions being

$$|T_2+1\rangle = (1/\sqrt{10})\{\sqrt{6}|+2, T_2-1\rangle - \sqrt{3}|+1, T_2 0\rangle + |0, T_2+1\rangle\}$$

$$|T_2 0\rangle = (1/\sqrt{10})\{\sqrt{3}|+1, T_2-1\rangle - 2|0, T_2 0\rangle + \sqrt{3}|-1, T_2+1\rangle\}$$

$$|T_2-1\rangle = (1/\sqrt{10})\{|0, T_2-1\rangle - \sqrt{3}|-1, T_2 0\rangle + \sqrt{6}|-2, T_2+1\rangle\}$$

From these functions it is clear that the matrix elements of both $\partial V/\partial Q_{x^2-y^2}$ and $\partial V/\partial Q_{xy}$ will only be equal to one-tenth of the corresponding matrix elements between $|T_2+1\rangle$ and $|T_2-1\rangle$ of $^5T_{2g}$ in O_h symmetry, so that for both ε_g and τ_{2g} coupling the respective coupling constants, C and B, will both be reduced by a factor of (1/10) and E_{JT}, with its quadratic dependence on the coupling constants, by a factor of (1/100).

The other cases arising, including $^4T_{1g}(t_{2g}^5e_g^2)$ of high-spin d^7 (and the excited $^4T_{1g}$ and $^4T_{2g}(t_{2g}^2e_g^1)$ of d^3), may readily be treated in an analogous fashion, but the ^2E_g ground states of d^9 and of low-spin d^7, together with the ^5E_g ground state of high-spin d^4, require further comment. In the former case the coupling of the spin, $S = \frac{1}{2}$, with the E orbital state leads to a single Γ_8 level. This however is readily shown to factorise into two independent doublets, both of which are only active in coupling to the ε_g mode, and which show identical Jahn-Teller behaviour to the parent ^2E_g state. In the latter case the ^5E_g level shows no first order splitting due to spin-orbit coupling, so that to a very good approximation all the O* components arising (A_1, A_2, E, T_1, and T_2) remain equi-energetic. Thus the result for ^5E_g in simple O_h symmetry, neglecting spin-orbit effects, remains perfectly satisfactory.

The above treatments, in both the weak and the strong field bases, have for simplicity been restricted to ground states which either constitute a single LS state or correspond to a single orbital occupancy. In those cases for which the ground state represents the *only* level of a given symmetry (for example $^2T_{2g}$ of d^1, ^5E_g of d^4, $^5T_{2g}$ of d^6, and ^2E_g of d^9) no problem arises, but for systems possessing excited states of the same symmetry as the ground level the possibility of significant mixing must arise unless the levels in question are substantially separated energetically. As an example one may treat the $^3T_{1g}$ ground state of d^2 in both the weak and the strong field formalism.

In the former case both the ground ^3F and the higher lying ^3P states of the d^2 configuration give rise to $^3T_{1g}$ components, and using Table A 19 of Griffith[10], with

Table 6. Jahn-Teller Coupling Constants for ML_6, O_h, Systems in the Strong Field Basis[a]

d^1	(t_{2g}^1)	$^2T_{2g} \otimes \varepsilon_g$ $^2T_{2g} \otimes \tau_{2g}$	$C = -(2/3)\sqrt{3}\dot{\pi}$ $B = +2\pi$
d^2	(t_{2g}^2)	$^3T_{1g} \otimes \varepsilon_g$ $^3T_{1g} \otimes \tau_{2g}$	$C = +(2/3)\sqrt{3}\dot{\pi}$ $B = -2\pi$
d^3	(t_{2g}^3)	$^4A_{2g}$	Jahn-Teller impotent
d^4	$(t_{2g}^3e_g^1)$ (t_{2g}^4)	$^5E_g \otimes \varepsilon_g$ $^3T_{1g} \otimes \varepsilon_g$ $^3T_{1g} \otimes \tau_{2g}$	$A = -(1/2)\sqrt{3}\dot{\sigma}$ $C = -(2/3)\sqrt{3}\dot{\pi}$ $B = +2\pi$
d^5	$(t_{2g}^3e_g^2)$ (t_{2g}^5)	$^6A_{1g}$ $^2T_{2g} \otimes \varepsilon_g$ $^2T_{2g} \otimes \tau_{2g}$	Jahn-Teller impotent $C = +(2/3)\sqrt{3}\dot{\pi}$ $B = -2\pi$
d^6	$(t_{2g}^4e_g^2)$	$^5T_{2g} \otimes \varepsilon_g$ $^5T_{2g} \otimes \tau_{2g}$	$C = +(2/3)\sqrt{3}\dot{\pi}$ $B = -2\pi$
	(t_{2g}^6)	$^1A_{1g}$	Jahn-Teller impotent
d^7	$(t_{2g}^5e_g^2)$	$^4T_{1g} \otimes \varepsilon_g$ $^4T_{1g} \otimes \tau_{2g}$	$C = -(2/3)\sqrt{3}\dot{\pi}$ $B = +2\pi$
	$(t_{2g}^6e_g^1)$	$^2E_g \otimes \varepsilon_g$	$A = +(1/2)\sqrt{3}\dot{\sigma}$
d^8	$(t_{2g}^6e_g^2)$	$^3A_{2g}$	Jahn-Teller impotent
d^9	$(t_{2g}^6e_g^3)$	$^2E_g \otimes \varepsilon_g$	$A = -(1/2)\sqrt{3}\dot{\sigma}$

[a] For d^4, d^5, d^6, and d^7 systems the results are listed for both the high-spin and the low-spin configurations

Table 7. Reduction Factors for the Jahn-Teller Coupling Constants and Stabilization Energies of ML_6, O^*, Systems in the Strong Field Basis

O_h State	O^* State	Reduction Factors (Coupling Constants)	Reduction Factors (E_{JT})
$^3A_{2g}$ $^4A_{2g}$ $^6A_{1g}$		No first order spin-orbit splitting; Jahn-Teller impotent	
^5E_g		No first order spin-orbit splitting; O_h result applies	
^2E_g	Γ_8	$A_\varepsilon = -A$	1
$^2T_{1,2g}$	Γ_8 Γ_8	$A_\varepsilon = \mp(1/2)\,C$ $A_\tau = +(1/3)\sqrt{3}\,B$	1/4 1/4
$^3T_{1,2g}$	$\Gamma_3(E)$ $\Gamma_4(T_1)$	$A = +(1/2)\,C$ $B = \mp(1/2)\,B$ $C = -(1/2)\,C$	1/4 1/4 1/4
	$\Gamma_5(T_2)$	$B = -(1/2)\,B$ $C = \pm(1/2)\,C$	1/4 1/4
$^4T_{1,2g}$	$\Gamma_8[3/2]$	$A_\varepsilon = \pm(2/5)\,C$ $A_\tau = -(4/15)\sqrt{3}\,B$	4/25 4/25
	$\Gamma_8[5/2]$	$A_\varepsilon = \mp(2/5)\,C$ $A_\tau = -(1/15)\sqrt{3}\,B$	4/25 1/100
$^5T_{2g}$	(J = 1) $\Gamma_5(T_2)$	$B = +(1/10)\,B$ $C = +(1/10)\,C$	1/100 1/100
	(J = 2) $\Gamma_3(E)$ $\Gamma_4(T_1)$	$A = -(1/2)\,C$ $B = -(1/2)\,B$ $C = +(1/2)\,C$	1/4 1/4 1/4
	(J = 3) $\Gamma_4(T_1)$	$B = -(1/2)\,B$ $C = 0$	1/4 0
	$\Gamma_5(T_2)$	$B = -(1/10)\,B$ $C = +(2/5)\,C$	1/100 4/25

$M_S = 1$ throughout, the two-electron wave functions may be written as

$$^3T_{1g}(^3F)\begin{cases} |+1\rangle = -(1/2\sqrt{10})\,\{5|-\overset{+}{1}, -\overset{+}{2}\rangle + 3|\overset{+}{2}, -\overset{+}{1}\rangle + \sqrt{6}|\overset{+}{1}, \overset{+}{0}\rangle\} \\ |\;\;0\rangle = +(1/\sqrt{5})\,\{|\overset{+}{2}, -\overset{+}{2}\rangle + 2|\overset{+}{1}, -\overset{+}{1}\rangle\} \\ |-1\rangle = -(1/2\sqrt{10})\,\{5|\overset{+}{2}, \overset{+}{1}\rangle + 3|\overset{+}{1}, -\overset{+}{2}\rangle + \sqrt{6}|\overset{+}{0}, -\overset{+}{1}\rangle\} \end{cases}$$

and

$$^3T_{1g}(^3P)\begin{cases} |+1\rangle = +(1/\sqrt{5})\,\{\sqrt{2}|\overset{+}{2}, -\overset{+}{1}\rangle - \sqrt{3}|\overset{+}{1}, \overset{+}{0}\rangle\} \\ |\;\;0\rangle = +(1/\sqrt{5})\,\{2|\overset{+}{2}, -\overset{+}{2}\rangle - |\overset{+}{1}, -\overset{+}{1}\rangle\} \\ |-1\rangle = +(1/\sqrt{5})\,\{\sqrt{2}|\overset{+}{1}, -\overset{+}{2}\rangle - \sqrt{3}|\overset{+}{0}, -\overset{+}{1}\rangle\} \end{cases}$$

The matrix elements within these $^3T_{1g}$ manifolds follow the standard complex orbital pattern (Table 1) and correspond to the coupling constants

$$(^3F)\quad B = +(6/5)\sigma - (9/5)\pi\ ; \qquad C = -(1/10)\sqrt{3}\dot{\sigma} + (2/5)\sqrt{3}\dot{\pi}$$
$$(^3P)\quad B = -(6/5)\sigma - (6/5)\pi\ ; \qquad C = -(2/5)\sqrt{3}\dot{\sigma} - (2/5)\sqrt{3}\dot{\pi}$$

In general however, for any component, Γ_m, of the ground state, and for any Jahn-Teller operator, Q_γ^Γ, we have for the ground state wave function and for the ensuing Jahn-Teller matrix element, α and β being the mixing coefficients,

$$|^3T_{1g}, \Gamma_m\rangle = \alpha|^3T_{1g}(^3F), \Gamma_m\rangle + \beta|^3T_{1g}(^3P), \Gamma_m\rangle \quad (\Gamma_m = 0, \pm 1)$$

and

$$\langle ^3T_{1g}, \Gamma_m|Q_\gamma^\Gamma|^3T_{1g}, \Gamma_n\rangle = \alpha^2\langle ^3T_{1g}(^3F), \Gamma_m|Q_\gamma^\Gamma|^3T_{1g}(^3F), \Gamma_n\rangle$$
$$+ \beta^2\langle ^3T_{1g}(^3P), \Gamma_m|Q_\gamma^\Gamma|^3T_{1g}(^3P), \Gamma_n\rangle + \alpha\beta\{\langle ^3T_{1g}(^3F), \Gamma_m|Q_\gamma^\Gamma|^3T_{1g}(^3P), \Gamma_n\rangle$$
$$+ \langle ^3T_{1g}(^3P), \Gamma_m|Q_\gamma^\Gamma|^3T_{1g}(^3F), \Gamma_n\rangle\}$$

so that, although the first two terms may easily be obtained from the values of B and C, the last two terms require the determination of various cross product matrix elements connecting ^3F and ^3P. These may however be found from the wave functions without too much trouble, using the matrix elements of Table 2. Note especially that the sum of the last two terms in the expression for the matrix element follows the same symmetry determined relationships as in the standard complex orbital T sets, and because of this the required coupling constants may still, in the mixed orbital basis, be calculated from simply the $\partial V/\partial Q_{z^2}$ (or $\partial V/\partial Q_{x^2-y^2}$) and the $\partial V/\partial Q_{xy}$ matrix elements. These latter prove to be

$$\langle ^3T_{1g}(^3F), \ 0|\partial V/\partial Q_{z^2}|^3T_{1g}(^3P), 0\rangle = -(1/5)\sqrt{3}\dot{\sigma} - (8/15)\sqrt{3}\dot{\pi}$$
$$\langle ^3T_{1g}(^3F), \pm 1|\partial V/\partial Q_{z^2}|^3T_{1g}(^3P), \pm 1\rangle = +(1/10)\sqrt{3}\dot{\sigma} + (4/15)\sqrt{3}\dot{\pi}$$
$$\langle ^3T_{1g}(^3F), \pm 1|\partial V/\partial Q_{x^2-y^2}|^3T_{1g}(^3P), \mp 1\rangle = +(3/10)\dot{\sigma} + (4/5)\dot{\pi}$$
$$\langle ^3T_{1g}(^3F), \pm 1|\partial V/\partial Q_{xy}|^3T_{1g}(^3P), \mp 1\rangle = \pm(9/10)i\sigma \pm (2/5)i\pi$$

the last element listed being of course Hermitian.

For the strong field basis set one may proceed in an exactly similar manner. Here the two states which may mix are the ground level $^3T_{1g}(t_{2g}^2)$ and the higher lying $^3T_{1g}(t_{2g}^1e_g^1)$. In the real orbital basis the relevant wave functions are, using Tables A 20 and A 24 of Griffith[10)], with $M_S = 1$,

$^3T_{1g}(t_{2g}^2)$

$$|x\rangle = |(\overset{+}{x}y)(\overset{+}{x}z)\rangle\ ; \qquad |y\rangle = |(\overset{+}{y}z)(\overset{+}{x}y)\rangle\ ; \qquad |z\rangle = |(\overset{+}{x}z)(\overset{+}{y}z)\rangle$$

and

$^3T_{1g}(t_{2g}^1e_g^1)$

$$|x\rangle = -\tfrac{1}{2}\{\sqrt{3}\,|(\overset{+}{z^2})(\overset{+}{y}z)\rangle + |(x^2 - y^2)(\overset{+}{y}z)\rangle\}$$
$$|y\rangle = +\tfrac{1}{2}\{\sqrt{3}\,|(\overset{+}{z^2})(\overset{+}{x}z)\rangle - |(x^2 - y^2)(\overset{+}{x}z)\rangle\}$$
$$|z\rangle = |(x^2 - y^2)(\overset{+}{x}y)\rangle$$

The coupling constants for these two $^3T_{1g}$ states are

$$^3T_{1g}(t_{2g}^2) \quad B = -2\pi\,; \quad C = +(2/3)\sqrt{3}\,\dot{\pi}$$
$$^3T_{1g}(t_{2g}^1e_g^1) \quad B = -\ \pi\,; \quad C = -\tfrac{1}{2}\sqrt{3}\,\dot{\sigma} - (2/3)\sqrt{3}\,\dot{\pi}$$

and the required cross product matrix elements are readily found, using Table 3, to be

$$\langle ^3T_{1g}(t_{2g}^2),\, x\,(y)\,|\,\partial V/\partial Q_{xy}\,|\,^3T_{1g}(t_{2g}^1e_g^1),\, y\,(x)\rangle = -(3/2)\,\sigma$$

In this case the sum of the last two terms in the matrix element expression follows the symmetry determined relationships of the standard real orbital T sets (Table 1), so that here too the coupling constants in the linear combination basis may be deduced using only the matrix elements of $\partial V/\partial Q_{z^2}$ (or $\partial V/\partial Q_{x^2-y^2}$) and $\partial V/\partial Q_{xy}$. Moreover, situations for d^x systems other than d^2 may be treated in an entirely analogous way in either the weak field or the strong field basis.

Finally it is appropriate briefly to consider the nature of the parameters, e_λ and $\partial e_\lambda/\partial R$, and how best these quantities may be determined. It is of course well known that for octahedral ML_6 systems ligand field theory derives only one orbital energy parameter, Δ (or 10 Dq), from the experimental data, representing the splitting of the d-orbital set into the e_g and t_{2g} levels. The angular overlap model readily shows this separation to be given by $\Delta = 3e_\sigma - 4e_\pi$, so that neither e_σ nor e_π can be separately established from the value of Δ. However, as shown by Smith[12], spectroscopic data from the lower symmetry tetragonal (D_{4h}) species, especially those of Cr(III), indicate that where the ligands are capable of π-bonding interaction the ratio of e_π to e_σ usually lies between about 0.15 and 0.20. In principle this ratio ought merely to reflect that of the squared group overlap integrals, S_π^2/S_σ^2, but, as pointed out by Smith[12], if this ratio is calculated simply for the ligand p orbitals its value will generally exceed that of e_π/e_σ, and it is here necessary also to include the ligand s orbital contributions. Nevertheless, the empirically determined e_π/e_σ ratios are sufficiently well defined to enable quite reasonably accurate estimates of E_{JT} to be made, as was shown by Bacci[1], and the uncertainty in the e_π/e_σ ratio does not constitute a major problem. Consequently where, as for the τ_{2g} bending mode coupling, terms in e_λ/R arise, there is no difficulty in deriving good estimates for the values of E_{JT}.

On the other hand the estimation of the $\partial e_\lambda/\partial R$ quantities entails somewhat more effort. There are effectively two methods of approach of which the simpler uses the unsophisticated electrostatic model. On this basis Dq, and hence e_λ, shows a $1/R^5$ distance dependence so that direct differentiation then indicates that at any given metal-ligand distance, R, $\partial e_\lambda/\partial R = -5\,(e_\lambda/R)$. In justification of this rather naive model it may be noted that Burns and Axe[13] found the strain dependence of Dq to be rather well described by the $1/R^5$ variation.

Table 8. Matrix Elements for the Calculation of the Jahn-Teller Coupling Constants for ML_6, O_h, Systems in the Strong Field Basis including State Mixing[a]

d^7 high-spin, $^4T_{1g}$: configurations mixed : – $(t_{2g}^5e_g^2)$, $(t_{2g}^4e_g^3)$

$(t_{2g}^5e_g^2)$	$\langle z\vert\ \partial V/\partial Q_{z^2}\ \vert z\rangle = -(1/2)\sqrt{3}\dot{\sigma} - (2/3)\sqrt{3}\dot{\pi}$
	$\langle x(y)\vert\ \partial V/\partial Q_{xy}\ \vert y(x)\rangle = -\pi$
$(t_{2g}^4e_g^3)$	$\langle z\vert\ \partial V/\partial Q_{z^2}\ \vert z\rangle = +(2/3)\sqrt{3}\dot{\pi}$
	$\langle x(y)\vert\ \partial V/\partial Q_{xy}\ \vert y(x)\rangle = -2\pi$
	$\langle (t_{2g}^5e_g^2)z\vert\ \partial V/\partial Q_{z^2}\ \vert (t_{2g}^4e_g^3)z\rangle = 0$
	$\langle (t_{2g}^5e_g^2)x(y)\vert\ \partial V/\partial Q_{xy}\ \vert (t_{2g}^4e_g^3)y(x)\rangle = -(3/2)\sigma$

d^4 low-spin, $^3T_{1g}$: configurations mixed : – (t_{2g}^4), $(t_{2g}^3[^2T_1]e_g^1)$, $(t_{2g}^3[^2T_2]e_g^1)$

(t_{2g}^4)	$\langle z\vert\ \partial V/\partial Q_{z^2}\ \vert z\rangle = -(2/3)\sqrt{3}\dot{\pi}$
	$\langle x(y)\vert\ \partial V/\partial Q_{xy}\ \vert y(x)\rangle = +2\pi$
$(t_{2g}^3[^2T_1]e_g^1)$	$\langle z\vert\ \partial V/\partial Q_{z^2}\ \vert z\rangle = +(1/2)\sqrt{3}\dot{\sigma}$
	$\langle x(y)\vert\ \partial V/\partial Q_{xy}\ \vert y(x)\rangle = 0$
$(t_{2g}^3[^2T_2]e_g^1)$	$\langle z\vert\ \partial V/\partial Q_{z^2}\ \vert z\rangle = -(1/2)\sqrt{3}\dot{\sigma}$
	$\langle x(y)\vert\ \partial V/\partial Q_{xy}\ \vert y(x)\rangle = 0$
	$\langle (t_{2g}^4)z\vert\ \partial V/\partial Q_{z^2}\ \vert (t_{2g}^3[^2T_1]e_g^1)\rangle = 0$
	$\langle (t_{2g}^4)z\vert\ \partial V/\partial Q_{z^2}\ \vert (t_{2g}^3[^2T_2]e_g^1)\rangle = 0$
	$\langle (t_{2g}^3[^2T_1]e_g^1)z\vert\ \partial V/\partial Q_{z^2}\ \vert (t_{2g}^3[^2T_2]e_g^1)\rangle = 0$
	$\langle (t_{2g}^4)x(y)\vert\ \partial V/\partial Q_{xy}\ \vert (t_{2g}^3[^2T_1]e_g^1)\rangle = +(1/4)\sqrt{6}\sigma$
	$\langle (t_{2g}^4)x(y)\vert\ \partial V/\partial Q_{xy}\ \vert (t_{2g}^3[^2T_2]e_g^1)\rangle = +(3/4)\sqrt{2}\sigma$
	$\langle (t_{2g}^3[^2T_1]e_g^1)x(y)\vert\ \partial V/\partial Q_{xy}\ \vert (t_{2g}^3[^2T_2]e_g^1)\rangle = +\sqrt{3}\pi$

d^5 low-spin, $^2T_{2g}$: configurations mixed : – (t_{2g}^5), $(t_{2g}^4[^3T_1]e_g^1)$, $(t_{2g}^4[^1T_2]e_g^1)$

(t_{2g}^5)	$\langle z\vert\ \partial V/\partial Q_{z^2}\ \vert z\rangle = +(2/3)\sqrt{3}\dot{\pi}$
	$\langle yz(xz)\vert\ \partial V/\partial Q_{xy}\ \vert xz(yz)\rangle = -2\pi$
$(t_{2g}^4[^3T_1]e_g^1)$	$\langle z\vert\ \partial V/\partial Q_{z^2}\ \vert z\rangle = -(1/2)\sqrt{3}\dot{\sigma} - (2/3)\sqrt{3}\dot{\pi}$
	$\langle yz(xz)\vert\ \partial V/\partial Q_{xy}\ \vert xz(yz)\rangle = -\pi$
$(t_{2g}[^1T_2]e_g^1)$	$\langle z\vert\ \partial V/\partial Q_{z^2}\ \vert z\rangle = +(1/2)\sqrt{3}\dot{\sigma} - (2/3)\sqrt{3}\dot{\pi}$
	$\langle yz(xz)\vert\ \partial V/\partial Q_{xy}\ \vert xz(yz)\rangle = -\pi$
	$\langle (t_{2g}^5)xy\vert\ \partial V/\partial Q_{z^2}\ \vert (t_{2g}^4[^3T_1]e_g^1)xy\rangle = 0$
	$\langle (t_{2g}^5)xy\vert\ \partial V/\partial Q_{z^2}\ \vert (t_{2g}^4[^1T_2]e_g^1)xy\rangle = 0$
	$\langle (t_{2g}^4[^3T_1]e_g^1)xy\vert\ \partial V/\partial Q_{z^2}\ \vert (t_{2g}^4[^1T_2]e_g^1)xy\rangle = 0$
	$\langle (t_{2g}^5)yz(xz)\vert\ \partial V/\partial Q_{xy}\ \vert (t_{2g}^4[^3T_1]e_g^1)xz(yz)\rangle = -(3/2)\sigma$
	$\langle (t_{2g}^5)yz(xz)\vert\ \partial V/\partial Q_{xy}\ \vert (t_{2g}^4[^1T_2]e_g^1)xz(yz)\rangle = -(1/4)\sqrt{6}\sigma$
	$\langle (t_{2g}^4[^3T_1]e_g^1)yz(xz)\vert\ \partial V/\partial Q_{xy}\ \vert (t_{2g}^4[^1T_2]e_g^1)xz(yz)\rangle = 0$

[a] The results for the d^2 $^3T_{1g}$ system are given in the text (Sect. 2)

Alternatively, one may start from the basic assumptions of the Angular Overlap Model by writing

$$E^*_\lambda = c_\lambda S^2_\lambda$$

where E* is the anti-bonding energy of the $\sigma(e_g)$ or $\pi(t_{2g})$ metal d orbitals, c_λ is a constant of proportionality, and S_λ is the appropriate metal-ligand group overlap integral. This may be rewritten in the form

$$E^*_\lambda = c_\lambda S^{*2}_\lambda F^2_\lambda$$

where S^*_λ and F_λ are respectively the radial and angular components of the overlap integral, S_λ. From the Angular Overlap Model one has

$$E^*_\lambda = e_\lambda F^2_\lambda \qquad \text{or} \qquad e_\lambda = c_\lambda S^{*2}_\lambda$$

Since however

$$S_\lambda = S^*_\lambda F_\lambda \qquad \text{then} \qquad e_\lambda = c_\lambda (S^2_\lambda/F^2_\lambda)$$

whence

$$\partial e_\lambda/\partial R = (c_\lambda/F^2_\lambda) \cdot (\partial S^2_\lambda/\partial R)$$

In principle one may now determine E^*_λ in terms of e_λ, so that c_λ and hence $\partial e_\lambda/\partial R$ may be found. More directly, for any given distance, R, one may write

$$c_\lambda/F^2_\lambda = e_\lambda/S^2_\lambda \qquad \text{so that} \qquad \partial e_\lambda/\partial R = (e_\lambda/S^2_\lambda) \cdot (\partial S^2_\lambda/\partial R)$$

In this way therefore the $\partial e_\lambda/\partial R$ quantities arising for the ε_g stretching mode coupling may be evaluated on a firmer theoretical basis. Nevertheless, the quantity $\partial S^2_\lambda/\partial R$, as a function of R, usually passes through rather a sharp maximum whose position is sensitive to the quality of the wave functions employed to calculate S_λ. Consequently, the resulting values of $\partial e_\lambda/\partial R$ are significantly dependent upon the basis set used and somewhat sensitive to the assumed metal-ligand distance, R.

The only other terms involved in the calculation of the E_{JT} values are of course the force constants, K_ε and K_τ, for the stretching (ε_g) and bending (τ_{2g}) modes respectively. A very detailed survey of the methods of derivation of these parameters from vibrational spectral data has been given by Labonville, Ferraro, Wall, and Basile[14)], which showed that generally K_ε can be rather accurately determined but that a significantly greater uncertainty attends the values of K_τ. Consequently, a rather greater uncertainty will arise in the estimation of E_{JT} for coupling to the τ_{2g} mode than for coupling to the ε_g vibration.

Nevertheless, limitations imposed by the fiducial limits of the force constants will apply irrespective of the method by which the coupling constants themselves have been derived, and the great advantage of the Angular Overlap Model is that it enables good estimates of E_{JT} to be deduced very simply and without the complexity of previous and more convoluted methods of calculating the coupling constants. (See for example Van

Vleck[15)] and Liehr and Ballhausen[16)].) Thus, as shown by Bacci[1)], the Jahn-Teller stabilization energy arising from t_{2g} orbital degeneracies (T states) will amount only to a few hundred cm^{-1} whereas those from e_g levels (E states) can easily amount to a thousand or more cm^{-1}. The greatest utility of the Angular Overlap Model lies however not so much in the absolute calculation of the magnitude of E_{JT} (although here it probably performs at least as well as its more complicated rivals), but in enabling simple comparisons to be made between the E_{JT} values to be expected for varying d^x systems in terms of a very limited number of parameters. It is therefore with this aspect of the method that the next Section will be primarily concerned.

3 Results

In Tables 4 and 5 are shown the results calculated for the Jahn-Teller coupling constants in the weak field scheme: the former contains those derived in O_h symmetry, using the $|LSM_LM_S\rangle$ basis set and the latter the constants found when spin-orbit effects are specifically included via the O^* $|LSJM_J\rangle$ basis. Similarly, Table 6 shows the coupling constants obtained in the strong field scheme in O_h symmetry whilst the reduction factors, both for the coupling constants and for E_{JT}, consequent upon the allowance for spin-orbit coupling (O^* symmetry) are listed in Table 7.

Neglecting for the moment the effects of spin-orbit coupling, it may be noted that where the ground level is actually Jahn-Teller active the O_h symmetry results are the same in the weak field and in the high-spin strong field bases for the d^1, d^4, d^6, and d^9 situations. This is not surprising however since these are just those systems for which the ground state is the only level of that particular irreducible representation. Nevertheless, for the remaining two cases, d^2 and high-spin d^7, the weak field and strong field results are markedly different. Thus in the strong field basis the σ-interacting propensities of the e_g set and the π-interacting tendencies of the t_{2g} set are sharply and clearly distinguished whilst in the weak field basis both σ- and π-terms contribute for either coupling. This however arises from the fact that in both cases there exist two levels of the same irreducible representation originating from differing LS states in the weak field scheme and corresponding to differing t_{2g} and e_g orbital occupancies in the strong field scheme. Thus for d^2 systems both the ground ^3F and the excited ^3P levels (weak field) and the ground (t_{2g}^2) and excited $(t_{2g}^1e_g^1)$ levels (strong field) give rise to $^3T_{1g}$ states so that in general the system is most aptly described by some linear combination of the $^3T_{1g}$ levels of ^3F and ^3P or of (t_{2g}^2) and $(t_{2g}^1e_g^1)$. Obviously, as $Dq \to 0$ the weak field ^3F description is adequate whilst as $Dq \to \infty$ the strong field (t_{2g}^2) description suffices, but frequently the coupling constants must be evaluated for a linear combination wave function as outlined in Sect. 2 (vide supra). In an exactly similar way the d^7 high-spin system must in general be described in terms of a linear combination of the $^4T_{1g}$ components of the ground level ^4F and the excited state ^4P (weak field) or of the ground state $(t_{2g}^5e_g^2)$ and excited state $(t_{2g}^4e_g^3)$ levels (strong field). As in the d^2 case this necessitates the determination of a number of cross product matrix elements, but these are readily found just as for the d^2 case. (See also Table 8.)

With this in mind, both the weak field and the high-spin strong field O_h treatments lead to essentially similar predictions concerning the extent of Jahn-Teller activity –

clearly no parallel with the low-spin strong field behaviour would be expected. Thus the d^3, d^5, and d^8 systems are inactive, the T orbital degeneracies of d^1, d^2, d^6, and d^7 but modestly active, and only for the d^9 and high-spin d^4 E orbital degeneracies are substantial E_{JT} values likely. This is of course in general agreement with the experimental evidence for the various transition series, so that the results of the calculations in the weak field $|LSJM_J\rangle$ basis therefore appear strikingly incongruous. Thus the use of the $|LSJM_J\rangle$ basis has brought about mixing of the σ- and π-contributions in all cases but more obviously the d^3 and d^8 ground states, hitherto predicted to be Jahn-Teller impotent, are now active, whilst the strongly active d^4 system is now predicted to be inactive. Clearly therefore the $|LSJM_MJ\rangle$ basis cannot constitute a particularly satisfactory description of the experimental situation but this is, fortunately, not hard to explain. Thus the adoption of the $|LSJM_J\rangle$ basis would assume that the spin-orbit coupling constant, ζ, is quite substantial with respect to the electron repulsion terms which in turn, in a weak field basis, ought to be substantial with respect to Dq. Experimentally however this situation does not really occur. In the 3d series for example ζ is initially small with respect to the electrostatic repulsion parameter, B (the Racah parameter), and although these quantities become of comparable magnitude towards the end of the series, ζ is never dominant. Thus only in the 4d and to a greater extent in the 5d series does ζ become substantially greater than B, and under these conditions Dq has also increased greatly so that the basic assumptions of the weak field model simply do not apply. [Moreover, it has been pointed out (C. K. Jørgensen, personal communication) that with strong spin-orbit effects there are really three ratios to consider: ζ/Δ, ζ/B, and Δ/B. Thus in a sense the use of the $|LSJM_J\rangle$ basis overemphasises B, and it would be interesting to treat, say, $5d^x$ systems (small ζ/Δ, large ζ/B and Δ/B) using the spin-orbit orbitals, γ_8 and γ_7 instead of t_{2g}, and γ_8 instead of e_g.]

The d^4 and d^6 situations however require some further comment. In the $|LSJM_J\rangle$ basis the various ^5D_J components are assumed to be significantly separated with $^5D_0$ (Γ_1) lying lowest and, naturally, Jahn-Teller impotent. Nevertheless, when Dq becomes at all appreciable the ^5D state splits into a ground ^5E_g level and an upper $^5T_{2g}$ level, and to the first order the former state is *not* split by spin-orbit coupling and consequently the simple $|LSM_LM_S\rangle$ results for O_h symmetry constitute a much better description of the system. Similarly, for d^6 systems in the $|LSJM_J\rangle$ basis the $^5D_4$ state lies lowest and the lowest lying component thereof (Γ_5) shows coupling constants which, although somewhat reduced with respect to the usual activity of T states, do not lead to totally negligible values of E_{JT}. Again therefore this is an unrealistic description of the system since, as shown by Van Vleck[11] and herein, one would expect E_{JT} here to be reduced to only one-hundredth of its value in the parent $^5T_{2g}$ state if and when spin-orbit coupling should be substantial.

Similarly the coupling constants calculated for the d^3 $(^4F_{3/2})$ and d^8 $(^3F_4)$ systems, although quite small, do not lead to totally negligible E_{JT} values as might be expected because of the strong correlation of these levels in the strong field limit with the orbitally non-degenerate state $\Gamma_8(^4A_{2g}, t_{2g}^3)$ and $\Gamma_5(^3A_{2g}, t_{2g}^6e_g^2)$ respectively. Once more then the $|LSJM_J\rangle$ basis proves to be less than adequate since in the strong field scheme all the Jahn-Teller matrix elements within these Γ_8 and Γ_5 levels vanish by virtue of spin-orthogonality, and a calculation for the d^3 case (vide infra), with specific allowance for spin-orbit mixing with excited states in the strong field scheme, predicts only very small values indeed of E_{JT}. It may be concluded therefore that the $|LSJM_J\rangle$ basis, except possibly as applied to *complete* LS state manifolds, is not a satisfatory basis for the description of ML_6 d-orbital systems but that, certainly for the 3d series, the ordinary weak field $|LSM_LM_S\rangle$ basis is reasonably adequate, allowing where appropriate for state mixing.

Turning now to the strong field results, the calculations neglecting spin-orbit coupling are entirely unremarkable, simply illustrating the complete separation of the σ- and π-contributions into the e_g and t_{2g} sets respectively. For the d^2 and high-spin d^7 cases allowance for state mixing may be made as previously described, and can also be applied with advantage in the low-spin d^4 and low-spin d^5 situations. (See Table 8 where the necessary matrix elements in the strong field basis are listed.) For the low-spin d^7 system with a $^2E_g\,(t_{2g}^6e_g^1)$ ground level state mixing is unlikely to be significant since only *doubly* excited ^2E_g states arise.

In the strong field O* basis the effects of spin-orbit coupling can very easily be determined as shown in Sect. 2, and may be incorporated either within a complete spin × orbital basis or within the separate O* components produced, assuming these to be well separated. With this latter assumption the extent of reduction of the Jahn-Teller coupling constants, relative to those of the parent O_h states, are shown in Table 7, together with the corresponding reduction factors for the magnitude of E_{JT}. Note that these values are entirely symmetry determined and depend only on the orbital symmetry of the given state and the value of the coupled spin, S.

Of the states listed in Table 7 those of A_1 and A_2 orbital symmetry remain inactive and require no further comment. The ^5E level shows no first order splitting due to spin-orbit coupling so that the simple O_h situation is essentially undisturbed whilst the Γ_8 level derived from a ^2E state shows no reduction in the value of E_{JT}. The results for the various T_1 and T_2 states however indicate that in these cases substantial diminution of Jahn-Teller activity may ensue when spin-orbit effects are large.

Thus for the Γ_8 ground level resulting for d^1, $^2T_{2g}\,(t_{2g}^1)$ the Jahn-Teller stabilization energy for both couplings is reduced to one-quarter of its value in the parent state; for the d^2 $^3T_{1g}\,(t_{2g}^2)$ ground state spin-orbit splitting leads to a $\Gamma_3(E)$ and a $\Gamma_5(T_2)$ level being almost equienergetic and lying lowest, but for all the possible couplings E_{JT} is again reduced to a quarter of its original value. For the low-spin d^4 ground state, $^3T_{1g}\,(t_{2g}^4)$, however, spin-orbit coupling leads to an inactive $\Gamma_1(A_1)$ component as the lowest lying level so that when ζ is large the Jahn-Teller activity will tend to be quenched completely. Moreover, for the low-spin d^5 situation a similar result ensues since in O* symmetry the lowest lying component of $^2T_{2g}\,(t_{2g}^5)$ is a Jahn-Teller impotent Kramers' doublet, Γ_7. The high-spin d^6 ground state, $^5T_{2g}\,(t_{2g}^4e_g^2)$, previously considered by Van Vleck[11)], also suffers a marked quenching of Jahn-Teller activity since the lowest lying O* component, $\Gamma_5(T_2)$ $(J = 1)$, exhibits only one-hundredth of the previous value of E_{JT}, whilst the high-spin d^7 ground state, $^4T_{1g}\,(t_{2g}^5e_g^2)$, yields another inactive Kramers' doublet, Γ_6, as its lowest O* level. In those cases for which ζ is large therefore E_{JT} will be subject to substantial quenching along the lines shown in Table 7, although a consideration of the full spin × orbital ground state manifold would be required to treat the situation in which ζ and E_{JT} were of comparable magnitude.

To summarise therefore either the weak field or the strong field basis may be used in O_h symmetry for those situations in which spin-orbit coupling is relatively small as long as allowance is made for state mixing in the cases for which this is likely to be significant. The weak field $|LSJM_J\rangle$ basis is however unsatisfactory for the treatment of those d^x systems in which spin-orbit coupling is either substantial or dominant. Nevertheless, under these circumstances all the required matrix elements within the O* ground state spin × orbital manifold may readily be found in the strong field basis, being symmetry determined, in terms of the simple O_h coupling constants.

A useful check upon the calculation of the various Jahn-Teller coupling constants is afforded in some cases by the application of the diagonal sum rule and its fulfillment is illustrated in Table 9 for two simple cases, namely the d^1 O* situation in both the weak and strong field schemes and the d^2 O_h case for the triplet (S = 1) states, again in both the weak and strong field bases. In the former case it is interesting to note that in the weak field scheme the σ- and π-contributions are scrambled together whereas in the strong field basis the separation of these terms between the e_g and t_{2g} orbital sets remains inviolate. Similarly, in the latter case, this same separation of e_g (σ) and t_{2g} (π) contributions, which is reflected in the strong field coupling constants, is destroyed in the weak field basis.

Finally, as an example illustrative of many of the points discussed above, it is instructive to calculate the magnitude of the Jahn-Teller stabilization energy, E_{JT}, for the ground state of a d^3 system. When spin-orbit coupling is neglected this system corresponds to the $^4A_{2g}$ component of ^4F in the weak field basis or to the $^4A_{2g}(t_{2g}^3)$ level in the strong field scheme: in both cases this orbital singlet will be Jahn-Teller impotent. When spin-orbit coupling is large however the weak field ground state becomes $^4F_{3/2}$ which yields a single Γ_8 level whose coupling constants are shown in Table 5. This level correlates quite strongly with the strong field Γ_8 O* level derived from $^4A_{2g}(t_{2g}^3)$, but whereas the latter is, to, the first order, inactive due to spin orthogonality, the weak field $^4F_{3/2}\,\Gamma_8$ level yields coupling constants which, although obviously small, may nevertheless produce quite appreciable E_{JT} values in some cases. As an example one may treat the $[ReF_6]^{2-}$ anion since in the 5d series a rather large value of ζ would be anticipated. In fact the experimental data are well fitted[17] using Dq = 3,280 cm^{-1} and ζ = 2,550 cm^{-1}, so that, assuming $e_\pi = 0.20\ e_\sigma$, one has e_σ = 14,909 cm^{-1} and e_π = 2,982 cm^{-1}. Taking the ionic radii of Re^{4+} and F^- as 0.68 Å and 1.36 Å respectively gives R = 2.04 Å, so that on the simple electrostatic model one has $\dot{\sigma} = -3.654 \times 10^{12}\ cm^{-2}$ and $\dot{\pi} = -0.731 \times 10^{12}\ cm^{-2}$. For the force constants for the ε_g and τ_{2g} modes respectively one may without great error assume the values listed[14] for the $5d^6$ anion, $[PtF_6]^{2-}$, these being 3.82 mdyne/Å and 0.52 mdyne/Å using the MOVFF model. (Multiply by 5.035×10^{20} to convert to cm^{-1}.) With these figures the $^4F_{3/2}$ ground state yields E_{JT} values for the $\Gamma_8 \otimes \varepsilon_g$ and $\Gamma_8 \otimes \tau_{2g}$ couplings of 70.5 cm^{-1} and 82.9 cm^{-1} respectively.

On the whole these stabilization energies, although small, seem rather too large for what is essentially an orbital singlet ground state and are indeed not so very much smaller than the E_{JT} values estimated by Bacci[1] for a t_{2g} orbital degeneracy. Since however it is known that the $|LSJM_J\rangle$ basis set is unsatisfactory for various other cases the alternative strong field O* calculation should prove more informative. Here the ground Γ_8 level remains inactive to the first order and Jahn-Teller activity is only introduced by allowing for spin-orbit mixing with excited Γ_8 levels. From Table A 34 of Griffith[10] it is seen that only two such levels in fact mix in this way, namely $^2T_{2g}(t_{2g}^3)$ and $^4T_{2g}[3/2](t_{2g}^2e_g^1)$, the spin-orbit matrix elements connecting them with the $^4A_{2g}(t_{2g}^3)$ ground level being $+\zeta$ and $-(1/3)\sqrt{15}\ \zeta$ respectively.

The wave functions for the $^2T_{2g}(t_{2g}^3)$ level are given by Griffith[10] in his Table A 24, whilst those for $^4T_{2g}(t_{2g}^2e_g^1)$ are easily constructed using his Table A 20. It is then readily shown that the $^2T_{2g}(t_{2g}^3)$ state is Jahn-Teller inactive for both ε_g and τ_{2g} coupling whereas

Table 9. Illustrations of the Diagonal Sum Rule for the Jahn-Teller Coupling Constants of ML_6, O_h and O^* Systems

(A) d^1, O^*		Strong Field			Weak Field
$^2T_{2g}(t_{2g}^1)$	$\Gamma_8 \otimes \varepsilon_g$	$A_\varepsilon = - (1/3)\sqrt{3}\dot{\pi}$	$^2D_{3/2}$	$\Gamma_8 \otimes \varepsilon_g$	$A_\varepsilon = - (1/5)\sqrt{3}\dot{\sigma} - (1/5)\sqrt{3}\dot{\pi}$
$^2E_g(e_g^1)$	$\Gamma_8 \otimes \varepsilon_g$	$A_\varepsilon = - (1/2)\sqrt{3}\dot{\sigma}$	$^2D_{5/2}$	$\Gamma_8 \otimes \varepsilon_g$	$A_\varepsilon = - (3/10)\sqrt{3}\dot{\sigma} - (2/15)\sqrt{3}\dot{\pi}$
	ε_g	$- (1/2)\sqrt{3}\dot{\sigma} - (1/3)\sqrt{3}\dot{\pi}$		ε_g	$- (1/2)\sqrt{3}\dot{\sigma} - (1/3)\sqrt{3}\dot{\pi}$
$^2T_{2g}(t_{2g}^1)$	$\Gamma_8 \otimes \tau_{2g}$	$A_\tau = + (2/3)\sqrt{3}\pi$	$^2D_{3/2}$	$\Gamma_8 \otimes \tau_{2g}$	$A_\tau = + (2/5)\sqrt{3}\sigma + (2/5)\sqrt{3}\pi$
			$^2D_{5/2}$	$\Gamma_8 \otimes \tau_{2g}$	$A_\tau = - (2/5)\sqrt{3}\sigma + (4/15)\sqrt{3}\pi$
	τ_{2g}	$+ (2/3)\sqrt{3}\pi$		τ_{2g}	$+ (2/3)\sqrt{3}\pi$

(B) $d^2(S = 1)$, O_h		Strong Field			Weak Field
$^3T_{1g}(t_{2g}^2)$	$\otimes\ \varepsilon_g$	$C = + (2/3)\sqrt{3}\dot{\pi}$	$^3T_{1g}(^3F)$	$\otimes\ \varepsilon_g$	$C = - (1/10)\sqrt{3}\dot{\sigma} + (2/5)\sqrt{3}\dot{\pi}$
$^3T_{2g}(t_{2g}^1e_g^1)$	$\otimes\ \varepsilon_g$	$C = + (1/2)\sqrt{3}\dot{\sigma} - (2/3)\sqrt{3}\dot{\pi}$	$^3T_{2g}(^3F)$	$\otimes\ \varepsilon_g$	$C = + (1/2)\sqrt{3}\dot{\sigma} - (2/3)\sqrt{3}\dot{\pi}$
$^3T_{1g}(t_{2g}^1e_g^1)$	$\otimes\ \varepsilon_g$	$C = - (1/2)\sqrt{3}\dot{\sigma} - (2/3)\sqrt{3}\dot{\pi}$	$^3T_{1g}(^3P)$	$\otimes\ \varepsilon_g$	$C = - (2/5)\sqrt{3}\dot{\sigma} - (2/5)\sqrt{3}\dot{\pi}$
	ε_g	$- (2/3)\sqrt{3}\dot{\pi}$		ε_g	$- (2/3)\sqrt{3}\dot{\pi}$
$^3T_{1g}(t_{2g}^2)$	$\otimes\ \tau_{2g}$	$B = - 2\pi$	$^3T_{1g}(^3F)$	$\otimes\ \tau_{2g}$	$B = + (6/5)\sigma - (9/5)\pi$
$^3T_{2g}(t_{2g}^1e_g^1)$	$\otimes\ \tau_{2g}$	$B = - \pi$	$^3T_{2g}(^3F)$	$\otimes\ \tau_{2g}$	$B = - \pi$
$^3T_{1g}(t_{2g}^1e_g^1)$	$\otimes\ \tau_{2g}$	$B = - \pi$	$^3T_{1g}(^3P)$	$\otimes\ \tau_{2g}$	$B = - (6/5)\sigma - (6/5)\pi$
	τ_{2g}	$- 4\pi$		τ_{2g}	$- 4\pi$

the $^4T_{2g}(t_{2g}^2e_g^1)$ state yields the coupling constants

$$B = +\pi ; \qquad C = -(1/2)\sqrt{3}\dot{\sigma} + (2/3)\sqrt{3}\dot{\pi} .$$

The contribution of the $^2T_{2g}(t_{2g}^3)$ level to the Jahn-Teller activity is therefore nil and any cross product terms vanish due to spin orthogonality so that only the $^4T_{2g}(t_{2g}^2e_g^1)$ state need be considered. Thus, by perturbation theory, one may to a good approximation write the Γ_m component ($m = \varkappa, \lambda, \mu, \nu$) of the Γ_8 d^3 O* ground state as

$$|\Gamma_8, \Gamma_m\rangle = |\Gamma_8, {}^4A_{2g}, \Gamma_m\rangle + (1/3)\sqrt{15}(\zeta/\Delta E)|\Gamma_8, {}^4T_{2g}[3/2], \Gamma_m\rangle$$

where $\Delta E = E(^4T_{2g}) - E(^4A_{2g})$. From Table 7 it is seen that for the $^4T_{2g}[3/2]\Gamma_8$ level E_{JT}, for either coupling, is reduced by a factor of (4/25). Moreover, from Table A 20 of Griffith[10] it is easily found that all the $\langle ^4A_{2g} \| ^4T_{2g}\rangle$ cross product terms vanish so that one may write

$$E_{JT}(\Gamma_8) = (15/9) \cdot (\zeta/\Delta E)^2 \cdot (4/25)\, E_{JT}(^4T_{2g})$$

The E_{JT} values for the coupling of $^4T_{2g}(t_{2g}^2e_g^1)$ to either the ε_g or the τ_{2g} mode may at once be found from the above expressions for C and B and from the parameters evaluated previously and prove to be 2,327 cm^{-1} for $^4T_{2g} \otimes \varepsilon_g$ and 90.7 cm^{-1} for $^4T_{2g} \otimes \tau_{2g}$. Thus one easily finds that for the Γ_8 ground state E_{JT} for the ε_g coupling is 2.25 cm^{-1} and for the τ_{2g} coupling 0.08 cm^{-1}. In both cases therefore the activity of the Γ_8 levels proves as expected to be insignificant: taking[14] $h\omega$ for the ε_g mode as 576 cm^{-1} and for the τ_{2g} mode as 210 cm^{-1}, the $E_{JT}/h\omega$ ratios are only 0.004 and 0.0004 respectively. Note however that this calculation is not altogether trivial since Γ_8 levels are in general especially sensitive to the operation of the Ham effect[18, 19] and that even $E_{JT}/h\omega$ ratios as small as 0.10 can result in appreciable quenching of angular momentum contributions and significant magnetic anomalies[20].

4 Conclusions

As noted throughout the calculation of the Jahn-Teller stabilization energies via the Angular Overlap Model, although not capable of the highest precision, does afford a satisfactory method for obtaining good estimates of the order of magnitude of E_{JT}. Furthermore, it produces results of comparable accuracy to those obtained by more intricate procedures whilst itself retaining the merit of extreme simplicity and may thus usefully be employed to deduce approximate values for the strength of the Jahn-Teller coupling in various cases as reflected by the ratio $E_{JT}/h\omega$.

The greatest utility of the Angular Overlap Model lies however in its ability to comprehend the Jahn-Teller activity in terms of a simple one-electron operator. The matrix elements of this operator must of course first be determined within the chosen basis set for the required symmetry and stoichiometry and the method is here applied to give a comprehensive coverage for d^x systems resulting from ML_6, O_h, species. In principle however the technique is applicable to any situation in which orbital degeneracy persists, thereby producing potentially Jahn-Teller active levels, for example the tetragonal d^x ML_4 or trans-$ML_4L'_2$ systems, both of D_{4h} symmetry. Moreover, it is generally possible to adopt either a weak or a strong field formalism, with appropriate allowance for the effects of state mixing, and if desired to include the effects of spin-orbit coupling in a very simple fashion. Most importantly, the technique allows an easy comparison to be made of the Jahn-Teller activities of any given set of electronic states.

5 References

1. Bacci, M.: Chem. Phys. Letters *58*, 537 (1978)
2. Bacci, M.: Chem. Phys. *40*, 237 (1979)
3. Warren, K. D.: Inorg. Chem. *19*, 653 (1980)
4. Warren, K. D.: ibid. *21*, 3467 (1982)
5. Bersuker, I. B.: Coord. Chem. Rev. *14*, 357 (1975)
6. Liehr, A. D.: J. Phys. Chem. *67*, 389 (1963)
7. Warren, K. D.: Chem. Phys. Letters *89*, 395 (1982)
8. Stevens, K. W. H.: Proc. Phys. Soc. (London) *A 65*, 209 (1952)
9. Hutchings, M. T.: Solid State Phys. *16*, 227 (1964)
10. Griffith, J. S.: The Theory of Transition Metal Ions, Cambridge University Press, London 1961
11. Van Vleck, J. H.: Physica *26*, 544 (1960)
12. Smith, D. W.: Structure and Bonding *35*, 87 (1978)
13. Burns, G., Axe, J. D.: J. Chem. Phys. *45*, 4362 (1966)
14. Labonville, P., Ferraro, J. R., Wall, M. C., Basile, L. J.: Coord. Chem. Rev. *7*, 257 (1972)
15. Van Vleck, J. H.: J. Chem. Phys. *7*, 79 (1939)
16. Liehr, A. D., Ballhausen, C. J.: Ann. Phys. (New York) *3*, 304 (1958)
17. Allen, G. C., Warren, K. D.: Structure and Bonding *19*, 105 (1974)
18. Ham, F. S.: Phys. Rev. *138*, A 1727 (1965)
19. Ham, F. S.: ibid. *166*, 307 (1968)
20. Warren, K. D.: Chem. Phys. Letters *99*, 427 (1983)

The Composition, Structure and Hydrogen Bonding of the β-Diketones

John Emsley

Department of Chemistry, King's College, Strand, London WC2R 2LS, U.K.

Proton transfer and hydrogen bonding are two aspects of the chemistry of hydrogen that respectively govern the behaviour and structure of many molecules, both simple and complex, from water to DNA. The β-dicarbonyls exhibit both of these features, and in ways which have singled them out for detailed study for many years. They provide the best known examples of keto $\leftrightharpoons$ enol tautomerism, with the advantage of slow proton transfer and high concentrations of the enol tautomers in most cases. These enols are stabilized by intramolecular OHO hydrogen bonds which at various times have been thought of as being centred, linear hydrogen bonds that are somehow incorporated into the delocalised π system to give "aromatic" systems. Research involving structural, spectroscopic and computational techniques has deepened our understanding of these compounds and changed our picture of them. The hydrogen bonding is surprisingly strong, surprising since it is neither centred, nor linear, nor involved in the ring's delocalized bonding, although it is certainly coupled to it. This review deals with controversies that have surrounded the β-dicarbonyls and discusses the current view that the enol tautomers are a double-minimum potential well with a low energy barrier. The review ends with a brief look at the β-thioxoketones, which provide an analogous system based on a heteronuclear SHO hydrogen bond in the enol or enethiol tautomer.

List of Abbreviations

The abbreviations which refer to β-dicarbonyl compounds are meant to refer to all those species which comprise the β-dicarbonyl system, i.e. the keto and the various enol tautomers.

AA	acetylacetone; 2,4-pentanedione; $MeCOCH_2COMe$
BA	benzoylacetone; $PhCOCH_2COMe$
αClAA	α-chloroacetylacetone; 3-choro-2,4-pentanedione; MeCOCH(Cl)COMe
DBM	dibenzoylmethane; $PhCOCH_2COPh$
DMAA	dimethylacetylacetone; 3,3-dimethyl-2,4-pentanedione; $MeCOCMe_2COMe$
^2H_2AA	deuteriated acetylacetone; $MeCOCD_2$-COMe
HFAA	hexafluoroacetylacetone; CF_3COCH_2-$COCF_3$
IR	infrared
MAA	methylacetylacetone; 3-methyl-2,4-pentanedione; MeCOCH(Me)COMe
MDA	malondialdehyde; propanedial; $HCOCH_2COH$; the end form of this is sometimes referred to as β-hydroxyacrolein or 3-hydroxy-2-propenal, HCOCH=C(OH)H.
NMR	nuclear magnetic resonance
R	Raman
TAA	thioacetylacetone; $MeCSCH_2COMe$
TAE	tetraacetylethane; $CH(COMe)_2CH$-$(COMe)_2$
TFAA	trifluoroacetylacetone; CF_3COCH_2-COMe
UV	ultraviolet

I. Introduction

β-Dicarbonyl compounds have, over the years, been of considerable interest to organic, inorganic and physical chemists. The organic chemists have made use of them chiefly as β-ketoesters in such reactions as the malonic ester synthesis, the Knoevenagel condensation, the Michael addition and the reverse Claisen condensation. The inorganic chemist on the other hand has found β-diketones to be useful chelating ligands, the acetylacetonate anion especially so. Physical chemists have used them to study the keto ⇆ enol equilibrium which is best exemplified by the β-diketones, but can be displayed by any molecule with the COCHXCO subunit (or the CHXCO subunit in theory). These aspects of the β-dicarbonyls are covered in advanced textbooks of the respective branches of chemistry. This review concentrates principally on the β-diketones and β-ketoaldehydes – their tautomeric composition, structure and bonding.

The β-dicarbonyls have been a source of controversy for decades. Early this century the argument centred round the possibility of tautomerism and the extent to which enolization could occur[1], Fig. 1(I). Later came the suggestion that acetylacetone (AA) complexes had benzenoid character around the chelate ring:[2] Fig. 1(II). This was subsequently refuted[3], but the idea of a π ring system continued when attention shifted to the hydrogen bonded enol tautomer which was reported to be strong and centred:[4] Fig. 1(III). The use of the terms enol chelate or proton chelate derives from this supposed analogy with metal chelates.

We shall be concerned with compounds of general formula $RCOCH_2COR'$, and their mono-α-substituted derivatives, where R and R′ can be H or any organic groups or organic framework such as part of a ring system. We thus encompass β-ketoaldehydes, including malondialdehyde (MDA) R = R′ = H, as well as β-diketones but not β-ketoesters (R = OR″). Attention will focus chiefly on the cis enol forms with their intramolecular hydrogen bonding, seeking to explain the nature of this unusual kind of hydrogen bond and showing the effect it has on the composition of the equilibrium, the structure, reactions and spectroscopic properties of these remarkable compounds. The final section of the review (VII) will deal with the β-thioxoketones in which a sulfur atom replaces one of the carbonyl oxygens. This change brings an extra degree of complexity

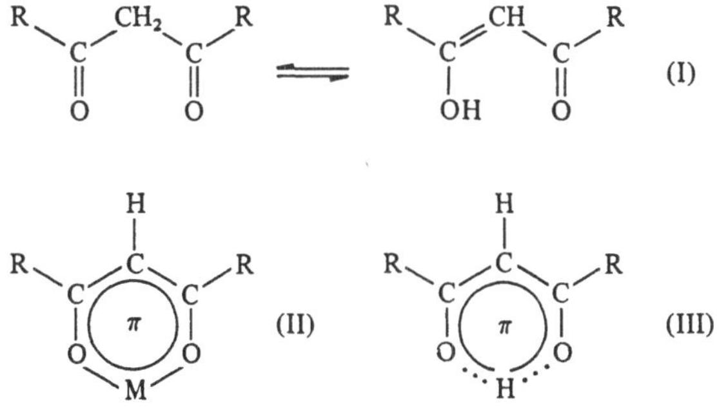

Fig. 1 (M=metal atom)

with the possibility of enethiol as well as enol tautomers. The intramolecular OHS hydrogen bonds in these are similar in many respects, including strength, to the OHO bonds of the β-diketone counterparts.

A. Tautomers and Conformers

The term keto ⇋ enol tautomerism implies two isomers and for most purposes this simple view is adequate. However for a symmetric β-diketone, $RCOCH_2COR$, or for MDA, $HCOCH_2COH$, there are three diketo conformers and six enol ones, assuming that is, that the skeletal framework of oxygens and carbon atoms is planar: Fig. 2. For an asymmetric β-diketone or β-ketoaldehyde this increases to four and eleven respectively since it is possible to distinguish the various conformers of RC(OH)=CHCOR′ from those of RCOCH=C(OH)R′. The same is true of symmetric β-thioketones, $RCSCH_2COR$.

There are several ways of naming these various conformers. They can be designated cis/trans, U/S/W or E/Z. But since in reality only a limited number of conformers will be encountered the terms cis and trans will be used to indicate conformers that have an intramolecular hydrogen bond and those that have not. The symmetry of the hydrogen bond, whether it be centred or non-centred, will be shown by using C_{2v} or C_s respectively. The U/S/W and Z,Z/Z,E/E,Z/E,E notation will be reserved for the relatively rare occasions when the molecular framework itself is under discussion.

Which conformers of Fig. 2 are important depends upon the type of β-dicarbonyl compound. For β-diketones two predominate to the virtual exclusion of the others: these are the cis diketo and the cis enol. Each has a sufficiently distinguished set of proton signals in the ^1H NMR spectrum to make integration a reliable measure of molar proportions, and hence K, [enol]/[keto], can be easily obtained.

The β-ketoaldehydes on the other hand can display appreciable contributions of the trans enol tautomers, detectable at room temperature by ^1H NMR spectroscopy in some cases[5, 6]. MDA in $CDCl_3$ is mainly trans[6], and in water and other hydroxylic solvents it is entirely trans[6, 7, 8]. Interconversion between the enol tautomers is slow on the NMR time scale. Enolization occurs chiefly at the formyl side, COH, but the energy difference between the two cis enol conformers is less than 1.5 kJ mol^{-1}[5]. The trans enol tautomer was deduced to be of the S conformer so that for the benzoyl derivative the two species in solution were A(i) and A(ii) with an enthalpy difference of 11 kJ mol^{-1} favouring the cis enol[5].

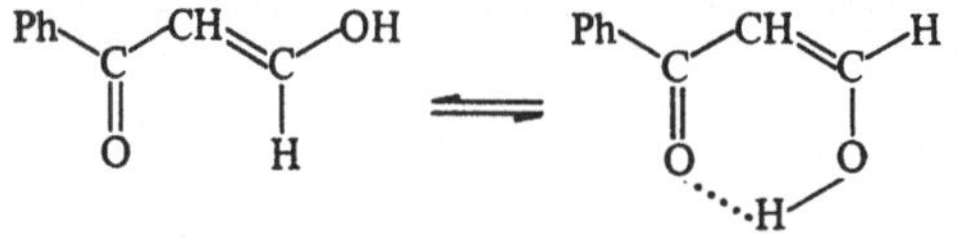

A(i) A(ii)

The trans A(i) makes a significant contribution even at − 80 °C. Some solvents that are both polar and basic, such as pyridine and DMF, increase the concentration of A(i), but neutral, non-polar solvents, such as CCl_4, favour A(ii) sometimes to the exclusion of the trans tautomer.

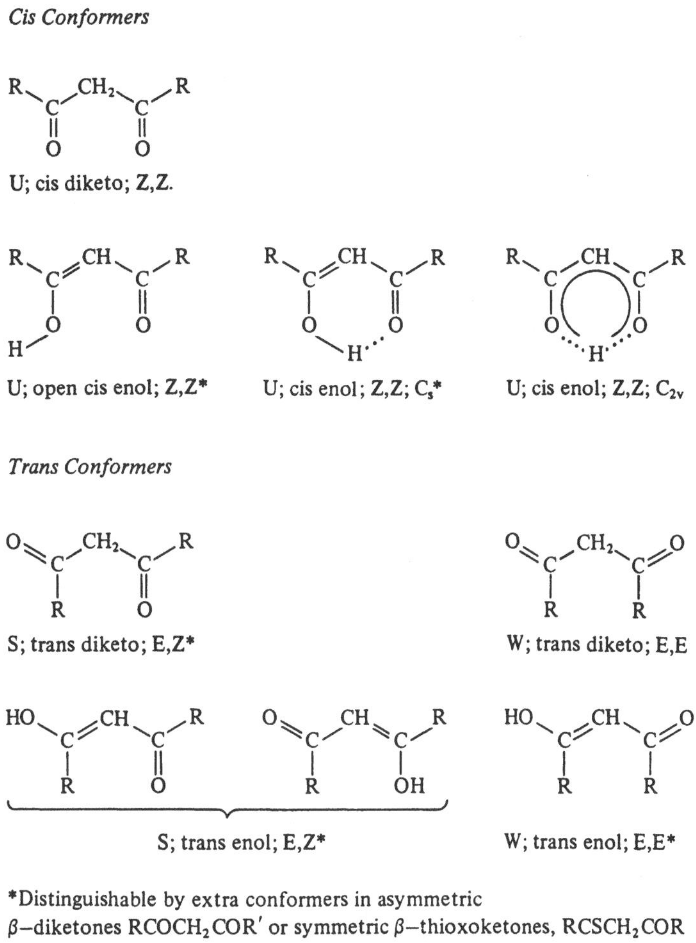

Fig. 2. The Conformers of a β-Diketone, $RCOCH_2COR$

The presence of a methyl group on the α carbon destabilizes the cis enol relative to the trans and the latter can then be observed even in neutral, non-polar solvents[5]. These effects of solvent and substituents on tautomerization are also principal factors in the enolization of the β-diketones – see below.

Within the β-diketone system three types of proton transfer can be distinguished:

(i) the relatively slow exchange between the vinylic CH and the enol OH sites, i.e. the keto ⇆ enol tautomerism, which takes hours, even days, to reach equilibrium,

(ii) the rapid exchange of the labile proton between OH groups on different molecules which is too fast to be distinguishable by ^1H NMR spectroscopy under normal conditions; and

(iii) the extremely rapid transfer of the proton from one oxygen of the cis enol tautomer to the other, i.e. the oscillation of the proton between the two minima of the potential energy well of the OHO hydrogen bond.

These three motions are effectively independent of one another since they differ in magnitude by at least 10^6 in each case, i.e. (i)/(ii) and (ii)/(iii).

The slow keto ⇆ enol proton transfer means separate signals in the NMR spectrum for the tautomers. The second exchange (ii) is responsible for the line broadening and loss of multiplet structure of the ^{1}H NMR signal of the enol proton. The third type of proton motion, (iii), is not resolvable by NMR so that ways around this have been sought in order to obtain a time-averaged analysis of the proton's location in the cis enol.

Despite the drawbacks NMR spectroscopy (^{1}H, ^{13}C, even ^{17}O) has been used more than any other technique in probing the riddle of the cis enol hydrogen bond.

In hydrocarbon solvents, β-diketones are predominantly (> 90%) enolized and these solutions have been subjected to flash photolysis, which causes photoisomerization to the diketo form[9]. Reversion of the diketone to the more stable cis enol was then followed by UV spectroscopy and pseudo first order rate constants at room temperature were in the range 14–68 × 10^{-3} s^{-1} (e.g. AA = 23 × 10^{-3} s^{-1}) with half-lives of several hours. The same research also reveals an alternative transformation on irradiation. In this the cis enol form is converted by rotation about a C–C or C=C bond into one of the possible trans enol isomers. These may then go on to the diketo form but mainly they revert very rapidly to the cis enol with rate constants of 0.1 to 70 s^{-1} (e.g. AA = 0.27 s^{-1})[9].

Though the proton transfer of the keto ⇆ enol equilibrium is only indirectly related to the hydrogen bond in the cis enol tautomer, nevertheless the two are linked, and the driving force of the former is attributed to the unusual strength of the latter. Chemists who probed the equilibrium almost invariably felt compelled to comment on the OHO bond, at least to the extent of discussing the chemical shift of the hydrogen bond proton, δ(OHO), and the factors affecting it. The composition of a β-dicarbonyl system is however a separate issue from the structure of the cis enol and the nature of the hydrogen bonding therein. This review will deal with these issues in this order.

II. The Keto ⇆ Enol Equilibrium Proton Transfer

In 1971 Kol'tsov and Kheifets were the authors of a review of all type of tautomerism and this is a useful summary of the intensive efforts that went into tackling the problem in the 1950s and 1960s[10]. The keto ⇆ enol equilibrium has three features of interest

(i) the composition at equilibrium,

(ii) the enthalpy (ΔH_{enol}) and entropy (ΔS_{enol}) of enolization, and

(iii) the rate of enolization or the rate of exchange.

Points (ii) and (iii) will be dealt with later in this review.

A. The % Enol at Equilibrium

The composition of a β-dicarbonyl system is usually expressed as the molar percentage of the enol tautomer at equilibrium, rather than as the equilibrium constant K ([enol]/[keto]). The amount of enol is influenced by a variety of factors:

(a) solvent;
(b) temperature;
(c) the presence of other species that are capable of hydrogen bonding;
(d) α-substituents;
(e) β-substituents; and, maybe,
(f) deuteriation.

Since most research has been carried out on AA, what follows will deal primarily with this β-diketone.

(a) Solvent Effects

The lower the polarity of the solvent, the higher the percentage of the enol tautomer. Table 1 lists the solvents for which data have been reported arranged in order of the

Table 1. Acetylacetone: Variation of composition at equilibrium with solvent and the chemical shift of the enol proton, δ(OHO)

Solvent	ε[a]	% cis enol tautomer	Ref.	δ(OHO) (ppm Me_4Si)	Ref.	Solvent type[b]
[gas phase	0	92	11]			
cyclo C_6H_{12}	2.0	97; 95	12; 13	15.6	16	neutral
dioxan	2.2	82	14	14.77	14	acceptor
CCl_4	2.2	95; 94	12; 13	15.05	14	neutral
C_6H_6	2.3	89	14	16.08	14	weak acceptor
Et_3N	2.5	100	15	15.5	16	strong acceptor
CS_2	2.6	94	14	15.13	14	neutral
Et_2O	4.3	95	14	15.05	14	acceptor
$CHCl_3$	4.8	87	15	15.07	14	weak donor
CH_3CO_2H	6.2	67; 73	14; 15	14.87	14	strong donor and acceptor
C_5H_5N	12.3	82	17			acceptor
Me_2CO	20.7	75; 78	13; 17	15.58	18	acceptor
EtOH	24.3	82	14	5.48[c]	14	acceptor and donor
neat	25.7	79; 81	13; 12; 14	15.40	14	–
MeOH	32.6	74	12; 14	4.98[c]	14	acceptor and donor
MeCN	36.2	62	14	15.57	14	weak acceptor
DMF	37.8	66	12			acceptor
DMSO	46.6	62; 60	12; 14	14.27	14	acceptor
H_2O	78.5	16	12	[c]		strong donor and acceptor

[a] Weast R.C. ed.: *Handbook of Chemistry and Physics,* 60 edn. 1980, E 56–E 58, CRC Press, Boca Raton, Florida, USA
[b] Solvent classified according to their hydrogen bonding abilities
[c] Proton exchange between AA and solvent

relative permittivity, ε, of the solvent. There is an approximately linear, inverse relationship between ε and the percentage enol, except for solvents such as dioxan and acetic acid whose polarity as solvents is not reflected in their ε values[19,20].

Another influence of the solvent on the keto $\leftrightarrows$ enol equilibrium should be the hydrogen bonding propensity of the solvent itself. Hydrogen bond donors should interact better with the keto tautomer while hydrogen bond acceptors may possibly compete for the enol proton's attentions. The only purely donor solvent is $CHCl_3$ and this has slightly less enol than expected for its low ε value. The purely acceptor solvents show more enol than expected.

Table 1 also lists the δ(OHO) values of AA in the various solvents. There is little correlation with either ε or the hydrogen bonding potential of the medium. The signal furthest downfield occurs in benzene solution and probably arises due to anisotropy effects of the benzene ring which is in close association with the enol tautomer[14].

Investigations on the solvent's effect on other β-diketones have not been carried out in any systematic fashion. Studies on malondialdehyde and acetylacetaldehyde, $MeCOCH_2COH$, are complicated by other tautomers; the cis enol forms are predominant in CCl_4, $CHCl_3$ and CH_2Cl_2 solutions, but in water the trans enol form was present[7]. For MDA there was no evidence of the diketo form even in water, but for acetylacetaldehyde there was 12% of the keto tautomer in $CHCl_3$ and 76% in water where it was present as the hydrated diol species.

(b) Temperature; Enthalpy and Entropy of Enolization

As the temperature increases the amount of the enol tautomer decreases. Published data is rather sparse. Investigations which observed β-diketones over a range of temperature were not all designed to measure changes in the enol : keto ratio, but to observe the upfield shift of δ(OHO), see later. Those which were designed to measure the enthalpy of enolization rarely gave the actual variation of composition with temperature but Table 2 shows the effect on AA and ^2H_2AA, although some of the data[21] should be treated with caution[23]. Table 3 gives the enthalpies and entropies of enolization. The discrepancy in the two values for MAA should be noted.

(c) Other Dissolved Species

Other hydrogen bond acceptors have been added to solutions of β-diketone systems to determine the effect their presence should have on the equilibrium. These are the

Table 2. Effect of temperature on the keto $\leftrightarrows$ enol equilibrium of AA and ^2H_2AA

AA	Temp./°C	−19	2.5	33	37.3
	% enol	89.5	86.4	79	78.2
	Ref.	21	21	22	21
^2H_2AA	Temp./°C	−15	0	26	50
	% enol	88	86	81	75
	Ref.	23	23	23	23

Table 3. Enthalpies and entropies of enolization of β-Diketones

β-Diketone	$\Delta H/\text{kJ mol}^{-1}$	$\Delta S/\text{J mol}^{-1} \text{ deg}^{-1}$	Ref.
AA	−11.9; −11.7	−27.9; −30.5	22; 13
MeCOCH(Cl)COMe	−24.8	−59.0	22
MAA	−55.6; − 8.9	−23.6; −30.5	22; 13
BA	−11.7	−16.3	13
DBM	−13.4	−14.2	13
HFAA	− 5.4	−25.5	13
MeCOCH(Ph)COMe	−10.5	−35.1	13

fluoride ion, water and triethylamine. All should encourage the enol to form trans conformers or open cis conformers by offering an alternative hydrogen bonding receptor atom.

The effect of the fluoride ion is dramatic and to the extent of posing more questions than it answers. Clark[24] showed that dissolved tetraalkylammonium fluorides converted AA entirely to the enol form in solvents such as $CHCl_3$ and even in DMF. Crystalline complexes, $R_4NF \cdot AA$, can be isolated and this is a useful method for dehydrating these fluoride salts which are notoriously hygroscopic. In the presence of F^-, AA can be easily alkylated at the α position with alkyl iodides[24].

More remarkably Clark also discovered that Bu_4NF or K(crown)F, when dissolved in MeCN containing AA, also gave 100% enol even when the mole ratio AA:F^- was 50:1[25]. In this system, and also with DBM and 2-acetylcyclopentanone, the hydrogen bonded proton's chemical shift was upfield at 2.6–4.4 ppm. This dramatic change may signify a very asymmmetric hydrogen bond formed after proton transfer to the F^-, i.e. $O^- \cdots H$–F, so that it enters the sphere of shielding of the fluoride ion.

On refluxing AA and KF in DMF for several hours intermolecular self condensation occurs Eq. (1)[26].

$$2CH_3COCH_2COCH_3 \xrightarrow[\text{DMF}]{F^-}$$

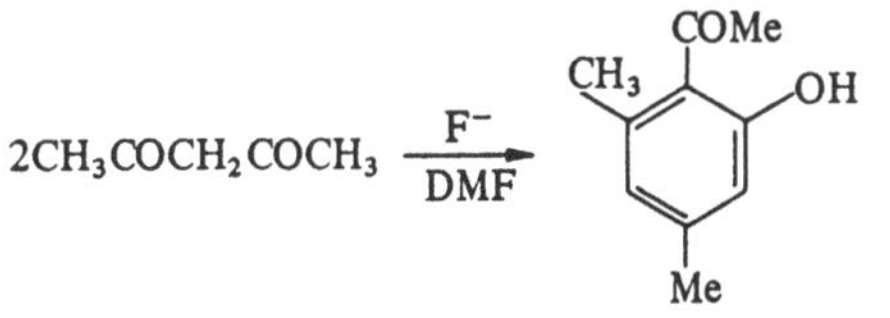

$$(1)$$

If water is added to AA dissolved in benzene it too increases the % enol to 93.1% at 5.5 °C[18]. This observation, the opposite to that when water is the bulk solvent for AA, is due to a 1:1 association of AA and H_2O. This association is thought to be between water and the enol tautomer; the equilibrium constant for the 1:1 complex was 2.4 (kg solvent mol^{-1})[2]. MAA and DMAA showed no association, but as they do and must exist only as the keto tautomers it was deduced that it was the enol tautomer that was the necessary species for interaction. This implies association via the enol OH as hydrogen bond donor.

Several diketones, AA, TFAA, BA and DBM, were dissolved in CHF_2Cl and Et_3N added[16]. At low temperatures (down to − 150 °C) two signals for the proton were distinguishable. One was identified as $\delta(OHO)$ arising from the cis enol tautomer, the other was $\delta(OHN)$ arising from proton transfer to the amine to give Et_3NH^+ which then was hydrogen bonded to the enolate anion. This association between AA and Et_3N has

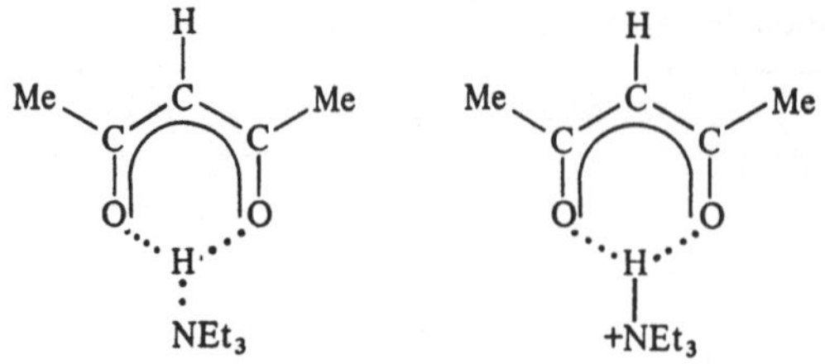

(I) (II) **Fig. 3**

no $\Delta\delta(^1H, ^2H)$ shift showing the hydrogen bond to be a normal, weak bond unlike the intramolecular bond of the cis enol[27]. Nevertheless the presence of the amine increased the percentage of the enol from 81.5% enol in neat AA to 88.5% when 0.10 mole fraction of Et_3N was present[27].

In another study of the AA/Et_3N system[28], this time also using ^{13}C and low temperature NMR, the 1:1 complex was postulated to involve a bifurcated hydrogen bond, Fig. 3(I), rather than proton transfer to the amine. On the other hand the complex formed between AA and Et_2NH, although still with a bifurcated hydrogen bond, did involve proton transfer, Fig. 3(II)[1].

(d) α-Substituents

Bulky alkyl substituents on the *α* carbon of AA leads to decreased amounts of cis enol. This is one deduction from the data in Table 4 but size alone is not the determining factor – some groups that are bulky, e.g. the chloro group, have the effect of increasing the cis enol. The bulky alkyl groups, *iso* propyl and *sec* butyl depressed the % enol almost to zero, but the enol tautomer that was present was there as the cis enol since δ(OHO) was 17.3 and 17.5 ppm respectively[36].

Table 4. The effect of *α*-substituents on the % enol tautomer of acetylacetone, MeCOCHXCOMe[a]

X	% enol tautomer	Ref.	X	% enol tautomer	Ref.
H	81	12, 13, 14	CH_2Ph	46	13
CH_3	28; 30	13; 30	Cl	92; 94	33; 30
C_2H_5	26	31	Br	46	30
$CH(CH_3)_2$	0.17	32	CN	100	34
$CH_2CH(CH_3)_2$	0.7	32	CO_2Me	100	35
$CH_2CH{=}CH_2$	42	13	SCH_3	100	33

[a] neat liquids

1 The interaction of AA and secondary (and primary) amines generally produces enaminones.
$MeCOCH_2COMe + R_2NH \rightarrow MeCOCH{=}C(NR_2)Me + H_2O$
The reaction with pyrrolidine is immediate production of the enaminone, the reaction with diethylamine is much slower[29]

In cyclic β-diketones such as 1,3-cyclopentanedione (B) bulky α-substituents favoured the keto form, but primary alkyl groups favour the enol[36].

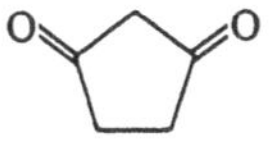

(B)

The presence of a bulky α group might be expected to force the carbonyl oxygens closer together thus strengthening any hydrogen bond between them and favouring the enol form, but at the same time there will be interaction with β-substituents that can only be alleviated by the molecule adopting a diketo conformation, S or W (E,Z or E,E).

A methyl group at the α-position depresses the % enol not only of AA as in Table 4, but also of BA from 98 to 4%, and DBM from 100 to 0%[13]. In water the % enol form of 0.2 M α-methyl acetylacetone is 2.79 ± 0.03[37].

(e) β-Substituents

Again there may be steric reasons for the % enol varying according to the nature of the β-substituents, but the relationship is not very clear – Table 5. Electron withdrawal is also a factor and this favours the enol tautomer.

β-substituents which could enhance the delocalization around the enol ring were noted also to give sharper signals in the ^{1}H NMR spectrum and this in turn was deduced as favouring a more symmetrical hydrogen bond[41]. In particular, phenyl groups in β positions should encourage centred hydrogen bonds. The effects of β-substituents are not only felt at the enol proton, the ring CH proton's chemical shift is also sensitive to changes and this too can be attributed to anisotropy changes due to changes in the π system[42].

Table 5. Effects of β-substituents on the % enol tautomer $R'COCH_2COR''$

R′	R″	% enol tautomer	solvent	Ref.
Me	Me	79; 81; 94	neat; CCl_4	13; 30; 17
Me	Ph	98; 91	CCl_4; $CDCl_3$	13; 38
Ph	Ph	100	CCl_4	13; 30
Me	CF_3	97; 100	CCl_4	13; 30
CF_3	CF_3	100	CCl_4	13; 30
Me	$C_6H_4pNMe_2$	77	$CDCl_3$	38
Me	C_6H_4pOMe	88; 95	$CDCl_3$; CCl_4	38; 39
Me	C_6H_4pMe	90; 96	$CDCl_3$; CCl_4	38; 39
Me	$C_6H_4pNO_2$	98; 100	$CDCl_3$; CCl_4	38; 39
Me	H	88	$CHCl_3$	7
CF_3	Ph	100	$CDCl_3$	40
Bu^t	Ph	100	$CDCl_3$	40
Bu^t	Bu^t	100	$CDCl_3$	40

(f) Deuteriation

Deuteriation can have a profound influence on the rate of enolization and on various properties of the OHO enol hydrogen bond but it should not affect the equilibrium constant, at least in theory[43]. Kinetic isotope effects on the keto → enol reaction, which is base catalysed, have been studied since the 1930s and for a variety of β-dicarbonyls[37, 43, 44]. For instance the rate of enolization of (C) has a kinetic isotope ratio, k_H/k_D of ca. 5[45]. The actual rates are 7.13 ± 10^{-6} dm^3 mol^{-1} s^{-1} in H_2O and 1.13×10^{-6} dm^3 mol^{-1} s^{-1} for the deuteriated form of (C) in D_2O. (The H form of (C) in D_2O had k = 5.08×10^{-6} dm^3 mol^{-1} s^{-1}). The enolization is catalysed by acetate ions and in the presence of this basic species the above rates were 44.4×10^{-3} (H form of (C) in H_2O), 6.72×10^{-3} (D form in D_2O) and 36.9×10^{-3} dm^3 mol^{-1} s^{-1} (H form in D_2O)[45].

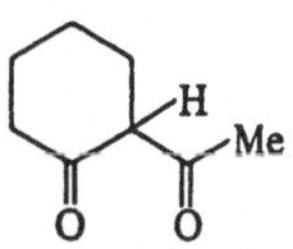

(C)

MDA has also been investigated and the tautomerization in this case involves trans enol[46]. The solvent was ethanol and in this medium at least, the mechanism of enolization was shown to involve intermolecular proton transfer[46].

The question of whether deuteriation can change the equilibrium constant has not been successfully answered. Early work seemed to say that it could, with one report showing that for AA the % enol at 37.3 °C was 78.2% whereas for $^2H_2$AA it was 80.9% at the same temperature[21]. However this experimental evidence was challenged on the grounds that insufficient time had been allowed for the deuteriated compound to establish equilibrium[23]. Given enough time, there was no measurable difference between the two systems and the enthalpy of enolization was determined to be ca. 10.5 kJ mol^{-1} for both AA and $^2H_2$AA[23].

The percentage of enol tautomer of deuteriated MAA, in a 0.25 M D_2O solution, was 2.12 ± 0.01, which is significantly less than that of the protonated form in H_2O at the same concentration, i.e. $2.79 \pm 0.03\%$[37]. The ratio of the two equilibrium constants was 0.76.

According to Leipert[27] in the AA system there are both centred (C_{2v}) and non-centred (C_s) enol tautomers and the ratio keto : enol C_{2v} : enol C_s is 37 : 55.1 : 7.9 in AA but 39.6 : 44.2 : 16.2 in $^2H_2$AA. This study relied on observing the changes brought about by increasing mole fractions of triethylamine being added to the AA and $^2H_2$AA. This idea of their being two enol tautomers with different hydrogen bonds has been challenged, and rightly so[47].

However there are aspects of Leipert's work that can be taken more seriously especially his observation of the effect of deuteriation on the chemical shift $\Delta\delta(^1H, ^2H)$. This isotopic shift has been instrumental in solving the problem of the hydrogen bond in AA cis enol tautomer – see Sect. III.

III. NMR Spectra of the β-Dicarbonyls

Nearly all the data of Sect. II was gathered from NMR spectra. In this section we turn our attention to the hydrogen bonding proton of the cis enol tautomer, looking firstly at the more usual properties of δ(OHO), line width, and coupling constants. The second part will cover the substituent effects on the ^{1}H and ^{13}C NMR chemical shift. The third part will deal with the isotopic shift $\Delta\delta(^{1}H, ^{2}H)$.

A. ^{1}H NMR Signal of the cis Enol Tautomers

The NMR signal of a proton participating in a strong hydrogen bond is generally at the downfield end of the ^{1}H NMR spectrum and is characteristically very broad[48)]. Such protons are involved in rapid exchange at room temperature between different sites and it requires extremely low temperatures to quench such exchange. Most data on β-dicarbonyls has been collected at ambient temperatures and the features which has been discussed as having most influence on the chemical shift δ(OHO) are α- and β-substituents. Table 6 arranges δ(OHO) in numerical order. Not included in this compilation are some derivatives of 2-C_4H_3S (thienyl)[41, 42)], and various para substituted phenyls[5, 38, 53)]. Larger ring cyclic β-ketoaldehydes have also been omitted[50)].

By comparison, the δ(OHO) of acetoacetic ester is to be found at 12.17 ppm and benzoylacetic ester at 12.83 ppm[30)]. This class of β-dicarbonyl derivative has the cis enol tautomer signals in the narrow range of 12–13 ppm, relatively insensitive to α and β and ester substituents.

Table 6 shows only one substance, CH_3COCH_2CHO, with signals extremely sensitive to solvent, δ(OHO) = 13.58 ppm (CCl_4)[7)] and 15.38 ppm (toluene)[5)]. Some sensitivity to solvent is to be expected and found[5)] but this is generally less than 1 ppm, even with solvents such as acetone which are not inert in hydrogen bonding terms. MDA has two widely different values for δ(OHO) reported in the literature[6, 16)].

The arrangement of Table 6 shows that steric factors predominate, especially of α-substituents. Perhaps only slightly less important are electronic effects, and curiously electron withdrawing groups, such as CF_3, at the β-positions actually increase the shielding of the cis enol proton.

An investigation into 3-substituted-2,4-pentadiones by ^{1}H NMR and IR spectroscopy concluded that substituents that could conjugate mesomerically with the enol ring's π delocalization had most effect[3)]. The electron withdrawal via this mechanism (D) was supported by Hückel MO calculations. The effect of the SMe group was explained by invoking C–S pπ-dπ bonding.

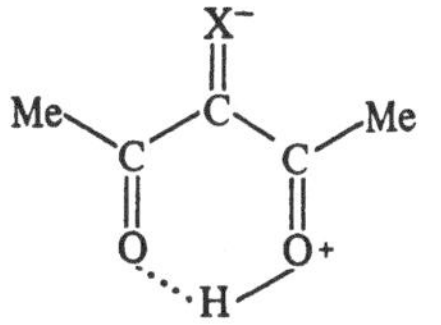

(D)

Table 6. The cis enol proton chemical shift, δ(OHO)/ppm Me_4Si, of β-diketone and β-ketoaldehyde enol tautomers of general formula R′COCH(R″)COR‴

δ(OHO)	R′	R″	R‴	solvent	Ref.
9.92[a]	H	H	H	$CDCl_3$	6
10.61	H	Me	H	$CDCl_3$	49
11.35	H	CH_2CH_2	CH_2	CCl_4	50
13.00	CF_3	H	CF_3	CCl_4	42
13.58	H	H	Me	CCl_4	7
14.10	H	Ph	H	CCl_4	51
14.24	Me	H	CF_3	CCl_4	42
14.27	H	$CH_2(CH_2)_2$	CH_2	CCl_4	50
14.3	CF_3	H	2-C_4H_3O[b]	CCl_4	41
14.33	Me	CH_2Ph	Me	neat	13
14.51	H	Me	Et	acetone	52
14.53	H	H	$C_6H_4pNO_2$	DMF	52
14.60	H	$CH_2(CH_2)_3$	CH_2	CCl_4	50
14.71	H	Me	Me	acetone	5
15.00	H	Me	Pr^i	acetone	5
15.06	CF_3	H	Ph	CCl_4	52
15.35	H	H	Bu^i	toluene	5
15.38	H	H	Me	toluene	5
15.40	2-C_4H_3O[b]	H	2-C_4H_3O[b]	CCl_4	41
15.40[c]	Me	H	Me	CCl_4	42
15.40	H	Ph	Me	CCl_4	51
15.55	Me	Cl	Me	CCl_4	33
15.67	Me	H	$C_6H_4pNO_2$	$CDCl_3$	38
15.8	Me	Br	Me	CCl_4	40
15.83	Me	Et	Me	neat	13
15.84	H	H	Bu^t	toluene	5
15.95	Me	H	Bu^t	CCl_4	52
16.27[d]	Me	H	Ph	$CDCl_3$	38
16.30	H	H	Ph	toluene	5
16.49	Bu^t	H	Bu^t	CCl_4	52
16.50	Me	Me	Me	neat	13
16.55	Me	Pr^n	Me	CCl_4	40
16.64	Ph	H	$C_6H_4pNO_2$	CCl_4	52
16.7	Ph	H	Bu^t	CCl_4	40
16.8	Me	Ph	Me	CCl_4	41
16.90	Me	CN	Me	CCl_4	33
17.00	Me	Me	Ph	neat	13
17.08	Me	SMe	Me	CCl_4	33
17.13	Ph	H	Ph	CCl_4	42
17.5	Me	Bu^s	Me	neat	32
18.62	Me	CO_2Me	Me	CCl_4	35

[a] Reported at 16.40 in CD_2Cl_2–$CFCl_3$ at 200 K[46];
[b] 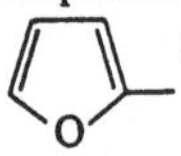

; [c] 15.83[33]; [d] 16.11[52]

The width of the OHO signal is broader than other signals in the spectrum and is very sensitive to traces of water. Nonhebel[40] showed that bulky substituents on α and β sites not only shifted δ(OHO) downfield but produced a sharper signal, both facts which he took to mean that a stronger hydrogen bond was formed as the two oxygen atoms were forced closer together. The stronger bond reduced exchange between the cis enol forms and this explained the sharper signal. However as we shall see in Sect. IV the changes in R(O··O) brought about by various groups at the α-positions are not very large although the range of values, 245–255 pm, spans a region in which it is known that OHO hydrogen bonds undergo profound changes, with respect to their potential energy surface, as the oxygens approach each other[54].

If the hydrogen bond is a double minimum, as now seems most likely, then in an asymmetric β-dicarbonyl or in a β-ketoaldehyde one minimum should be of lower energy than the other and the proton will thus show a preference for one oxygen over the other.

Attempts have been made to show which cis enol tautomer is preferred. Para-substituted BA offers a way of fine-tuning the electron density at one carbonyl and this has shown that in BA derivatives the proton prefers to reside on the benzoyl oxygen[38,51,55]. This preference does not carry over into formylbenzoylmethane, $PhCOCH_2CHO$, and it has been concluded that in this substance the minima are roughly of equal depth so that there is a 50 : 50 benzoyl : formyl distribution[53]. However in para-substituted derivatives of this compound the ratio could vary from 56 : 44 for p-nitrobenzoyl to 40 : 60 for p-dimethylaminobenzoyl. These ratios were calculated from coupling constants, J(enol-formyl) being 0 Hz when the proton is at the benzoyl side of the molecule, and ca. 13 Hz when located on the formyl oxygen.

The strongest preference for one cis enol tautomer was found in (E), (+)-hydroxylmethylenementhone, in which the form shown accounts for 75% of the cis enol[56]. This again was deduced from the observed coupling constant of 2.0 Hz and assuming J(enol-formyl) was 11 Hz when the proton resided on the formyl oxygen.

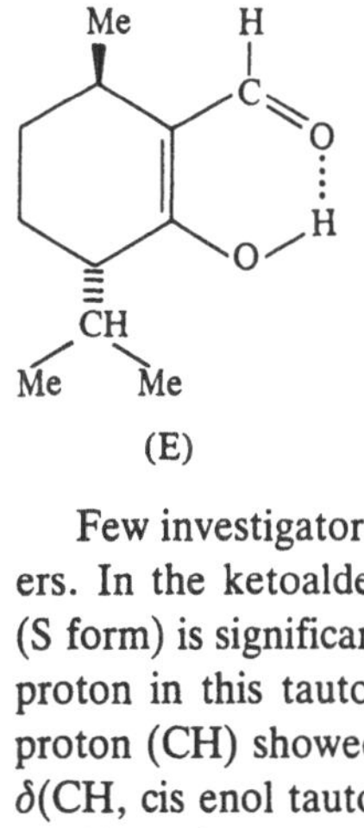

(E)

Few investigators have postulated contributions to δ(OHO) from trans enol tautomers. In the ketoaldehyde, $PhCOCH_2CHO$, it has been stated that the trans tautomer (S form) is significant under certain conditions and the chemical shift of the OH group's proton in this tautomer is 13.55 pm compared to 15.9 for the cis enol[5]. The formyl proton (CH) showed the most marked difference: δ(CH, trans enol tautomer) = 12.0, δ(CH, cis enol tautomer) = 4.6 ppm.

Since cis ⇆ trans enol conversion is likely to be indistinguishable by ^{1}H n.m.r. under normal conditions this is yet another factor that should be borne in mind when discussing the data of Table 6, although for β-diketones it is unlikely to be significant because of the increased steric repulsion that the trans forms experience compared to the cis tautomers. However this steric factor serves to stabilize the trans enol tautomers of ketoaldehydes.

Another attempt to distinguish the enol group from the carbonyl group in cis enol tautomers made use of ^{17}O NMR spectroscopy. No firm answer could be reached in most cases because of the rapid proton transfer between the two sites under the conditions of

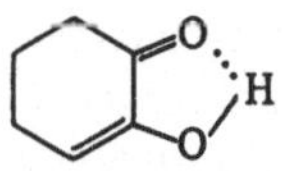

(F)

the experiment[57]. For BA it was deduced that the proton prefers the benzoyl side and for (F) two separate ^{17}O signals were in fact observed since in this molecule proton transfer is slow due to the molecular rearrangement that must synchronise with it. This early work, using enriched ^{17}O samples, was carried out at 27 °C, so that using modern FT methods and low temperature probes might be expected to produce a clearer picture from ^{17}O NMR spectroscopy.

The next-but-one atoms to the hydrogen bonding centre are carbons and ^{13}C NMR spectroscopy has thrown up some interesting results, and at loggerheads to current ideas about the nature of this bond. ^{13}C NMR, and ^{1}H NMR, chemical shifts can be reduced to components that are calculable from so-called substituent effects.

B. Substituent Effects

An ingenious proof that the proton of the hydrogen bond was not a double minimum was advanced by Shapet'ko. This was based on his discovery[52] that the chemical shift $\delta(OHO)$ could be resolved into two components that were related solely to the β-substituents X and Y of β-diketone, which he depicted in a manner (G) that implies that

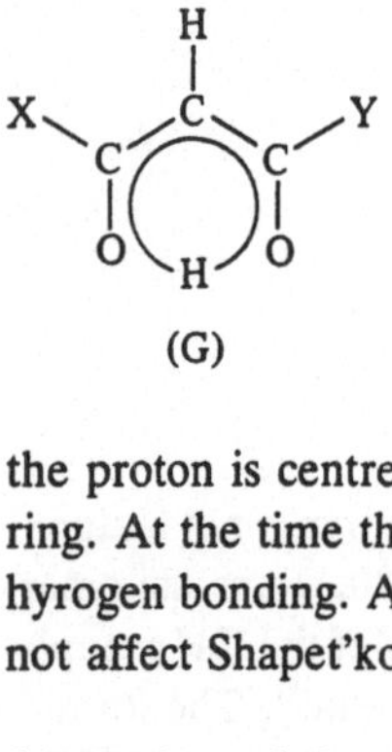

(G)

the proton is centred and is involved in the delocalized π bonding around the cis enol ring. At the time the theory was proposed (1972) there was evidence in favour of such hyrogen bonding. Although we may no longer support such an interpretation this does not affect Shapet'ko's evidence of a linear relationship (Eq. 2)

$$\delta(OHO) = \Delta(X) + \Delta(Y) \quad (2)$$

where Δ refers to the increment apportioned to each of the β-substituents and calculated from $\Delta(Me) = 7.70$ ppm for AA, for which compound $\delta(OHO) = 15.40$ ppm. Using this as the basis, then Δ values can be obtained from other asymmetric β-diketones and the validity of Δ checked by computing a value for $\delta(OHO)$ for yet other symmetric β-diketones. Thus CF_3COCH_2COMe permits $\Delta(CF_3)$ to be calculated as 6.54 ppm and this in turn predicts $\delta(OHO)$ for $CF_3COCH_2COCF_3$ to be 13.08 ppm which compares very well with the value of 13.00 ppm in Table 6. Table 7 lists the β-substituent value for other groups.

Table 7. β-Substituent Factors, Δ, for β-Diketones ^{1}H NMR δ(OHO)[a]

Substituent	Δ/(ppm)	Substituent	Δ/(ppm)
Me	7.70	2-C_4H_3O	7.73
CF_3	6.54	2-C_4H_3S	8.05
Ph	8.46	$C_6H_4pNO_2$	8.20
Bu^t	8.25	2,4,6-$Me_3C_6H_2$	7.78

[a] Based on chemical shifts of β-diketones in CCl_4 solution at room temperature[52)]

Shapet'ko's logic in support of a centred hydrogen bond would seem to be sound. He argued that for a non-centred hydrogen bond then Eq. (3)

$$\delta(\mathrm{OHO}) = p_X\delta_X + p_Y\delta_Y \tag{3}$$

would apply where p_X is the probability of the proton being in the potential well whose minimum is nearer the X carbonyl and δ_X is the chemical shift of such a proton. Assuming the proton spends most time in the deeper well then Eq. (3) should not be reducible to Eq. (2), and yet it is. The conclusion would appear to be that the proton is in a single minimum potential well.

Shapet'ko found linear relationships based on ^{13}C NMR chemical shifts, not only for the β carbons but also for the α carbons Eqs. (4), (5) and (6)[58)].

$$\delta C_\alpha = 100.6 + \Delta(X)_\alpha + \Delta(Y)_\alpha \tag{4}$$

$$\delta C_\beta = 191.5 + \Delta(X)_n + \Delta(Y)_d \tag{5}$$

$$\delta C_\beta = 191.5 + \Delta(Y)_n + \Delta(X)_d \tag{6}$$

Here $\Delta(X)_\alpha$ is the substituent factor attributable to group X when calculating the chemical shift of the α carbon, δC_α. For calculating the chemical shifts of the two β carbons $\delta C_{\beta X}$ and $\delta C_{\beta Y}$ it is necessary to include a substituent factor for the group on the β carbon in question, i.e. the near group (n), and one for the group on the distant β carbon (d).

Table 8. β-Substituent Factors, Δ_α, Δ_n and Δ_d for ^{13}C NMR Chemical Shifts of α- and β-carbons of β-Diketones[a]

Substituent	Δ_α/(ppm)	Δ_n/(ppm)	Δ_d/(ppm)
Me	0.0	0.0	0.0
CF_3	−4.1	−14.9	+3.5
Ph	−4.2	−8.5	+2.0
Pr^i	−1.5	+9.0	−0.2
Bu^t	−5.1	+8.8	+0.4
2-C_4H_3S	−2.9	−12.0	−5.1
2,4,6-$Me_3C_6H_2$	+1.4	−2.1	+2.1
CO_2Me	+1.5	−24.2	+7.3

[a] Based on chemical shifts of β-diketones in CCl_4 solution or as neat liquids[58)]

Thus for each X substituent there are three substituent factors $\Delta(X)_a$, $\Delta(X)_n$ and $\Delta(X)_d$ and these values are given in Table 8.

Again Shapet'ko's interpretation is that electron redistribution in the enol chelate is the sole influence in operation and that there is no tautomeric equilibrium between the two cis enol species. Since this conflicts with other evidence that the OHO hydrogen bonding is non-centred, all that can be said is that Shapet'ko's substituent correlations show a type of hydrogen bond that behaves for NMR purposes as if it were centred. Shapet'ko uses the term quasisymmetrical when he uses another method of investigating these β-diketones.

Replacing the proton of the cis enol tautomer with a BF_2 group gives (H). The effect of this on the ^{13}C NMR chemical shifts of the carbon atoms of the chelate ring is small e.g. $\delta C_\alpha = 191.2$ and $\delta C_\beta = 100.4$ for AA enol chelate and $\delta C_\alpha = 192.9$ and $\delta C_\beta = 102.2$ ppm for the BF_2 chelate of AA[59]. Similar small changes are noted for a variety of β-diketones[59, 60]. Shapet'ko concludes that since very little perturbation of electron density around the ring occurs on going from the proton to the BF_2 chelate, which must be centred, then the proton chelate is probably centred as well – or at least behaves as if it were.

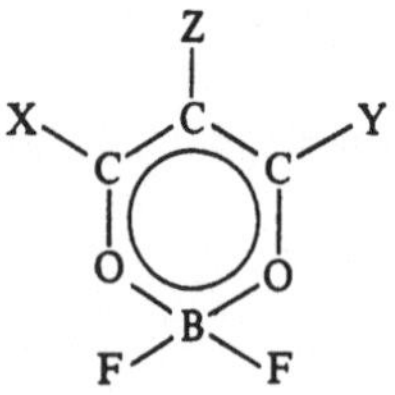

(H)

Not all β-diketone arrangements remain unperturbed however. Compound (J) and other esters show significant ^{13}C NMR chemical shift changes in going to the BF_2 chelate and it is deduced that in (J) the H in the enol tautomer is located near the carbonyl shown [60]. This result proves that Shapet'ko's method is not insensitive to the proton's location. However in the light of what follows (Sect. III.C) it will be seen that the potential energy well of the cis enol tautomer hydrogen bond must be a double minimum.

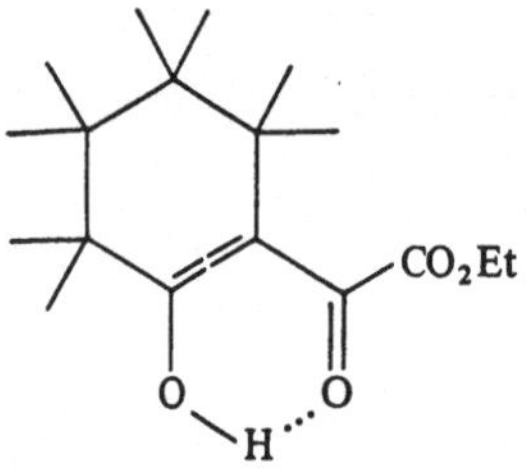

(J)

^{13}C NMR has recently been used to obtain equilibrium constants for the L (left) or R (right) configurations of the mixed methyl/trifluoromethyl/phenyl β-diketones[61] Fig. 4. $K_{L/R}$ was deduced to be 0.42 for TFAA ($X = CF_3$, $Y = Me$) 0.68 for BA ($X = Ph$, $Y = Me$), and 1.08 for $PhCOCH_2COCF_3$. Other compounds such as vinyl ketones, in solution

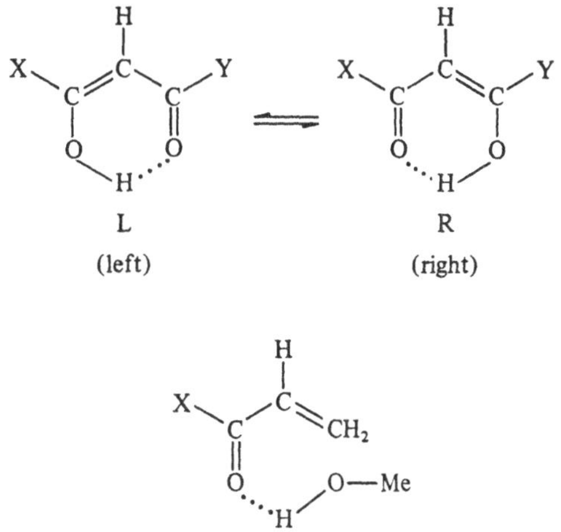

Fig. 4

with strong hydrogen bond donors, e.g. methanol, were used to simulate the grouping (I) in Fig. 4 and thus obtain $\delta(^{13}C)$ values which approximate to the carbonyl half of configuration R in Fig. 4. This work[61] shows the proton residence times for the double minimum potential well are dependent upon the nature of the β-substituents X and Y, as expected.

The ^1H NMR spectrum of MDA in $CFCl_3$–CD_2Cl_2 (1 : 1 mix) at temperatures lower than 200 K showed δ(OHO) at 16.40 ppm[46], which is at variance with an earlier value of 9.92[6]. Moreover the signal was a triplet due to coupling with the formyl hydrogen and J_{HOCH} was 6.1 Hz. The signal was not broadened in any way except by viscosity effects, and the molecule was behaving as if it had C_{2v} symmetry at least on the NMR times scale. The upper limit for the $C_s \rightarrow C_{2v} \rightarrow C_s$ interconversion was calculated as 25 kJ mol^{-1}. The significance of this will become apparent.

C. Proton-Deuterium NMR Isotopic Shift $\Delta\delta(^1H, ^2H)$

The change in the chemical shift between the resonance of the hydrogen bonding proton in the ^1H NMR spectrum and that of the hydrogen bonding deuteron in the ^2H NMR spectrum has been shown to be a good indication of the type of potential energy well in which the proton or deuteron is confined. The quantity $\Delta\delta(^1H, ^2H)$ corresponding to this isotopic difference in chemical shifts is defined as $\delta(^1H)$-$\delta(^2H)$ so that a positive value of $\Delta\delta(^1H, ^2H)$ means an upfield shift on substituting deuterium for hydrogen, and a negative value indicates a downfield shift. The former implies a double minimum potential well, the latter means a single minimum or a double minimum of very low barrier.

The shape of the well is determined solely by the heavy nuclei and the distance between them[62]. Where the hydrogen atom spends its time is determined by the depths of the wells and the height of the barrier between them. The ^2H isotope sits deeper within a particular well than does ^1H; and the ^3H isotope deeper still.

If both heavy atoms of the hydrogen bond are the same element in the same molecular environment then there are basically five kinds of potential energy well they can

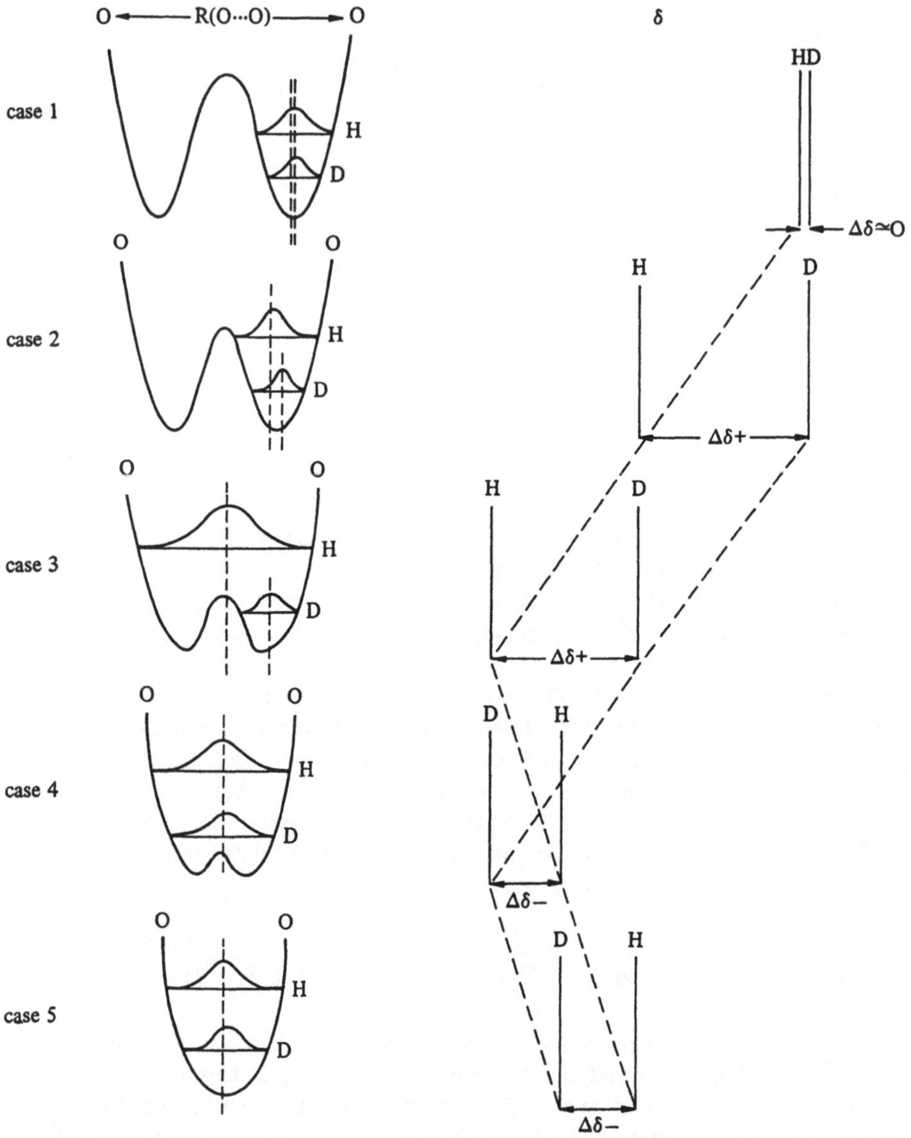

Fig. 5. Potential energy diagrams of O···O hydrogen bonds and their associated ^1H and ^2H (D) NMR chemical shift differences, Δδ

produce. These are shown in Fig. 5 for an O···O combination of the kind we are concerned with in the β-dicarbonyls.

The weak hydrogen bond, case 1, is a double minimum – high barrier in which the proton's motion is approximately harmonic and the deuteron's more so. Their motions are centred around the same point so that their shielding is roughly the same and thus there is no change in δ on going from one to the other. The vast majority of hydrogen bonds are of this weak variety.

As the oxygen atoms get closer and the barrier falls, the proton's motion becomes more anharmonic than the deuteron's, and the centres of their motions diverge (case 2). It may even be the case that the proton's movement is free of the barrier while that of the deuteron is still constrained to one well (case 3). In both these cases there will be an upfield shift of $\Delta\delta(^1H, ^2H)$ on deuteriation of the hydrogen bond because the deuteron remains in the shielded location while the proton is free to move into the deshielded region of space between the two oxygens.

As the oxygens move even closer we come to case 4 in which both ^1H and ^2H are above the barrier and again their time averaged locations coincide. However now their chemical shifts in their NMR spectra are not coincident because the proton's wider amplitude results in more overall shielding by the electron clouds of the oxygen atoms than the deuteron experiences. Isotopic substitution of such hydrogen bonds as cases 4 and 5 will result in a downfield shift, i.e. a negative value for $\Delta\delta(^1H, ^2H)$, proof that the hydrogen bond is very strong with a potential well that is either of very low barrier, flat bottomed or even a single minimum, as in case 5. As far as the proton is concerned case 3 is also this type of hydrogen bond. As the two oxygens approach each other in cases 3, 4 and 5 the shielding of the proton will increase and δ(OHO) move upfield in the ^1H NMR spectrum whereas the shielding of the deuteron will first decrease (case 3 to 4) but then increase (4 to 5), as shown in Fig. 5.

This picture of the chemical shift and isotopic effects also explain the somewhat puzzling fact that the downfield limits of δ(OHO) at ca. 20 ppm do not represent the strongest of hydrogen bonds. This situation indicates case 3 type bonds.

For O···O hydrogen bonds the changeover from weak to strong to very strong occurs in the range 255–230 pm of R(O··O)[48]. As shown in Sect. IV below the R(O··O) distances in β-diketones lie in the region 255–245 pm suggesting these will fall in either the case 2 or 3 categories so that $\Delta\delta(^1H, ^2H)$ values should be positive and large as indeed they are. Because of their molecular dimensions the β-diketones fall in a region of the hydrogen bond spectrum where there are relatively few examples, as yet. Luckily they are capable of providing compounds that cover this interesting range of R(O··O), and it remains to be seen if there are relationships between $\Delta\delta(^1H, ^2H)$ and other bond parameters, although Table 9 offers little hope with its wide variations.

The discovery of the so-called anomalous isotope effect of the enol protons of AA, MAA, BA and DBM was made by Chan et al.[63] who interpreted their results as evi-

Table 9. ^1H and ^2H NMR shift, $\Delta\delta(^1H, ^2H)$ of β-Diketone Enol Tautomers' Hydrogen Bonds

β-Diketone	$\Delta\delta(^1H, ^2H)$	Refs.
AA	+0.58; +0.50; +0.61; +0.62	63–65, 27
BA	+0.42; +0.64; +0.67	63–65
DBM	+0.45; +0.72	63, 65
MAA	+0.45	63
HFAA	+0.30	66
AA + C_4H_5N[a]	+0.74	27
$Bu^tCOCH_2COBu^t$	+0.59	64

[a] Mole fraction pyrrole = 0.55

dence of there being two hydrogen bond states for the enol tautomer a lower energy centred (C_{2v}) and a higher energy non-centred (C_s) one. Curiously the significance of $\Delta\delta(^1H, ^2H)$ and this approach to hydrogen bonding was not followed up for many years, but then there was a spate of activity[27, 64–67] the results of which are tabulated in Table 9. Although there is some discrepancy in these results, all agree on there being substantial upfield shifts in all cases, including the mixed AA-pyrrole system[27], and the conclusion is that all these cis enol tautomers have a double minimum hydrogen bond[47, 67].

For some of the β-diketones $\Delta\delta(^1H, ^3H)$ has been measured: AA = +0.83; BA = +0.93; DBM = +1.10[65]. These are larger than $\Delta\delta(^1H, ^2H)$ since the tritium nucleus lies even deeper in the potential energy well than the deuteron.

This information of Table 9 combined with that of Table 6, δ(OHO) values, shows that with β-diketones we have a variety of double minimum hydrogen bonds of cases 2 and 3. Thus HFAA would appear to be case 2 judging by its δ(OHO) of 13.00 whereas DBM with δ(OHO) at 17.13 is clearly case 3. This is supported by the difference in the R(O··O) distances: HFAA has the longest enol hydrogen bond, 255 pm[68], whereas DBM has one of the shortest, 246 pm[69], see below. With the reporting of a simple but accurate method for measuring $\Delta\delta(^1H, ^2H)$ the use of this technique in studying hydrogen bonds should become routine[70].

Further support for the non-centred enol hydrogen bond has come from ^{13}C and ^2H spin-lattice relaxation times of AA and $^2H_2$AA from which deuterium quadrupole coupling constants are derived. These are explained in terms of a double minimum potential well[67]. These same workers also discuss the broadness of the δ(OHO) signal and conclude that it is due to bond lengthening rather than contributions from trans enol forms. The minimum line width for AA achieved by exhaustive drying was 0.5 Hz.

IV. Molecular Structures of β-Dicarbonyls

For over 20 years the structures of β-dicarbonyl compounds have been determined by X-ray, electron and neutron diffraction and microwave spectroscopy. Agreement as to how the data should be interpreted has led to centred and non-centred hydrogen bonds being proposed for the same substance. In almost all cases the cis enol is the stable tautomer, but in 1971 Karle et al.[71] reported the structure of AA, in both this and the diketo form, by electron diffraction.

The diketo tautomer adopts the U (Z,Z) conformation with the oxygen atoms 277 pm apart and the two carbonyl groups at a dihedral angle of 48.6° to each other. The enol tautomer was planar and the intramolecular hydrogen bond distance, R(O··O) was only 238 pm. The hydrogen bond was linear and symmetric as would be expected for such a short bond[71]. A bond this short was unheard of for a neutral molecule, and still is; but we have reason to believe that this information about AA is unsound.

Other electron diffraction studies on AA[72], TFAA[72] and HFAA[68] gave R(O··O) values of 251.9, 251.4 and 255.1 pm respectively. In these determinations the enol form was assumed to be the only one present in the gas phase and the rings were assumed to be planar. Since the ring bond parameters showed C_{2v} symmetry it was also assumed the

Table 10. R(O··O) values of hydrogen bonds

R(O··O)/pm	Type of hydrogen bond	Examples
300[a]–270	weak or normal; case 1[b]	H_2O; salt hydrates; ROH
270–255	medium; case 2	$(RCO_2H)_n$
255–245	strong; case 3	$H(RCO_2)_2^-$; other acid salts[c]
245–235	very strong; case 4	$H_5O_2^+$; H(maleate)$^-$; dioxime complexes
<235	extremely strong; case 5	$H(OH)_2^-$

[a] A hydrogen bond is defined as one that is shorter than the sum of the Van der Waals radii of the two heavy atoms, in this case 150 pm; Bondi, A.: J. Phys. Chem. *68*, 441 (1964)
[b] See Fig. 2
[c] For other examples such as $H(NO_3)_2^-$ etc. see Refs. 48, 74–76

proton of the hydrogen bond lay on this axis i.e. was centred. As we shall see this is only true on a time averaged basis – the hydrogen bonds are non-centred – although these R(O··O) values are more in keeping with those derived by X-ray and neutron diffraction methods.

Since these determinations were made in 1971–72 very short hydrogen bonds have turned up in scores of systems[48]. Few OHO bonds are shorter than 240 pm and the shortest of all is that of the $[H(OH)_2]^-$ ion which is 229 pm[73]. Several compilations of X-ray and neutron diffraction data on R(O··O) show a range of values that can be correlated with the various kinds of hydrogen bond[48, 74–76], as in Table 10.

A. X-Ray and Neutron Diffraction Data

Most structural determinations are based on X-ray diffraction data, which is not the best technique for spotting hydrogen atoms and especially those of strong hydrogen bonds. By their very nature these are located in regions of space that are low in electron density and consequently are poor at deflecting X-ray radiation. Information about such hydrogen bonds has to be inferred from other geometrical features of the molecule.

Among the first β-diketones to be investigated by X-ray methods were the m-chloro[77] and m-bromo[78] derivatives of DBM. These proved to have short hydrogen bond lengths, R(O··O), of 247.5 and 246.4 pm respectively and by virtue of the symmetry of C–O and C–C bonds around the enol ring the proton was deemed to be centrally placed even though it could not be located. Similar conclusions were drawn about the intramolecular hydrogen bonds of tetraacetylethane in which R(O··O) = 242.4 pm[79].

Usnic acid, found in lichens, has three different cis enol units within the one molecule, and one of these is very short[80]. In order to locate the protons the X-ray determination was done on a crystal at − 110 °C. The unit cell contains two usnic acid molecules whose dimensions for the cis enol OHO bonds differ slightly – see Fig. 6. The shortest hydrogen bond shows that the proton is not centred in either molecule – in one it is nearer the acetyl carbonyl, in the other nearer the ring carbonyl.

Later, more sophisticated X-ray analysis gave a better idea of the position of the proton, and several structural determinations carried out using this diffraction method have been able to describe the cis enol hydrogen bond in terms not only of R(O··O) but

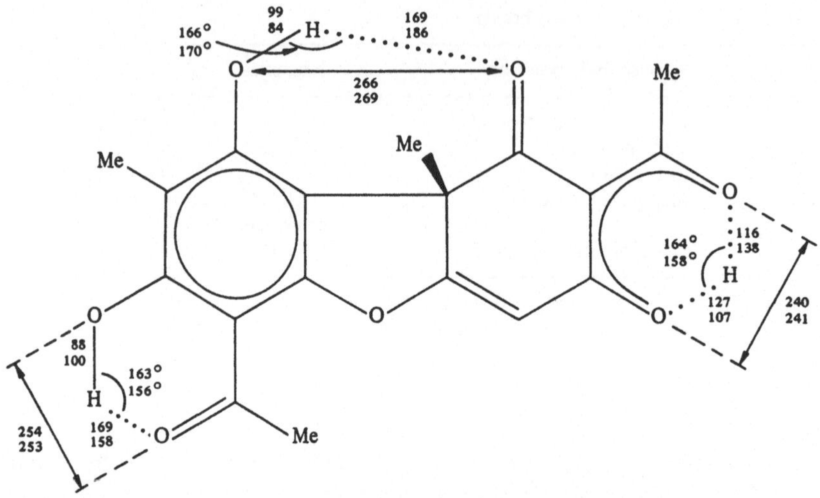

Fig. 6. The hydrogen bonds of usnic acid[80]. The upper and lower figures for each bond length and bond angle refer to the two usnic acid molecules of the unit cell

also of R(O–H), R(H···O) and ∠OHO. The results of these investigations are given in Table 11. In all cases the proton is non-centred although the neutron diffraction of BA shows that the hydrogen atom is only 4 pm off-centre[84], although this is only an average because of the large thermal motion of this proton (20–30 pm r.m.s amplitude of vibration).

In some determinations the symmetry of the remainder of the ring has led to postulations of a symmetrical hydrogen bond as in p-bromo-benzoylacetone[87] [R(O··O) = 248.1 pm] and p-nitro-benzoylacetone[88] [R(O··O) = 245.7 pm], even though the enol proton was not located. Yet there are good theoretical grounds for expecting a symmetrical skeleton and enhanced hydrogen bonding with the phenyl derivatives[89].

The symmetry of the ring is obvious in the many metal methylacetonate complexes that have been determined and the average ring bonds and angles from such compounds are shown in Fig. 7(I)[90]. By comparison the rings of AA (II) and 3,3′-trithioacetyl-

Table 11. The hydrogen bond parameter of β-diketones $RCOCH_2COR'$

R	R′	R(O··O)/ppm	∠OHO	R(O–H)/ppm	R(O··H)/pm	Method	Ref.
Me	Me	253.5	141	103	166	X-ray	81
Me	Me[a]	246.1	165	117	139	X-ray/neutron	82
Me	Ph	249.8	151	118	140	X-ray	83
Me	Ph	248.5	153	123.5	131.9	neutron	84
Ph	Ph	246.0	154	122	128	X-ray	85
Ph	Ph	246.3	155	116.1	136.2	neutron	86

[a] In 3,3′-trithiobisacetylacetone

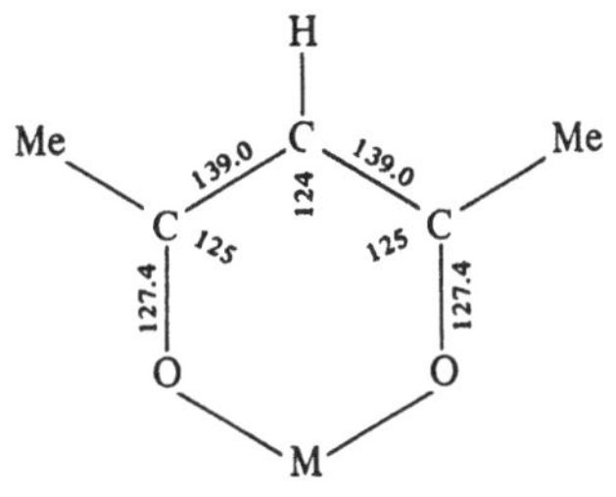

(I)[90]

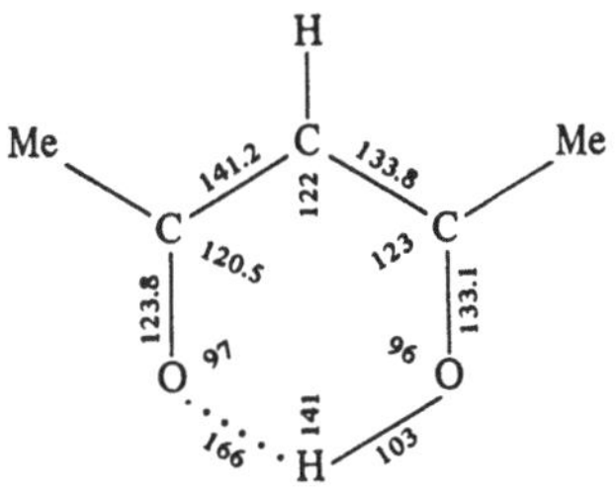

(II)[81]

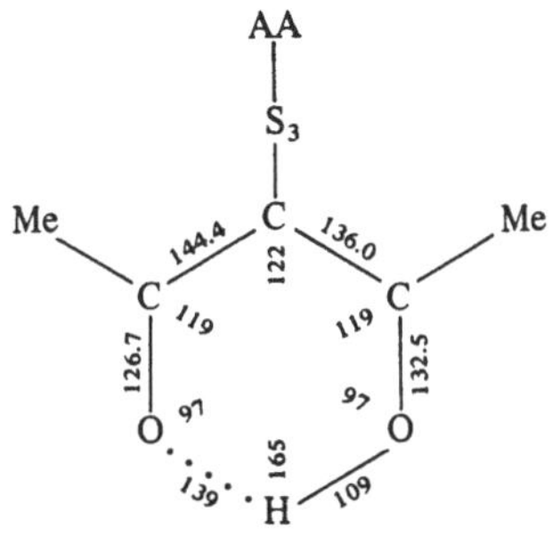

(III)[82]

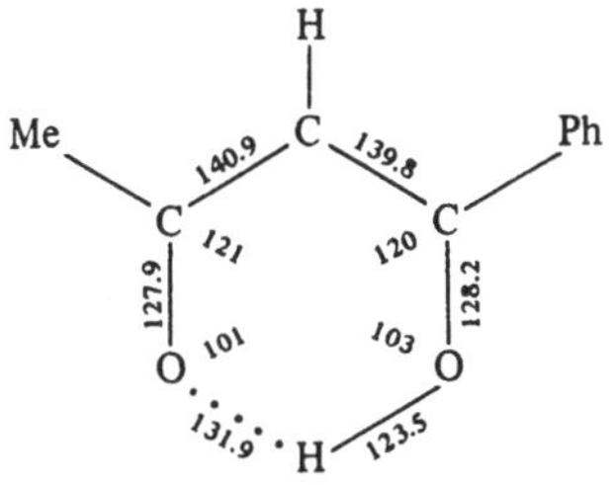

(IV)[84]

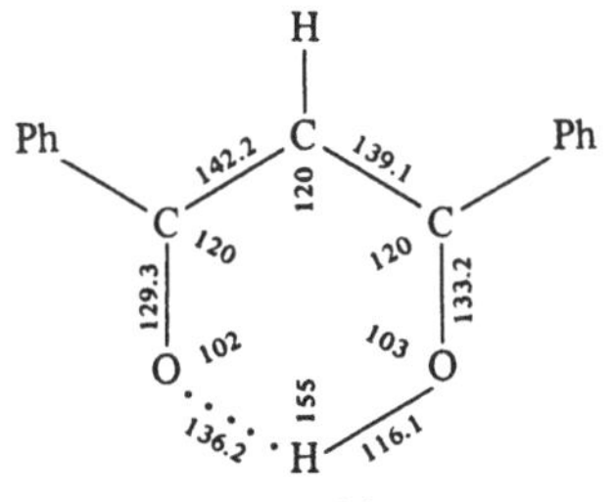

(V)[86]

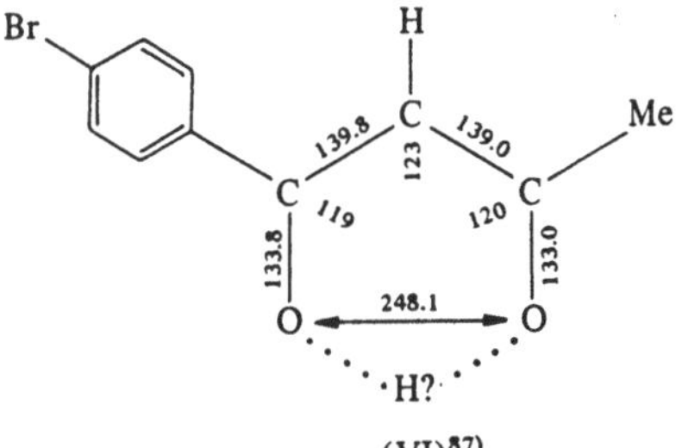

(VI)[87]

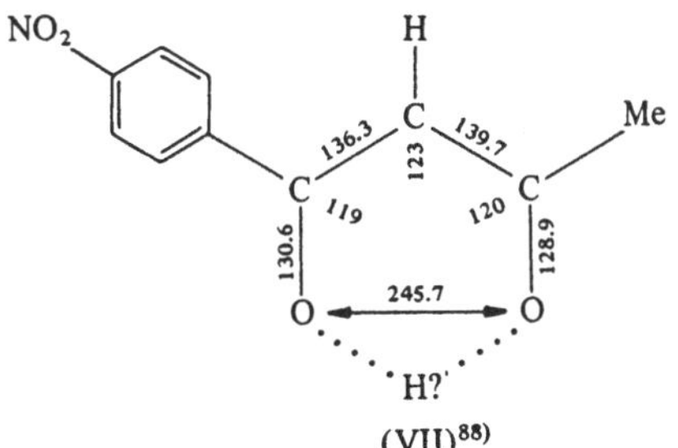

(VII)[88]

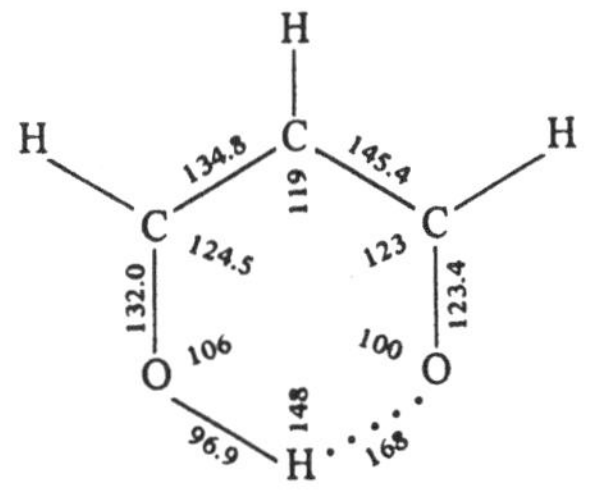

(VIII)[93]

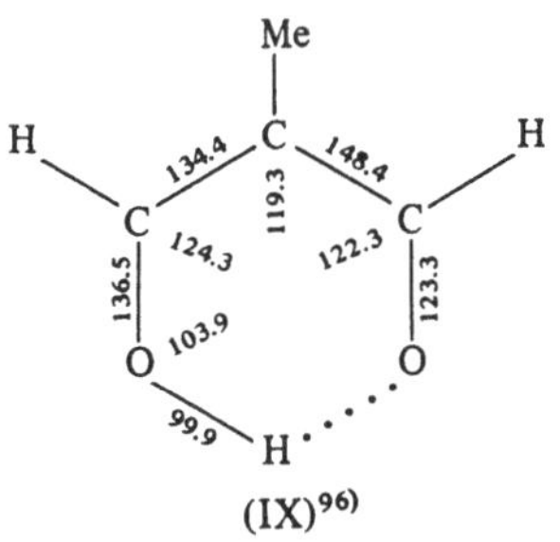

(IX)[96]

Fig. 7. Ring bonds and angles of β-diketone compounds. Bond lengths in pm

acetone (III) are not as delocalized, with alternating bond lengths around the ring. There is less obvious a difference between the enol and keto halves of the BA (IV) and DBM (V) rings although in the former the enol proton seems to incline to the benzoyl oxygen as previous NMR studies detected. The symmetry of the p-bromo-benzoylacetone (VI) is remarkable and on the basis of the position of the enol proton in (IV) and (V) one is tempted to predict a centred hydrogen bond. Almost the same is true of (VII).

AA, with its m.p. of − 23 °C, fortuitously turned up as a solvate in the crystalline phase of the drug complex diphenylhydantoin/9-ethyladenine[81]. Although the AA molecules were claimed to be free of other intermolecular attractions in this crystal, the structure revealed a hydrogen bonding proton that was 36 pm above the plane of the cis enol ring. This suggests that not all lattice forces are insignificant and the asymmetry of (II) in Fig. 3 may likewise be exaggerated especially the long R(O··H) distance which differs very much from that of (III).

Deuteriated β-diketones have not been subjected to crystal structure determinations so far, yet this would seem to be the next logical step since there are good theoretical grounds for expecting significant changes in R(O··O) if the hydrogen bond is deuteriated[91]. Double minimum hydrogen bonds should increase in length on deuteriation whereas very strong hydrogen bonds of the single minimum should contract. Within the limits of experimental error it appears that very strong hydrogen bonds are unaffected by deuteriation[92].

B. Microwave Spectroscopy Etc

This has been used to elucidate the structure of the simpler β-dicarbonyls MDA[93,94] and triformylmethane $HC(CHO)_3$[95]. MDA was shown to exist as the planar cis enol conformer with R(O··O) = 255.3 pm, in keeping with the double minimum potential energy well calculated by *ab initio* methods for this molecule – see next section. The microwave spectrum of trideuteriated MDA showed a non-tunnelling molecule with C_s structure (non-centred) and R(O··O) = 257.4 pm for the deuteriated bond[94]. The full structure is shown in Fig. 7 (VIII). Tunnelling is quenched in the trideuteriated molecule but not in protonated MDA. The tunnelling energy separation, observed in the far IR spectrum, is 21 cm^{-1}[94].

The microwave spectrum of methylmalondialdehyde shows the molecule is hydrogen bonded in the gas phase with a non-centred proton, R(O–H) = 99.9 pm, as in Fig. 7 (IX)[96].

The microwave spectra of both the ^1H and ^2H forms of triformylmethane again proved the structure to be a cis enol, and again with a non-centred hydrogen bond[95]. The

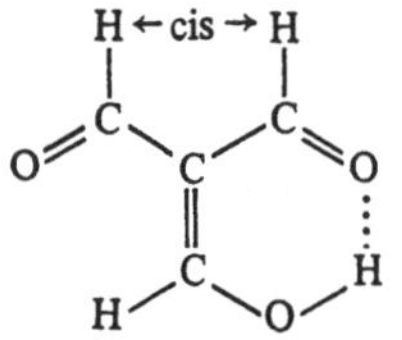

(K)

whole framework of the molecule was planar with the proton of the nonhydrogen bonded formyl group orientated cis with respect to the formyl proton of the acceptor carbonyl as in (K). Moments of inertia could not decide between cis and trans in the protonated spectrum but the deuteriated isomer showed it to be cis and also proved the hydrogen bond was not centred. There was no evidence in the spectrum for the triketo tautomer[95)].

X-ray photoelectron spectroscopy has been used to probe the hydrogen bonding on the basis that C_{2v} structures would show a single ionization for its equivalent oxygens while C_s should give two[46, 97)]. HFAA has two O_{1s} binding energies, showing it is C_s, as does MDA. AA itself gives an asymmetric doublet attributed to overlapping of one of the peaks with the keto form. (DMAA for which only the keto form is possible gives a single peak[97)].)

Malondialdehyde is unique in being a small molecule yet having an intramolecular hydrogen bond that persists in the vapour phase monomer. Its electronic spectrum shows two intense absorption transitions at wavelengths below 1800 Å[98)]. The one at 2630 Å (ν_{max}, 38,000 cm^{-1}) has no vibrational fine structure and arises from $\pi^* \leftarrow \pi$. The other, at 1959.5 Å (ν_{max}, 51,030 cm^{-1}), has moderate vibrational structure and is the ν_{00} band, $\pi^* \leftarrow \pi$[98)]. The most intense part of the spectrum lies at 3630–2950 Å and has many vibronic components with a progression of bands at 185 cm^{-1} intervals[99)]. The band system is due to $\pi^* \leftarrow n$ transitions in the molecule and the authors conclude that the electronic spectral evidence for MDA is consistent with a planar cis enol arrangement.

Malondialdehyde is unique in another way. Because it is a molecule of only nine atoms it is the chosen model for theoretical calculations on a β-dicarbonyl type system.

V. Theoretical Calculations and Hydrogen Bond Energies

The information which theoretical calculations can supply about a system resembles experimental information in that both are capable of being improved, given better facilities, and both are capable of being misinterpreted if the researcher desires to prove a particular point. Just as experimental evidence on the symmetry of the enol hydrogen bond has "proved" it to be centred or non-centred at different times so theoretical evidence has been forthcoming to support both concepts of this bonding. The final picture of the bonding, obtained from a fully and simultaneously optimized structure of MDA, and employing the 9 s5 p/4 s2 p/4 s (C/O/H) basis set, shows that there is an asymmetric C_s hydrogen bond of the double minimum, moderate barrier type[100)].

Calculations which produce geometries of molecules can be checked against the actual molecule in many cases, and this is so for MDA. Calculations which produce molecular energies are less easy to check against an experimental yardstick, but it is precisely this lack of data which justifies the effort put into such computations. At the end of this Section we shall bring together a variety of calculated and measured energies for MDA and other β-dicarbonyls and attempt to answer the question: how strong is the hydrogen bond of a cis enol tautomer?

A. Theoretical Calculations

The earliest theoretical work on a model molecule for the β-diketones naturally chose MDA and CNDO/2 caluclations showed a double minimum potential energy well for the cis enol tautomer[101]. Similar CNDO/2 results for AA suggested a double minimum, low energy barrier ($\Delta E = 2$ kJ mol^{-1}) potential well[102]. Ring angles were assumed to be 120° at the three carbon atoms. The result was challenged and for AA and TFAA a symmetrical strong hydrogen bond system was forecast on the basis of INDO calculations[103]. For AA R(O··H) was computed to be 117.4 pm with ∠OHO 152°, and TFAA gave a similar but slightly asymmetric hydrogen bond with the proton fractionally nearer the CF_3 group, R(O··H) 118.6, 116.7 pm[103]. The R(O··O) distances were 227.9 pm (AA) and 228.4 pm (TFAA) which are indicative of a very strong, single minimum bond (case 5). Both these values underestimate the actual hydrogen bond length by ca. 25 pm.

Using these INDO optimized geometries and an STO-3 G minimal basis set the energy of MDA was computed to be -262.13910 hartrees for this C_{2v} bond[104]. Other calculations showed that the barrier height is the most sensitive function of the internuclear distance between the receptor atoms[62]. Expanding the R(O··O) distance caused a double minimum system at R(O··O) 238 pm with $E = -262.13710$ hartrees with energy barrier 4.6 kJ mol^{-1}. These authors also calculated the energy differences between the open cis enol and the ring cis enol tautomer, C_s, (97.4 kJ mol^{-1}) and between the latter and the centred cis enol, C_{2v}, (13.4 kJ mol^{-1})[62].

Ab initio MO-LCAO-SCF calculations on the cis enol tautomer of MDA were done with a 4s2p/2s basis set on geometry optimized structures for the C_{2v} and C_s forms[104]. The results showed the latter to be 48 kJ mol^{-1} more stable than the former. The computed bond distances are given in Table 12. In a second paper correlation effects were included and the barrier height reduced to 42 kJ mol^{-1} [105]. A slightly different set of geometrical parameters and a minimal STO-3 G basis set gave an energy difference between C_{2v} and C_s of 27.6 kJ mol^{-1} [106]. This geometry optimization has however been criticized[100].

4-31 G calculations on MDA have also been performed[107], the details of which are given in Table 12, to determine the frequency of tunnelling. This is calculated to be 7×10^{11} s^{-1} ($\nu = 2.4$ cm^{-1}) and can be explained by assuming that only the enol proton moves ($C_s \rightarrow C_{2v} \rightarrow C_s$) while the other atoms of MDA remain in their average positions[107]. The computed barrier height seems rather large (67 kJ mol^{-1}).

During the investigation of sigmatropic rearrangements in various systems Radom also covered the cis enol tautomer of DMA using various basis sets, Table 12[108]. The proton transfer across the barrier was 40 kJ mol^{-1} [4-31 G with configuration interaction (CI)] which is to be compared with a barrier of 260 kJ mol^{-1} for proton transfer in 1,3-pentadiene (L) via a similar cyclic intermediate.

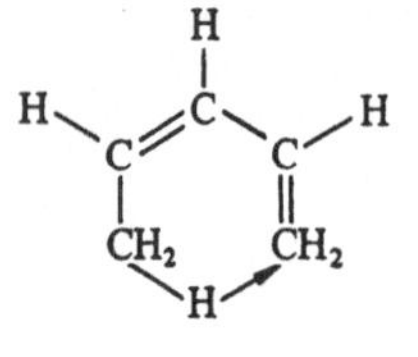

(L)

Table 12. Calculated hydrogen bond distances and total energies of the centred (C_{2v}) and non-centred (C_s) forms of MDA[a]

	C_{2v}	C_s	
R(O··O)/pm	229[100, 104, 109]	260[104], 256[106], 269[100]	
R(O··H)/pm	118.2[104], 116.8[106] 136.0[107], 119.1[100]		
R(O–H)/pm		97.5[104], 100[106], 93.5[107] 95.9[100]	
R(O···H)/pm		165[106], 179.9[107], 191.6[100]	
Basis set	Energies/hartrees	Energies/hartrees	ΔE^b/kJ mol^{-1}
STO-3 G[108]	−262.14532	−262.15587	28
STO-3 G[109]	−262.15577	−262.14531	28
4-31 G[108]	−262.21689	−265.23331	43
4-31 G[107]	−265.20177	−265.22723	67
4-31 G + CI[108]	−265.22368	−265.23922	40
6-31 G[109]	c	c	45
4 s2 p/2 s[104]	c	c	48
9 s5 p/4 s2 p/4 s[100]	−266.00124	−266.01472	36

a C_{2v}= 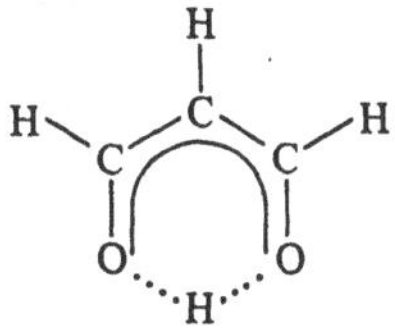 C_s= 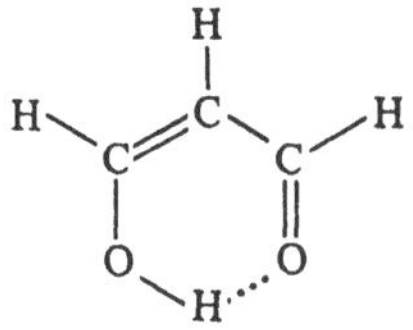

b Energy barrier to transition $C_s \rightarrow C_{2v}$
c Data not given

Calculations on MDA with semiempirical methods (INDO and CNDO/2) for C_{2v} and C_s conformers gave almost coincidental potential wells whereas STO-3 G shows the C_{2v} to have a single minimum but of higher energy than C_s by 44.5 kJ mol^{-1} (6-31 G basis set)[109].

The results of the most comprehensive theoretical analysis so far, by Bicerano et al.[100], with total geometry optimization and with a very large basis set are given in Table 12. The full bond details are shown in Fig. 8 and can be compared with the actual bond lengths and angle obtained from microwave spectroscopy in Fig. 7 (VIII). The agreement with the more stable C_s arrangement is quite good but the unusually long computed value for R(O··O) of 268.5 pm would suggest a weak, normal type (case 1) hydrogen bond in MDA whereas the observed value of 255 pm[94] shows it to be much stronger and certainly case 2. The final word on the theoretical approach to MDA has still to be written. A satisfactory meeting of experiment and theory may yet be achieved, as it has for the archetypal very strong hydrogen bond in HF_2^-[110].

B. The Energetics of Tautomerism and the Enol Hydrogen Bond

Whereas the energy of a hydrogen bond AHB can be defined with respect to AH + B (or A + HB if this energy is lower[111]), this definition assumes there to be no other interaction between AH and B before they hydrogen bond. When AH and B are part of the same molecule this simple definition must be qualified since there may be many conformers and energy levels against which the hydrogen bond, or bonds, may be defined. For β-dicarbonyls there is also the added complication of the enolization process itself and one can imagine different energy levels for all the nine tautomers of Fig. 1.

The only certainty in the ordering of these energy levels is that the cis enol, C_s form, is the lowest according to the theoretical calculations on MDA, and it would seem to be a reasonable assumption that the same is true of AA and most other β-dicarbonyls. One experimentally determined energy difference that can be related to this is the enthalpy of enolization ΔH_{enol} which has been accurately measured for AA in the gas phase as -8.9 kJ mol^{-1}[112]. For the liquid phase the value is -11.9 kJ mol^{-1}[13, 22, 112].

In solution the solvent has an effect, the range being from -7.5 in DMSO to -12.5 kJ mol^{-1} in cyclohexane[12]. Theoretically derived values for ΔH_{enol} for AA are -7.7[113] and -16 kJ mol^{-1}[114].

For other β-dicarbonyls there is a wide variation in ΔH_{enol} values: MDA, -29 kJ mol^{-1}, calculated[114]; α-chloroacetylacetone, -24.8 kJ mol^{-1}, measured[22]; α-methylacetylacetone, -25.5[13] or -5.56 kJ mol^{-1}[22] both measured values; BA, -16.3 kJ mol^{-1}, measured[13]; and DMB, -14.2 kJ mol^{-1}, measured[13].

From the point of view of this review the most important energy level difference relative to the cis enol C_s is the cis enol C_{2v}, the symmetrical or chelate tautomer. Several calculated values for this quantity, ΔE, the energy barrier to the transfer of the proton from one oxygen to the other, have been obtained and are listed in Table 12. Of these the most acceptable should be that of 36 kJ mol^{-1} computed with the largest basis set[100]. ^1H NMR studies[46] however on MDA place an upper limit of ca. 25 kJ mol^{-1} for the effective barrier to interconversion of the two C_s forms via the C_{2v}.

Unfortunately neither ΔH_{enol} nor ΔE tell us very much about the hydrogen bond energy, which we can define as the energy of the open cis enol relative to the hydrogen bonded cis enol C_s. The literature supplies three values for this quantity, all calculated for MDA, as -18.4[114], -33.1[115] and -97.4 kJ mol^{-1}[62]. From our knowledge of the strength of other OHO hydrogen bonds[48] the first of these is too low, lower even than

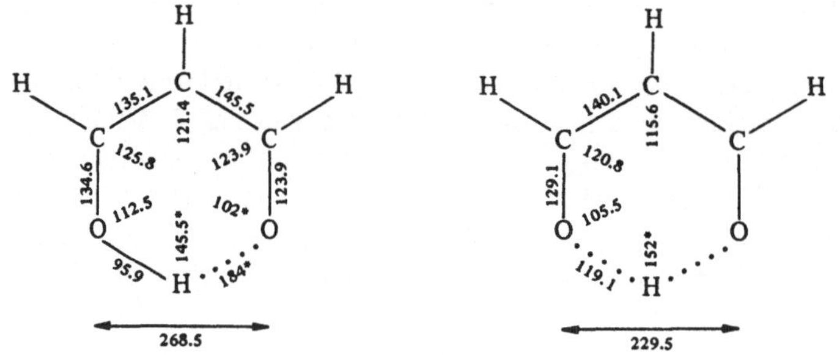

C_s — Lower energy state

C_{2v} — Transition state

*Calculated assuming a planar molecule around the hydrogen bond.

Fig. 8. Theoretically calculated structures of malondialdehyde enol tautomer[100]. Bond lengths in pm

most weak OHO bonds, and the second value of -33.1 kJ mol^{-1} still puts it in the category of weak hydrogen bonds. In calculating this, however, the R(O··O) value was taken as 296.4 pm (not optimized) which is 40 pm too long[115]. Very strong hydrogen bonds are generally >100 kJ mol^{-1} [110], the strongest being that of HF_2^- measured as 163 kJ mol^{-1} [116]. The third value of -97.4 kJ mol^{-1} thus seems most reasonable even though it is a little high for a case 2 or case 3 hydrogen bond. The average of these three values is 50 kJ mol^{-1} and this result has been used in compiling Fig. 9.

Figure 9 also contains the energy difference between a trans enol tautomer and the cis enol (C_s), -23 kJ mol^{-1} [115], which is surprisingly less than any of the energy differences

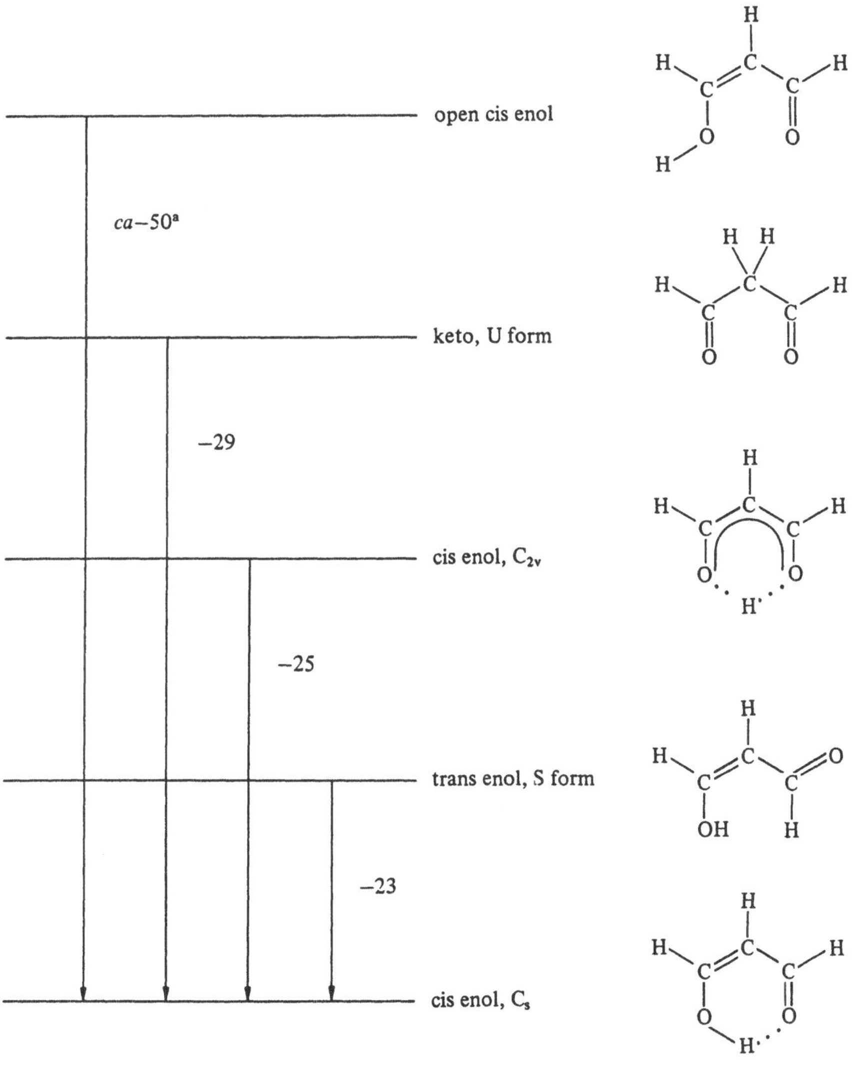

ᵃaverage of range −18 to −97 kJ mol^{-1}, see text

Fig. 9. The relative energies (kJ mol^{-1}) of the tautomers of MDA

discussed above but again suffers from the same criticism regarding the overlong R(O··O) distance in the cis enol tautomer. In view of Fig. 9 which shows that a fairly large input of energy is required to transform the trans enol to the cis open enol (>25 kJ mol^{-1}), it is somewhat surprising that more β-diketones do not show appreciable quantities of trans enol. However it does not follow that the same relative ordering of energies for MDA will be found with other β-dicarbonyls.

The change in hydrogen bond energy of an intramolecular bond that is part of a delocalized π system, over that without such delocalization, has been investigated by Kopteva and Shigorin[117]. By comparing systems of type (N) and (P) they showed an energy difference of ca. 60 kJ mol^{-1} between the two.

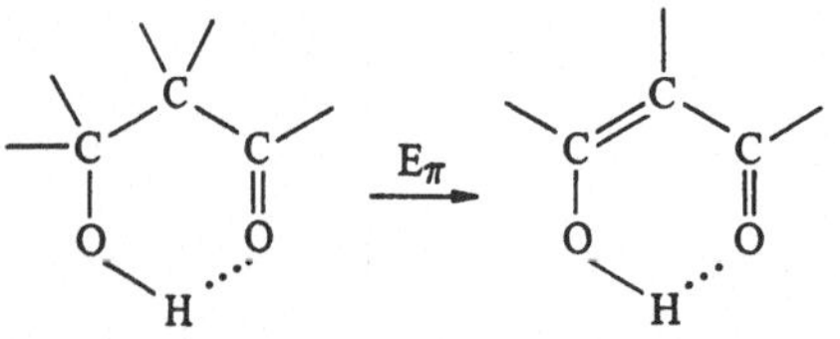

(N) (P)

Kreevoy and coworkers have devised a method for obtaining the shape of the potential energy well of a hydrogen bond[118, 119]. The method is based on isotopic fractionation factors, φ, defined as the ratio of D to H in the compound in question, compared to the D to H ratio of the surrounding medium. For HX ⇌ DX in H_2O–D_2O then ϕ_X is given by Eq. (8).

$$\phi_X = \frac{[DX]}{[HX]} \cdot \left(\frac{H}{D}\right)_{aq} \tag{8}$$

The method was tested on a variety of hydrogen bonded species including very strong hydrogen bonds such as HF_2^- and $H(CF_3CO_2)_2^-$ [119], and ϕ_X was shown to be related to potential energy function V, Eq. (9),

$$V = Ax^4 + Bx^2 \tag{9}$$

which describes one-dimensional curves of the double minimum type used in representing hydrogen bonds.

Isotopic redistribution of HFAA with Bu^{t}OH/Bu^{t}OD in acetonitrile was measured and the concentrations determined spectroscopically at 2650–3000 nm[120]. A value of φ, 0.53, for the hydrogen bonding proton of HFAA was obtained. The potential energy diagram matching this is shown in Fig. 10.

A unique piece of information which this approach provides is the "true" barrier height above the minima. In the case of HFAA this is 2870 cm^{-1}, equivalent to 34 kJ mol^{-1}. The zero point energy levels of H and D above the minima are 1370 and 1040 cm^{-1} or 16 and 12 kJ mol^{-1} respectively. In terms of the energies required to reach the top of the barrier these are 18 and 22 kJ mol^{-1} respectively, which compare well with the 25 kJ mol^{-1} for the proton form of MDA (Fig. 9), proposed as the upper limit for this molecule[46]. This comparison is not inappropriate as both HFAA and MDA have the

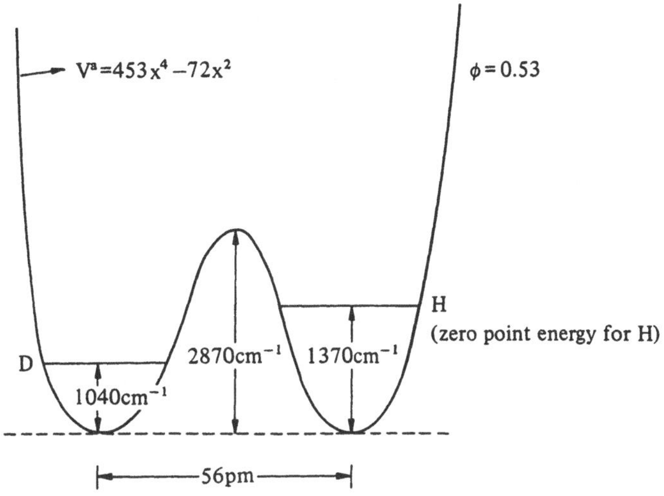

^{a}V in units $(cm^{-1})/10^3$

Fig. 10. The potential energy well of the enol tautomer hydrogen bond of HFAA[120)]

same R(O··O) of 255 pm[68, 96)], although the value for HFAA was obtained from an electron diffraction determination which showed the hydrogen bond to be linear and symmetric, so it may be unreliable.

VI. Vibrational Spectra of β-Dicarbonyls

The infrared (IR) and Raman (R) spectra of β-diketones show the vibrational absorptions of both the keto and enol forms if the spectra are run as neat liquids or in solvents. The keto form of AA shows two carbonyl stretching frequencies, ν(C=O), at 1727 and 1707 cm^{-1}, both very strong bands in the IR, with R counterparts at 1719 (polarized) and 1697 (depolarized) cm^{-1}. These are the in-phase and out-of-phase stretching modes of a molecule in which the dihedral angle between the two groups is ca. 90°[121)]. Needless to say it is not the keto form which has commanded the attention of spectroscopists but the cis enol form; and although a great deal of effort has been put into analysing the spectra of the β-diketones a definitive assignment of all vibrations of AA is still awaited.

For studying a system containing hydrogen bonds no techniques are better than those of vibrational spectroscopy, at least in theory[122)]. Each component of the equilibrium should be observable, and by means of isotopic substitution it should be possible unambiguously to assign all the motions which involve the hydrogen atoms. Correlations of IR band changes, Δν, and band intensities with hydrogen bond lengths, energies and chemical shifts enables a great deal of information to be got from the vibrational spectrum that cannot be obtained in other ways[122)].

The vibrational spectra of the following β-diketones and their deuteriated analogues have been published: AA, R & IR of the gas[123], CCl_4 solution[124], neat[121], solid[125], and MeCN solution[125]; BA[126]; DBM[126]; TFAA[126]; HFAA[123, 126]; TAE[126]; α-chloro-AA, neat[127], gas[126], solid[126]; α-methylthio-AA[127]; α-cyano-AA[124].

It is the vibrational modes of the cis enol ring which can reveal information about the hydrogen bonding. If ν(C=O) and ν(C–O) can be distinguished, or ν(C=C) and ν(C–C), then the clear assumption would be that the hydrogen bond was non-centred. The question then to be answered is the nature of the barrier of the potential well. Figure 11 shows the vibrations in which we are most interested. Of the 3n-5, i.e. 13, modes of a six-membered ring seven are significant, the rest are in-plane and out-of-plane ring deformations not involving the hydrogen atom.

The most comprehensive analysis of β-diketones[125, 126] proves the C_s symmetry is the correct one for the β-diketones involving CH_3, Ph and CF_3 groups. The band assignments are given in Table 12. The effect of deuteriation of the olefinic proton is also noticeable on the modes ν(C=C) and ν(C–C), especially the latter, but it is the effect of deuteriation on the hydrogen bond which has far reaching implications.

The isotopic ratio ν_H/ν_D for OHO hydrogen bonds can vary from ca. 1.35, the expected ratio, down to ca. 1.00 which in fact means that isotopic substitution has no effect at all[128]. Novak who investigated this anomaly linked it to the O··O internuclear distance, noting that the nil isotope effect was reached as R(O··O) approached 250 pm. Below this bond length the isotope effect reappeared and returned to its normal value of ca. 1.35 at 244 pm. This span of R(O··O) values is that of the β-diketones in their cis enol tautomers but Wood et al.[125, 126] found only ν_H/ν_D values of 1.34–1.38 (Table 13) for

Table 13. Vibrational band frequencies for β-diketones etc. (cm^{-1})[a]

β-Diketone[b]	ν_{as}(OH)[c]	ν_H/ν_D	γ(OH)	ν(C=O)[d]	ν(C–O)[d]	ν(C=C)[e]	ν(C–C)[e]
AA	2750	1.36	957	1625	1460	1600	1298
$^2H_2$AA	2020		707	1635	1390	1544	1082
BA	2650	1.35	952	1605	1458	1580	1274
$^2H_2$BA	1960		720	1602	1296	1520	1072
DBM	2620	1.34	975	1560	1470	1560	1282
$^2H_2$DBM	1950		725	1565	1322	1504	1045
TFAA	2900	1.37	892	1660	1427	1600	1277
$^2H_2$TFAA	2120		650	1657	1390	1530	1044
HFAA	3000	1.38	856	1692	1440	1632	1289
$^2H_2$HFAA	2180		642	1683	1405	1598	1012
AA^- [f]				1575	1380	1520	1282
AA(calcd)[g]				1622	1411	1600	1216

[a] Data from Refs. 121, 123–126
[b] $^2H_2$ refers to the deuteriated compounds
[c] from gas phase spectra
[d] from IR spectra of gas phase, neat liquids or solutions
[e] from R spectra likewise
[f] Ref. 121
[g] Ref. 111, this also calculated δ(OH) as 1457 cm^{-1}

C_s C_{2v}

Hydrogen bond vibrations

$\nu_{as}(OH)$ $\nu_{as}(OHO)$

$\delta(OH)$ $\delta(OHO)$

$\gamma(OH)$ $\gamma(OHO)$

out of plane   out of plane

Ring skeletal vibrations

$\nu(C=O)$ $\nu_s(C \dddot{} O)$

$\nu(C-O)$ $\nu_{as}(C \dddot{} O)$

$\nu(C=C)$ $\nu_{as}(C \dddot{} C)$

$\nu(C-C)$ $\nu_s(C \dddot{} C)$

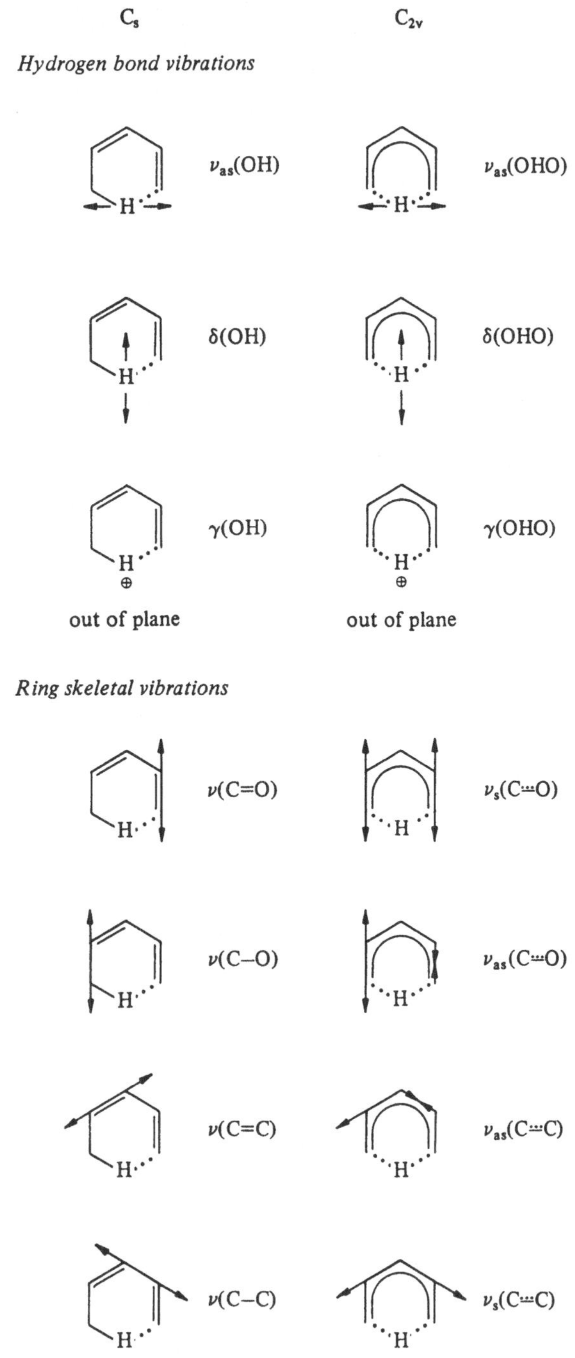

Fig. 11. Vibrational modes of the *cis* enol tautomer ring

compounds in this range. Ignoring the R(O··O) vs ν_H/ν_D relationship – on the grounds that the hydrogen bonds are not linear – means that these OHO bonds are either strong or very weak. Wood decided that the latter was the more likely and that the potential well is a double minimum with a high barrier and that if it were free of molecular constraints it would normally have R(O··O) in excess of 260 pm. It is the opinion of this reviewer that the former choice would be more appropriate in view of the other evidence of R(O··O), δ(OHO), $\Delta\delta(^1H, ^2H)$ etc. In other words the IR and R evidence supports strong hydrogen bonding – a double minimum but with a low barrier, at least for the proton if not for the deuteron[2].

The change in the OH stretching frequency on formation of a hydrogen bond, $\Delta\nu$(OH), is the most notable feature of the vibrational spectum. As R(O··O) shortens, the hydrogen bond strengthens and $\Delta\nu$(OH) increases. A direct relationship between $\Delta\nu$(OH) and bond energy was commonly used to estimate the latter quantity[129)] but the method has not stood the best of time and has fallen into disuse. Even so the $\Delta\nu$(OH) could be used to estimate the bond energy (open cis enol to C_s cis enol) if the IR band corresponding to the former were known. Early work on AA, TFAA, etc. as a neat liquid and in MeCN solution detected bands at 3600–3500 cm^{-1} that were assigned to trans enol[130)] and indeed the free OH of this and the open cis enol tautomers would be expected to be ca. 3600 cm^{-1}. This being so then $\Delta\nu$(OH) for the β-diketones in Table 13 varies over the range 600–1000 cm^{-1}, and consequently ΔE over the range of ca. 30–50 kJ mol^{-1} [129)]. The upper limit, for DBM, corresponds to the quantity averaged in the previous section. The order of bond strength, based on ν_{as}(OH) is TAE > DBM > BA > AA > TFA > HFAA. The lowest ν_{as}(OH) is that of TAE at 2540 cm^{-1} [125)].

The second most surprising feature of the IR bands of hydrogen bonds is their broadness and intensity. For the ν_{as}(OH) bands the intensity is not particularly strong but their broadness is, and the width at half band height of this band in AA is nearly 1000 cm^{-1}. On deuteriation the bands become significantly narrower[126)] and for ^2H_2AA is 400 cm^{-1}. Intensity is of little use in studying hydrogen bonds because they overlap so many other modes in most spectra. It has, however, been noted that intensity decreases as the symmetry and strength increases, and for the hydrogen maleate ion ν_{as}(OH) is indistinguishable from the background noise. This particular OHO bond is very strong and R(O··O) is 239 pm[131)]. The intensity in the β-diketones is surprisingly less than expected, which again suggests these have strong OHO bonds.

The bending modes of the hydrogen bond should be above 1000 cm^{-1} for the in plane mode δ(OH), and below this for the out of plane mode γ(OH). The in plane mode was calculated to fall at 1457 cm^{-1} [123)] and a band at 1460 cm^{-1} was allotted to this vibration. Other analysts discounted this assignment[125, 132)] in favour of its being ν(C–O). Both δ(OH) and δ(OD) remain unidentified in the β-diketones and it seems reasonable to infer that they are so because this vibration is coupled to some other mode.

The out-of-plane bending mode, γ(OH), predicted to come at 945 cm^{-1} in AA[123)], is identifiable by its broadness in the region 1000–860 cm^{-1}. It shifts to lower wavenumbers on deuteriation (Table 13) and is R inactive. Often the deuteriated mode is observable

---

2 Wood[126)] also deduces that the barrier is high enough to prevent tunnelling which would otherwise cause splitting of ν(C=O) etc. Also he finds no tunnelling frequency ν_τ in the far IR. But see below

when the protonated band is hidden by other vibrations and this prompted Ogashi[133] to use γ(OD) rather than γ(OH) in illustrating the close linear relationship between this and the ^{1}H NMR δ(OHO) for 14 β-dicarbonyls. Again an order of bond strengths for the β-diketones is obtained in agreement with that above based on ν_{as}(OH). A close link between the hydrogen bond and the π conjugation around the ring was also inferred[133].

Theoretical treatment of the behaviour of the hydrogen bonded proton in MDA gave the following computed values[134]: ν_{as}(OH) 3094; δ(OH) 2039; γ(OH) 944 and 791 cm^{-1}. For the γ(OH) vibrations there was coupling with C–H bending (944 cm^{-1}) or with C–C–C bending (791 cm^{-1}). The optimized geometries of the C_s and C_{2v} forms of MDA are significantly different from those determined by Bicerano[100] – Fig. 8 – so that this theoretical approach may not be accurate in its predictions of frequencies but the work has two novel features. It predicts δ(OH) much higher than previously expected to that this mode may be obscured in other β-diketones. The carbonyl bands in the spectra may cover δ(OH) were it to be in the 1600–1800 cm^{-1} region. The authors[134] also put forward the suggestion that the proton moves in an ellipse. The proton transfer rate from one well to the other is calculated to be 9.77×10^{13} s^{-1}, the barrier height being less than 5 kJ mol^{-1} [134].

A. Proton Tunnelling in Double Minimum Potential Energy Hydrogen Bonds

Waves can tunnel through an energy barrier as can particles, such as electrons, which have significant wave nature. And so can the proton. The "tunnel effect" is the term used to describe the mechanism by which a particle can escape through a potential energy barrier, a barrier which in chemical terms should restrict it because its height exceeds the kinetic energy of the particle. Quantum mechanics provides the escape route via the wave properties which are significant for the smallest of particles. Since there is a finite probability of occurrence of the wave within the barrier, and beyond it, then the particle has a probability of existence beyond the barrier and thus provides a tunnel through the barrier for its escape[135]. Heavier particles such as the deuteron have much less chance of tunnelling and this can lead to anomalous isotope effects[135].

In discussing certain β-diketones tunnel effects are important. Theoretical and experimental work on MDA has dealt with the problem. For MDA the tunnelling motion is of large amplitude, and complicated the structural determination by microwave spectroscopy[94, 107]. The tunnelling energy of separation of the proton was calculated to be 26 ± 10 cm^{-1} and a far IR band near 21 cm^{-1} was assigned to this. MDA can exchange the proton between the two oxygen atoms by different routes – the reaction coordinate path or the direct transfer path. Both are symmetric but with different energy barriers of 48.5 and 66.7 kJ mol^{-1} respectively and different interminimal distances of 45.7 and 87.7 pm respectively[107]. The corresponding tunnelling frequencies are 2.27×10^{10} and 7.11×10^{10} Hz, the latter corresponding to 2.4 cm^{-1}, which is less than the experimentally estimated value.

The role of symmetry in the tunnelling of the proton between a double minimum pair of potential energy wells is crucial. If the initial and final states are indistinguishable tunnelling will occur but a slight change in structure which destroys this symmetry will

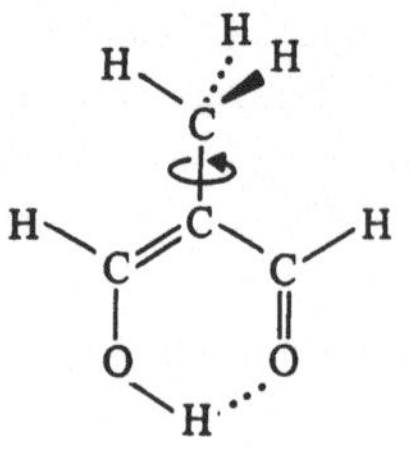

(Q)

quench tunnelling through all but the lowest energy barriers[136]. The rotation of the α-methyl group in methylmalondialdehyde (Q) is sufficient to do this.

It was shown by Sanders[96] in his microwave investigation of (Q) that proton tunnelling (inversion) and methyl rotation (torsion) are strongly coupled. The methyl group can be symmetric with respect to the rest of the molecule when one of its CH bonds is in a plane perpendicular to the principal plane of the molecule, which is the plane of the ring. Rotation from this geometry destroys the symmetry [as in (Q)] and this reduces the rate of proton tunnelling[137]. This explains why the methyl rotation and proton transfer are so coupled.

Another way of disturbing the symmetry of MDA very slightly, so as only partially to quench the tunnel effect, is to substitute ^{13}C or ^{18}O isotopes into the molecule[94]. Substituting a deuteron in place of the proton is a more dramatic change which quenches tunnelling by doubling the mass of the tunnelling particle. This course of action was necessary in order to determine accurately the structure of MDA itself[94].

Naphthazarin, Fig. 12 (I), and its mono-methyl and di-methyl derivatives, Fig. 12 (II), (III), (IV), are β-dicarbonyls that have been used to study the influence of symmetry on tunnelling[138]. They have no option but to exist in the cis enol configurations. A great deal of work has been done on the naphthazarins because of their structural relationship to certain anti-cancer antibiotics[139]. Debate has turned on the nature of the hydrogen bonds – are they centred or non-centred?

The ^1H NMR spectra of naphthazarin itself and the 2,6- and 2,7-dimethyl derivatives show only one signal for the ring protons δ(C–H) and this was interpreted as evidence for fast interconversion between the two tautomers[140]. For naphthazarin the single line is only resolved at − 190 °C in the solid state[139]. An early interpretation of the ^1H NMR data for the monomethyl and other monosubstituted derivatives favoured a dominant tautomer rather than a centred hydrogen bond[141], and recent SCF calculations[138] have confirmed that the non-centred bond is of lower energy than the centred structure. The calculated energy levels of the barrier, ΔE, are given in Fig. 12.

The monomethyl naphthazarin, (II), shows different CH environments, a singlet at 7.19 and a quartet at 6.89 ppm, with J_{HCCMe} = 1.4 Hz[138]. The alternative tautomer (R) was calculated to be only 3.8 kJ mol^{-1} higher in energy than the form shown in Fig. 12 (II) yet for this derivative no proton exchange occurs, compared to naphthazarin and the dimethyl derivatives, which have rates of exchange of $2 - 4 \times 10^4$ Hz. The rate of exchange for their deuteriated derivatives is $1 - 18 \times 10^3$ Hz.

The calculated R(O–H) values for the non-centred naphthazarin is 99.5 pm and 120.4 pm for the centred arrangement. The energy barriers are higher than in the β-diketones but structural and energy considerations do not serve to explain tunnelling in

Compound	δ(OH)/ppm	δ(CH)/ppm	ΔE/kJ mol^{-1}
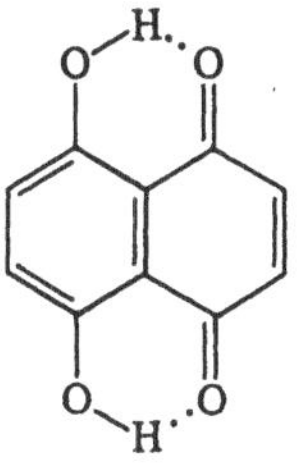 (I)	12.43	7.13[a]	117.1
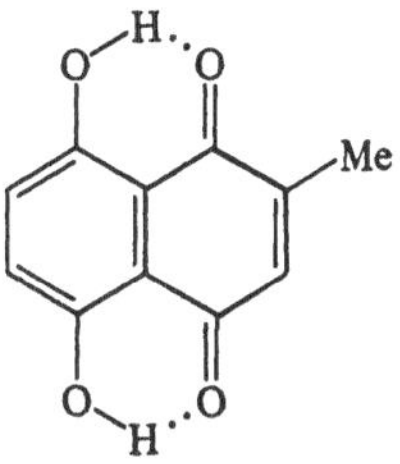 (II)	12.41 12.53	7.19 6.89	117.5
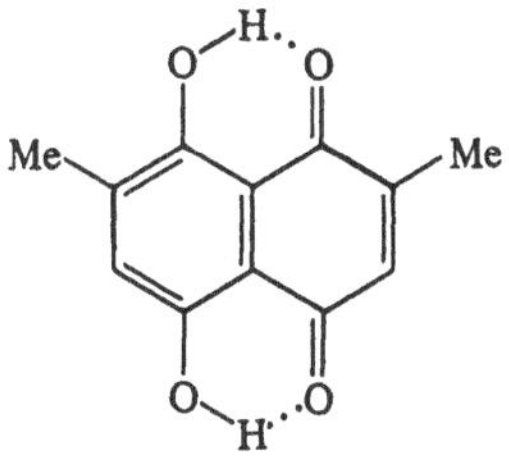 (III)	12.48 12.96	6.96	114.2
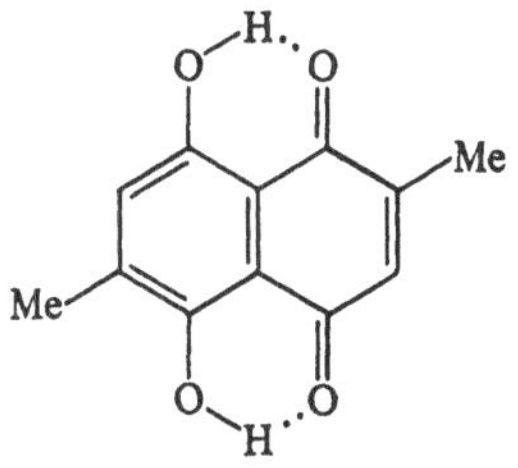 (IV)	12.72	6.96	113.7

[a]Unresolved even at −90°C

Fig. 12. Naphthazarin and its methyl derivatives

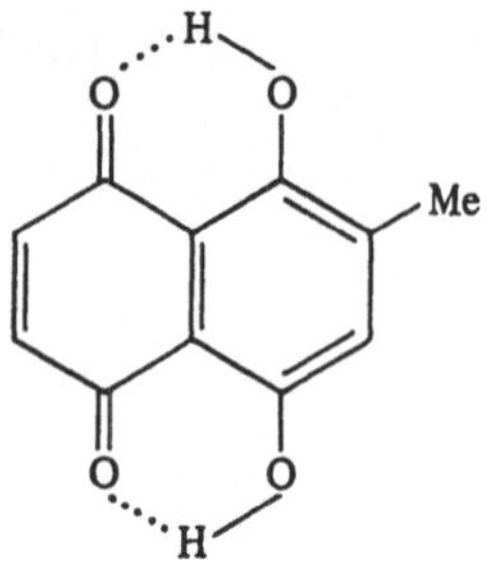

(R)

(I), (III) and (IV) but not (II). The loss of symmetry is put forward as the explanation, and this quenches tunnelling sufficiently to resolve the ring CH protons in the ^{1}H NMR spectrum.

VII. The *β*-Thioxoketones

Neither a symmetric hydrogen bond, nor proton tunnelling is possible for the intramolecular tautomers of the *β*-thioxoketones. Nevertheless these analogues have an added dimension of their own – the possibility of the cyclic tautomer being either the enethiol or enol forms, (S) or (T), with possibly an equilibrium mixture of the two. Nor are OHS hydrogen bonds necessarily weaker than OHO bonds although a distinction can be made between O–H···S and O···H–S with the latter weaker than the former[142]. Early work by Duus and Lewesson[143] showed by ^{1}H NMR and IR spectroscopy that only the enethiol forms (S) were present in the *β*-thioxoketone systems TAA (where X = Y = Me), and for X = Y = Ph and X = Me, Y = Ph. For TAA the *β*-diketone tautomer was also detected.

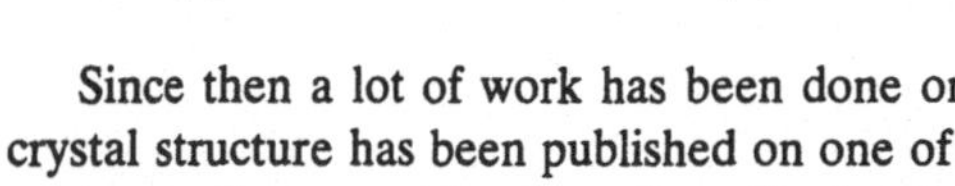

enethiol (S) enol (T)

Since then a lot of work has been done on *β*-thioxoketones and recently an X-ray crystal structure has been published on one of them which proves conclusively that the proton prefers to reside not on the sulfur but the oxygen atom – at least in the solid state. Figure 13 gives this structure[144]. The chemical shift of the hydrogen bonding proton, δ(OHS), is 11.77 ppm[144] which is similar to that of TAA, 11.48 ppm.

That the enol form (T) is the more stable was predicted by CNDO calculations on monothio analogues of MDA[145] which showed the O–H··S hydrogen bond to be twice as

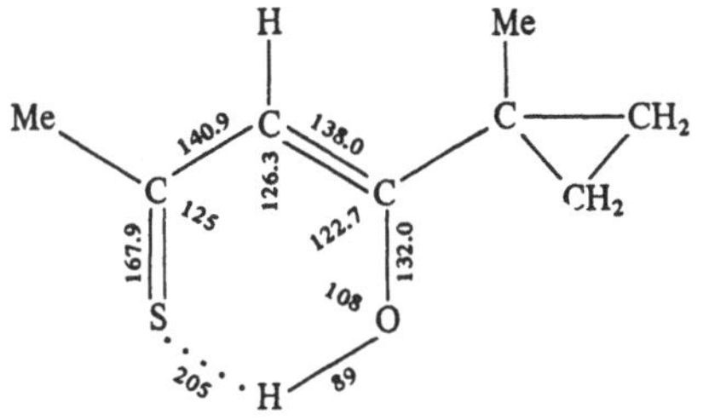

Fig. 13. X-ray crystal structure of 1-(1-methylcyclopropyl)-3-thioxobutan-1-one[144]. Bond lengths in pm

strong as the O···H–S alternative. The calculated energy differences and geometries of this compound in its open and closed cis enol and enethiol tautomers is shown in Fig. 14.

TAA was shown by low temperature UV studies at 95 °K to be the enol form[146]. Irradiation at this temperature gave the enethiol tautomer. At room temperature TAA has UV bands at 296 nm due to π,π^* of the enethiol form and at 354 nm due to π,π^* of the enol form. The former disappears on cooling but reappears at 288 nm on irradiation (the blue shift of 8 nm being due to the different temperature).

Other workers[147] have suggested TAA consists of the enethiol and a trans tautomer (U) which is the unidentified species reported by earlier investigators[148]. Matrix IR showed a band at 2507 cm^{-1} which was assigned to the enethiol ν_{as}(S–H···O) tautomer whereas the trans conformer (U) had a sharp band at 2558 cm^{-1} assigned to ν(S–H). The

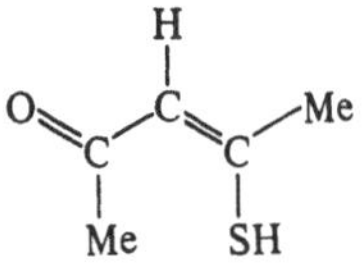

(U)

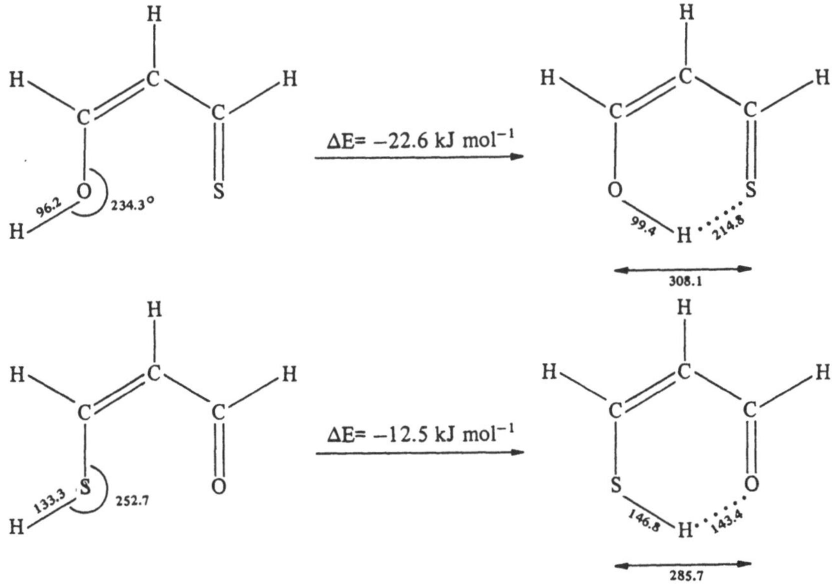

Fig. 14. Calculated hydrogen bond parameters and energies of monothiomalondialdehyde[145]. Bond lengths in pm

cis enethiol and trans enethiol tautomers could be interconverted by UV[147]. X-ray photoelectron spectroscopy has been used to record the O_{1s} and S_{2p} ionization spectra of TAA and other β-thioxoketones. This shows that both the enol and enethiol tautomers are detectable and for TAA the ratio of these is 61 : 39% respectively[149].

β-Thioxoketones with aromatic β-substituents show an equilibrium mixture of 4 : 1 of the enol : enethiol conformers[150]. Structures of some five and six-membered 2-acylcycloalkanethiones and 2-thioacylcycloalkanones have been shown by ^{1}H NMR, IR and UV spectroscopy also to exist as equilibrium mixtures of the enol and enethiols which interconvert very rapidly by intramolecular chelate proton transfer[151].

VIII. Conclusion

The hydrogen bonding of the cis enol tautomers of β-diketones is strong (> 50 kJ mol^{-1}) but not very strong (> 100 kJ mol^{-1}), short (245–255 pm) but not very short (< 240 pm), non-centred and non-linear. The proton of this OHO bond finds itself in a double minimum potential energy well separated by an energy barrier that is not much higher than its ground state vibrational energy level and through which it tunnels easily if symmetry requirements are met. The height of the barrier is governed by R(O··O) and with certain β-substituents it may be very low indeed, so that it approximates to a symmetrically centred hydrogen bond. Needless to say this hydrogen bond is the key factor in determining a lot of the chemistry of these compounds including the keto ⇆ enol equilibrium. In the spectrum of hydrogen bonds the cis enol tautomers occupy a niche intermediate between the vast number of weak, or so-called normal, hydrogen bonds and the growing number of very strong hydrogen bonds. As yet they remain the most thoroughly researched examples of their kind, but a growing number of other intramolecular hydrogen bonded molecules probably fall into this class such as tropolone [46,152], 3-hydroxyflavone[153], 6-hydroxy-2-formylfulvene[46,154], 9-hydroxyphenalenone[46,155], and phenylazoresorcinol[156]. The likely symmetry of the hydrogen bonding in these and other potentially strong systems has been discussed[89].

IX. References

1. For instance, Meyer, K.: Ber. dt. Chem. Ges. *44,* 2718 (1911); ibid. *45,* 2843 (1912)
2. Calvin, M., Wilson, K. W.: J. Am. Chem. Soc. *67,* 2003 (1945)
3. Holm, R. H., Cotton, F. A.: ibid. *80,* 5658 (1958)
4. Williams, D. E., Dumke, W. L., Rundle, R. E.: Acta Cryst. *15,* 627 (1962)
5. Noy, R. S., Grindin, V. A., Ershov, B. A., Kol'tsov, A. I., Zubko, V. A.: Org. Mag. Res. *7,* 109 (1975)
6. Bothner-By, A. A., Harris, R. K.: J. Org. Chem. *30,* 254 (1965)
7. George, W. O., Mansell, V. G.: J. Chem. Soc. B 132 (1968)
8. Yoffe, S. T., Petrovsky, P. V., Goryunov, Ye. I., Yershova, T. V., Kabachnik, M. I.: Tetrahedron *28,* 2783 (1972)
9. Veierov, D., Bercovici, T., Fischer, E., Mazur, Y., Yoger, A.: J. Am. Chem. Soc. *99,* 2723 (1977)
10. Kol'tsov, A. I., Kheifets, G. M.: Russ. Chem. Revs. *40,* 773 (1971)
11. Strohmeier, W., Hühne, I.: Z. Naturforsch. *7b,* 184 (1952)

12. Spencer, J. N., Holmboe, E. S., Kirschenbaum, M. R., Firth, D. W., Pinto, P. B.: Can. J. Chem. *60,* 1178 (1982)
13. Allen, G., Dwek, R. A.: J. Chem. Soc. B 161 (1966)
14. Rogers, M. T., Burdett, J. L.: Can. J. Chem. *43,* 1516 (1965)
15. Reeves, L. W.: ibid. *35,* 1351 (1957)
16. Kol'tsov, A. I., Golubev, N. S., Milevskaya, I. S.: J. Gen. Chem. USSR *50,* 2025 (1980): transl. Zh. Obshch. Khim. *50,* 2504 (1980)
17. Kondo, G., Takemoto, T., Ikenone, T.: J. Chem. Soc. Japan, Ind. Chem. Sect. *68,* 1404 (1968)
18. Kato, M., Watarai, H., Suzuki, N.: Can. J. Chem. *55,* 1473 (1977)
19. Reichardt, C.: Angew. Chem. Int. Edn. Engl. *1,* 29 (1965)
20. Gutman, V.: ibid. *9,* 843 (1970)
21. Thompson, D. W., Allred, A. L.: J. Phys. Chem. *75,* 433 (1971)
22. Rogers, M. T., Burdett, J. L.: ibid. *70,* 939 (1966)
23. Kol'tsov, A. I., Ershov, B. A.: J. Org. Chem. USSR *11,* 440 (1975); transl. Zh. Org. Khim. *11,* 450 (1975)
24. Clarke, J. H.: J. Chem. Soc. Perkin I, 1744 (1977)
25. Clarke, J. H.: J. Chem. Soc. Perkin II, 1326 (1978)
26. Clarke, J. M., Miller, J. M.: J. Chem. Soc. Perkin I, 2063 (1977)
27. Leipert, T. K.: Org. Mag. Res. *9,* 157 (1977)
28. Raban, M., Yamamoto, G.: J. Org. Chem. *42,* 2549 (1977)
29. Emsley, J., Kuroda, R., Parker, R. J.: unpublished results
30. Burdett, J. L., Rogers, M. T.: J. Am. Chem. Soc. *86,* 2105 (1964)
31. Mollin, Yu. N., Ioffe, S. T., Zaev, E. E., Solov'eva, E. K., Kuchneva, E. E., Voevodskii, V. V., Kabachnik, M. I.: Izv. Akad. Nauk SSSR Ser. Khim. 1556 (1965)
32. Yoffe, S. T., Fedin, E. I., Petrovskii, M. I., Kabachnik, M. I.: Tet. Lett. 2661 (1966)
33. Yoshida, Z., Ogoshi, H., Tokumitsu, T.: Tetrahedron *26,* 5691 (1970)
34. Wirzchowski, K. L., Shugar, D., Katritski, A. R.: J. Am. Chem. Soc. *85,* 827 (1963)
35. Forsen, S., Nilsson, M.: Acta Chem. Scand. *14,* 1333 (1960)
36. Houk, K. N., Davis, L. P., Newkomo, G. R., Duke Jr., R. E., Nauman, R. V.: J. Am. Chem. Soc. *95,* 8365 (1973)
37. Dahlberg, D. B., Long, F. A.: ibid. *95,* 3825 (1973)
38. Sardella, D. J., Heinert, D. H., Shapiro, B. L.: J. Org. Chem. *34,* 2817 (1969)
39. Gindin, E. A., Emelina, E. E., Ershov, B. A., Kloze, G., Kol'tsov, A. I., Shapet'ko, N. N.: Zh. Org. Khim. *5,* 1890 (1969)
40. Nonhebel, D. C.: Tetrahedron *24,* 1869 (1968)
41. Lintvedt, R. L., Holtzclaw Jr., H. F.: Inorg. Chem. *5,* 239 (1966)
42. Lintvedt, R. L., Holtzclaw Jr., H. F.: J. Am. Chem. Soc. *88,* 2713 (1966)
43. Bell, R. P.: The Proton in Chemistry, 2 edn., London, Chapman & Hall 1973
44. Long, F. A., Watson, D.: J. Chem. Soc. 2019 (1958)
45. Riley, T., Long, F. A.: J. Am. Chem. Soc. *84,* 522 (1962)
46. Brown, R. S., Tse, A., Nakashima, T., Haddon, R. C.: ibid. *101,* 3157 (1979)
47. Robinson, M. J. T., Rosen, K. M., Workman, J. D. B.: Tetrahedron *33,* 1655 (1977)
48. Emsley, J.: Chem. Soc. Revs. *9,* 91 (1980)
49. Shapet'ko, N. N., Kessenikh, A. V., Skoldinov, A. P., Protopopova, T. K.: Teor. Eksper. Khim. *2,* 757 (1966)
50. Garbisch, E. W.: J. Am. Chem. Soc. *85,* 1696 (1963)
51. Forsen, S., Nilsson, M.: Arkiv Kemi *19,* 41 (1963); ibid. *20,* 41 (1963)
52. Shapet'ko, N. N.: Org. Mag. Res. *5,* 215 (1973)
53. Gindin, V. A., Chipum, I. A., Ershov, B. E., Kol'tsov, A. I.: ibid. *4,* 63 (1972)
54. Poplett, I. J. F., Sabir, M., Smith, J. A. S.: J. Chem. Soc. Faraday Trans 2 *77,* 1651 (1981)
55. Lowe, J. U., Fergusson, L. N.: J. Org. Chem. *30,* 3000 (1965)
56. Shapet'ko, N. N., Radushnova, I. L., Bogachev, Yu. S., Berestova, S. S., Potapov, V. M., Kiryuschkina, G. V., Talebarovskaya, I. K.: Org. Mag. Res. *7,* 540 (1975)
57. Gorodetsky, M., Luz, Z., Mazur, Y.: J. Am. Chem. Soc. *89,* 1183 (1967)
58. Shapet'ko, N. N., Berestova, S. S., Lukovkin, G. M., Bogachev, Yu. S.: Org. Mag. Res. *7,* 237 (1975)

59. Berestova, S. S., Shapet'ko, N. N., Shigorin, D. N., Medvedeva, V. G., Skoldinov, A. P., Plakhina, G. D., Andreichikov, Yu. S.: Theoret & Exper. Chem. *15,* 449 (1979); transl. of Teoret. Eksp. Khim. *15,* 575 (1979)
60. Shapet'ko, N. N., Berestova, S. S., Medvedeva, V. G., Skoldinov, A. P., Andreichikov, Yu. S.: Doklady Acad. Nauk USSR *234,* 566 (1977); transl. of Doklady Acad. Nauk USSR *234,* 876 (1977)
61. Lazaar, K. I., Bauer, S. H.: J. Phys. Chem. *87,* 2411 (1983)
62. Isaacson, A. D., Morokuma, K.: J. Am. Chem. Soc. *97,* 4453 (1975)
63. Chan, S. I., Lin, L., Clutter, D., Dea, P.: Proc. Nat. Acad. Sci. *65,* 816 (1970)
64. Shapet'ko, N. N., Bogachev, Yu. S., Radushnova, I. L., Shigorin, D. N.: Doklady Acad. Nauk USSR *231,* 1085 (1976); transl. of Doklady Acad. Nauk USSR *231,* 409 (1976)
65. Altman, L. A., Laungani, D., Gunnarsson, G., Wennerström, H., Forsén, S.: J. Am. Chem. Soc. *100,* 8264 (1978)
66. Gunnarsson, G., Wennerström, H., Egan, W., Forsén, S.: Chem. Phys. Letts. *38,* 96 (1976)
67. Egan, W., Gunnarsson, G., Bull, T. E., Forsén, S.: J. Am. Chem. Soc. *99,* 4568 (1977)
68. Andreassen, A. L., Zebelman, D., Bauer, S. H.: ibid. *93,* 1148 (1971)
69. Jones, R. D. G.: Acta Cryst. *B32,* 1807 (1976)
70. Evans, D. F.: J. Chem. Soc. Chem. Comm. 1226 (1982)
71. Lowrey, A. H., George, C., D'Antonio, P., Karle, J.: J. Am. Chem. Soc. *93,* 6399 (1971)
72. Andreassen, A. L., Bauer, S. H.: J. Mol. Struct. *12,* 381 (1972)
73. Abu-Dari, K., Raymond, K. N., Freyberg, D. P.: J. Am. Chem. Soc. *101,* 3688 (1979)
74. Emsley, J., Jones, D. J., Lucas, L.: Revs. Inorg. Chem. *3,* 105 (1981)
75. Joswig, W., Fuess, H., Ferraris, G.: Acta Cryst. *B38,* 2798 (1982)
76. Olovsson, I., Jönsson, P.-G.: The Hydrogen Bond, Vol. II, p. 393, Amsterdam, North-Holland Publishing Co. 1976
77. Engelbretson, G. R., Rundle, R. E.: J. Am. Chem. Soc. *86,* 574 (1964)
78. Williams, D. E., Dumke, W. L., Rundle, R. E.: Acta Cryst. *15,* 627 (1962)
79. Schaefer, J. P., Wheatley, P. J.: J. Chem. Soc. (A) 528 (1966)
80. Norrestam, R., von Glehn, M., Wachtmeister, C. A.: Acta Chem. Scand. *28B,* 1149 (1974)
81. Camerman, A., Mastropaolo, D., Camerman, N.: J. Am. Chem. Soc. *105,* 1584 (1983)
82. Power, L. F., Jones, R. D. G.: Acta Cryst. *B27,* 181 (1971)
83. Semmingsen, D.: Acta Chem. Scand. *26,* 143 (1972)
84. Jones, R. D. G.: Acta Cryst. *B32,* 2133 (1976)
85. Hollander, F. J., Templeton, D. H., Zalkin, A.: ibid. *B29,* 1552 (1973)
86. Jones, R. D. G.: ibid. *B32,* 1807 (1976)
87. Jones, R. D. G.: ibid. *B32,* 301 (1976)
88. Jones, R. D. G.: J. Chem. Soc. Perkin Trans 2 513 (1976)
89. Haddon, R. C.: J. Am. Chem. Soc. *102,* 1807 (1980)
90. Lingafelter, E. C., Braun, R. L.: ibid. *88,* 2951 (1966)
91. Singh, T. R., Wood, J. L.: J. Chem. Phys. *50,* 3572 (1969)
92. Emsley, J., Jones, D. J., Kuroda, R.: J. Chem. Soc. Dalton Trans 2141 (1981)
93. Rowe, W. F. Jr., Duerst, R. W., Wilson, E. B.: J. Am. Chem. Soc. *98,* 4021 (1976)
94. Baughaun, S. L., Duerst, R. W., Rowe, W. F., Smith, I., Wilson, E. B.: ibid. *103,* 6296 (1981)
95. Turner, P. H.: J. Chem. Soc. Faraday Trans 2 *76,* 383 (1980)
96. Sanders, N. D.: J. Mol. Spectrosc. *86,* 27 (1981)
97. Brown, R. S.: J. Am. Chem. Soc. *99,* 5497 (1977)
98. Seliskar, C. J., Hoffman, R. E.: Chem. Phys. Letts. *43,* 481 (1976)
99. Seliskar, C. J., Hoffman, R. E.: J. Am. Chem. Soc. *99,* 7072 (1977)
100. Bicerano, J., Shaefer III, H. F., Miller, W. H.: ibid. *105,* 2550 (1983)
101. Pople, G. A., Santry, D. P., Segal, G. A.: J. Chem. Phys. *44,* 3289 (1966)
102. Schuster, P.: Chem. Phys. Letts. *3,* 433 (1969)
103. Gordon, M. S., Koob, R. D.: J. Am. Chem. Soc. *95,* 5863 (1973)
104. Karlström, G., Wennerström, H., Jönsson, B., Forsén, S., Almlöf, J., Roos, B.: ibid. *97,* 4188 (1975)
105. Karlström, G., Jönsson, B., Roos, B., Wennerström, H.: ibid. *98,* 6851 (1976)
106. Del Bene, J. E., Kochenour, W. L.: ibid. *98,* 2041 (1976)

107. Fluder, E. M., de la Vega, J. R.: ibid. *100,* 5265 (1978)
108. Bouma, W. J., Vincent, M. A., Radom, L.: Int. J. Quantum Chem. *14,* 767 (1978)
109. Catalán, J., Yánez, M., Fernández, J. I.: J. Am. Chem. Soc. *100,* 6917 (1978)
110. Emsley, J., Overill, R. E., Parker, R. J.: J. Chem. Soc. Faraday 2 *79,* 1347 (1983)
111. Emsley, J., Overill, R. E.: Chem. Phys. Letts. *65,* 616 (1979); Bouma, W. J., Radom, L.: ibid. *64,* 216 (1979)
112. Melia, T. P., Merrifield, R.: J. Applied Chem. *19,* 79 (1969)
113. Bouma, W. J., Radom, L.: Australian J. Chem. *31,* 1649 (1978)
114. Noack, W.-E.: Theoret. Chim. Acta *53,* 101 (1979)
115. Carlsen, L., Duus, F.: J. Chem. Soc. Perkin Trans 2 1081 (1980)
116. Larson, J. W., McMahon, T. B.: J. Am. Chem. Soc. *104,* 5848 (1982)
117. Kopteva, T. S., Shigorin, D. N.: Russ. J. Phys. Chem. *48,* 312 (1974); transl. Zhur. Fizich. Khim. *48,* 532 (1974)
118. Kreevoy, M. M., Liang, T. M., Chang, K.-C.: J. Am. Chem. Soc. *99,* 5207 (1977)
119. Kreevoy, M. M., Liang, T. M.: ibid. *102,* 3315 (1980)
120. Kreevoy, M. M., Ridl, B. A.: J. Phys. Chem. *85,* 914 (1981)
121. Ernstbrunner, E. E.: J. Chem. Soc. A 1558 (1970)
122. Hadzi, D., Bratos, S.: The Hydrogen Bond, Vol. II, Chap. 12, p. 567, Amsterdam, North-Holland Publishing Co. 1976
123. Ogoshi, H., Nakamoto, K.: J. Chem. Phys. *45,* 3113 (1966)
124. Wierzchowski, K. L., Shugar, D.: Spectrochim Acta *21,* 943 (1965)
125. Tayyari, S. F., Zeegers-Huyskens, Th., Wood, J. L.: ibid. *35A,* 1289 (1979)
126. Tayyari, S. F., Zeegers-Huyskens, Th., Wood, J. L.: ibid. *35A,* 1265 (1979)
127. Ogoshi, H., Yoshida, Z.: ibid. *27A,* 165 (1971)
128. Novak, A.: Structure and Bonding *18,* 177 (1974)
129. Sherry, A. D.: The Hydrogen Bond, Vol. III, Chap. 25, p. 1199, Amsterdam, North-Holland Publishing Co. 1976
130. Bratoz, S., Hadzi, D., Rossmy, G.: Trans. Faraday Soc. *52,* 464 (1956)
131. Hussain, M. S., Schlemper, E. O., Fair, C. K.: Acta Cryst. *B36,* 1104 (1980)
132. Mecke, R., Funck, E.: Z. Elektrochem. Ber. Buns. *60,* 1124 (1956)
133. Ogoshi, H., Yoshida, Z.: J. Chem. Soc. Chem. Comm. 176 (1970)
134. Kato, S., Kato, H., Fukui, K.: J. Am. Chem. Soc. *99,* 684 (1977)
135. Bell, R. P.: The Tunnel Effect in Chemistry, London, Chapman & Hall, 1980
136. de la Vega, J. R.: Acc. Chem. Res. *15,* 185 (1982)
137. Burch, J. H., Fluder, E. M., de la Vega, J. R.: J. Am. Chem. Soc. *102,* 4000 (1980)
138. de la Vega, J. R., Busch, J. H., Schauble, J. H., Kunze, K. L., Haggert, B. E.: ibid. *104,* 3295 (1982)
139. Shiau, W.-I., Duesler, E. N., Paul, I. C., Curtin, D. Y., Blann, W. G., Fyfe, C. A.: ibid. *102,* 4546 (1980) and refs. therein
140. Bratan, S., Strohbusch, F.: J. Mol. Struct. *61,* 409 (1980)
141. Moore, R. E., Scheuer, P. G.: J. Org. Chem. *31,* 3272 (1966)
142. Snyder, W. R., Schreiber, H. D., Spencer, J. N.: Spectrochim. Acta *29A,* 1225 (1973)
143. Duus, F., Lawesson, S.-O.: Arkiv Kemi *29,* 127 (1968)
144. Nørskov-Lauritsen, L., Carlsen, L., Duus, F.: J. Chem. Soc. Chem. Comm. 496 (1983)
145. Carlsen, L., Duus, F.: J. Chem. Soc. Perkin Trans 2 1080 (1980)
146. Carlsen, L., Duus, F.: ibid. 2 1532 (1979)
147. Gebicki, J., Krantz, A.: J. Chem. Soc. Chem. Comm. 486 (1981); J. Am. Chem. Soc. *103,* 4521 (1981)
148. Duus, F., Anthonsen, J. W.: Acta Chem. Scand. *31,* 40 (1977)
149. Jørgensen, F. S., Brown, R. S., Carlsen, L., Duus, F.: J. Am. Chem. Soc. *104,* 5922 (1982)
150. Carlsen, L., Duus, F.: J. Chem. Soc. Perkin Trans. 2 1768 (1980)
151. Duus, F.: J. Org. Chem. *42,* 3123 (1977)
152. Alves, A. C. P., Hollas, J. M.: Mol. Physics *23,* 927 (1972); ibid. *25,* 1305 (1973)
153. Woolfe, G. J., Thistlethwaite, P. J.: J. Am. Chem. Soc. *103,* 6916 (1981)
154. Pickett, H. M.: ibid. *95,* 1770 (1973)
155. Rossetti, R., Rayford, R., Haddon, R. C., Brus, L. E.: ibid. *103,* 4303 (1981)
156. Hibbert, F., Simpson, G. R.: ibid. *105,* 1063 (1983)

Author-Index Volumes 1–57